S0-EDJ-151

The Encyclopedia of WILD LIFE

The Encyclopedia of

WILD LIFE

Published by Salamander Books Limited
LONDON

A Salamander Book
Published by Salamander Books Ltd.,
Salamander House,
27 Old Gloucester Street,
London WC1N 3AF,
United Kingdom.

ISBN 0 86101 096 5

Fourth impression 1981

Distributed in the United Kingdom
by New English Library Ltd.

Filmset and reproduced by
Photoprint Plates Ltd, Rayleigh,
Essex, England

Printed in Belgium
by Henri Proost & Cie, Turnhout.

All correspondence concerning the content of
this volume should be addressed to
Salamander Books Ltd.

Many of the maps and drawings in this book are based on material originally appearing in 'The Living World of Animals' published by The Reader's Digest Association Limited, and in 'Book of British Birds' published by Drive Publications Limited.

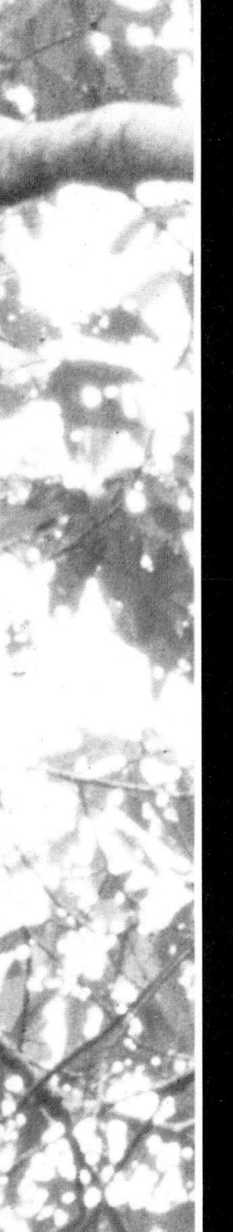

Editors: Eve Harlow and Iain Parsons

Original Editorial Plan: Cathy Kilpatrick

Art Director: Tod Slaughter,
Arka Cartographics

Cartography: Arka Cartographics

Drawings: Andrzej Bielecki

The publishers would like to thank the numerous contributors to this volume, particularly:

Professor A. d'A. Bellairs
Professor of Vertebrate Morphology
University of London at St. Mary's
Hospital Medical School:
The Introduction to Reptiles
The Evolution of Reptiles
Turtles and Tortoises
Alligators and Crocodiles
Lizards
Snakes
Linda Bennett:
Cuckoos
Woodpeckers
Dr. Valerie Brown
Zoology Dept., Royal Holloway College:
The Arthropods
Michael Chinery:
Insects
M. J. Everett:
Birds of Prey
Owls
Dr. J. F. D. Frazer:
The Introduction to Amphibians
The Evolution of Amphibians
Newts and Salamanders
Frogs and Toads
Dr. Ray Gambell and
Sidney G. Brown
Institute of Oceanographic Sciences:
Whales
Howard Ginn, M.A., Cantab.
British Trust for Ornithology:
The Introduction to Birds
Nicholas Hammond:
Perching Birds
Fowls
Flightless Birds
John Hannah:
Sea-Cows, Manatees and Dugongs
T. P. Inskipp
Parrots
Pigeons
Cathy Kilpatrick:
The Evolution of Invertebrates
The Single Celled Animals
Worms
The Introduction to Fish
The Evolution of Fish
Dr. P. Morris
The Introduction to The Encyclopedia
The Introduction to Invertebrates
The Introduction to Mammals
Bats
Marion Nixon, B.Sc., Ph.D.
University College, London:
Bony Fish
The Evolution of Mammals
David Le Roi, M.A., B.Sc., F.I.A.L.:
Primates
Dorothy Rook:
Goatsuckers
Kingfishers and Swifts
Cranes and Rails
Gareth J. Thomas:
The Evolution of Birds
Swimmers and Waders
Gulls and Auks
A. E. Vines, B.Sc., M.I.Biol.
Molluscs
Gnawing Mammals
Dogs, Bears and Relatives
Peter Ward, B.Sc.
Producer BBC School Nature Series:
Echinoderms
Elephants

Reptiles

Birds

Mammals

INTRODUCTION

There are about a million species of animals known to Science. They inhabit almost every part of the world, from the tropical forests to deserts and the polar ice caps; from mountain tops to the permanent blackness of the ocean depths. In fact, animals seem to live everywhere that environmental conditions do not contrive to make life seem impossible. Some overcome even this limitation and can survive in scalding springs, on permanent ice, in pools of oil or in similarly unlikely and inhospitable places.

Wherever they live, all animals need food, derived directly or indirectly, from plants; but few other factors are universally essential. So great is the diversity and adaptability of animal life that many species can do without what might be considered biological necessities.

The Mystery of Life

Many creatures can live without oxygen, like the worms that inhabit the smelly black mud at the bottom of stagnant ponds and estuaries. Certain fish and arthropods spend their entire lives in the pitch darkness of caves or the deep sea, without the need or the ability to use light to find their way about. Although all animals need water as a constituent of their bodies, many can go without drinking; some female Desert Rats can produce milk for their young without ever swallowing a drop of water. Certain simple animals such as snails, can become dormant in times of drought and may remain alive but inactive for years, awaiting the next rainfall. Paradoxically, some of the animals best adapted to live without water, are in fact aquatic. A few little midges for instance, that live in puddles in tropical Africa, are killed when the water is dried up by the fierce sun, but their eggs can survive dehydration indefinitely and will develop normally within minutes of being dampened. Using special biological mechanisms certain arctic fish can swim in water colder than the freezing point of their own blood and the special internal heating and cooling mechanisms of mammals and birds also permit fairly normal activity over a wide range of environmental temperatures; but for other animals heat and cold are often crucial in determining if and where they will live.

The size of animals is also affected by other physical factors, notably gravity. The bigger an animal is, the more support it needs for its weight. With aquatic forms this is less important because they are supported by the water. Jellyfish for example, live in the sea and can be over 6 ft (2 metres) in diameter but on land even the smallest species collapses, totally impotent. For land animals body support is provided by some form of skeleton, of which there are two basic types: the external (exoskeleton), and the internal (endoskeleton).

The exoskeleton is like a medieval suit of armour, a set of jointed tubes and boxes round the outside of the animal. It provides support for the body, attachments for muscles and also protection for the soft tissues. An exoskeleton is found in over three-quarters of all the known animal species; it is very efficient and can be very strong in proportion to its weight, (a Flea for example, can jump 130 times its own height, a feat that requires an immensely strong skeleton to withstand the forces involved). Exoskeletons reach their greatest degree of sophistication among the arthropods, (notably insects and crustaceans), where complicated peg and socket joints help overcome the basic clumsiness of the system. The strength of the exoskeleton permits saving of weight, enabling the insects to become the only group of invertebrates capable of flight. The biggest disadvantage of having the skeleton outside the body is that it impedes growth and must be cast off whenever the animal needs to increase its size. Arthropods have to moult their armour regularly as they grow and in doing so throw away their protection against predators. They can and do hide away during this period of vulnerability, but still cannot escape the force of gravity. Shedding the exoskeleton, especially for those that live on land, means loss of support and the animal is almost helpless until its new skeleton hardens. Without its skeleton, the terrestrial arthropod is just as helpless as the Jellyfish would be on land, and perhaps this major drawback of the exoskeleton is one of the main factors limiting the size of arthropods. The heaviest insect is the Goliath Beetle weighing only 3 oz (95 g) although some crabs may reach a weight of 40 lbs (18 kg).

An endoskeleton can grow with the animal and is less of a limitation on maximum size. Internal skeletons are best developed among the vertebrates, supporting the largest animal that ever lived, the Blue Whale. This species may exceed 98 feet (30 m) in length and weigh over 150 tons (tonnes). Part of a whale's immense bulk is supported by the surrounding water. Land animals are much smaller, the largest being the African Elephant weighing up to about 5 tons (tonnes). Here again maximum size is limited by a combination of physical and biological

factors. If an elephant were to be much larger, it would need such massive bones to hold up just its own skeleton that it could not move or support the extra weight of its muscles and intestines. On the other hand, certain giant Dinosaurs were more than 60 feet (20 m) long and were twice as heavy as an Elephant; others may have weighed 10 times as much, but how they managed it is something of a mystery.

The importance of body temperature

The problem of size in animals relates to the crucial importance of temperatures. A rise of 10 °C doubles the speed of chemical reactions and since all biological activity is the result of chemical changes, temperature affects digestion, respiration, muscular movement and the functioning of nerves, and indeed every aspect of an animal's mechanism. This is important to a cold blooded animal, which may need to bask in the sun to keep its body 'ticking over', and especially important to large animals. Without a warm body an animal would suffer from having a slow brain (since that is comprised of nerves whose activity depends upon temperature), from the long delay in finding out what was going on in the furthest parts of its body and from the even longer pause before he could do anything about it! The larger the animal, the more serious these problems become.

Only the mammals and birds have evolved the ability to maintain both high and constant body temperatures. The constancy permits all parts of the body to function most efficiently and reduces the effects of temperature changes in the environment. The relatively high temperature (35 °–40 °C), conveys a range of advantages. These 'warm blooded' animals can digest food faster, move more rapidly and achieve a much faster rate of energy utilisation in the body. Their growth is faster too; a baby bird may multiply its weight a hundred times in a matter of days. In a cold blooded frog, signals in the nerves travel at only 60 feet (20 m), per second; in a warm blooded mammal the chemical and physical processes are speeded up by the higher temperature and additional biological refinements ensure that the nerves conduct information 5 or 6 times as fast. Taken together, these effects mean greater activity, faster responses and very significant advantages over other animals, although higher temperatures do mean that adequate body insulation is essential; in the form of fur, feathers and fat.

Sedentary animals

Although speed and movement seem important, even characteristic, features of animals, it is worth remembering that many animals do not move at all, but become fixed to a solid object early in life and remain in the same place until they die. Corals, Barnacles and various burrowing shellfish are familiar examples, many of which do not survive if dislodged from their base. Since they do not crop vegetation or actively pursue prey, these sedentary animals are filter feeders with complex mechanisms for removing tiny food particles from the medium in which they live. Casting away the great animal advantage of mobility does have three particular biological benefits. One is that they save energy by not having to seek food, another is that they escape competition from more conventional feeders. The third advantage is that they are independent of the food content of their substrate; a cow or a snail must live and move on a surface abounding in vegetable food, whereas a filter-feeding cockle can burrow in sterile sand and a barnacle can adhere to the bottom of a boat. Both obtain food from the surrounding water which must, in any case, pass through the animal's body for respiratory purposes.

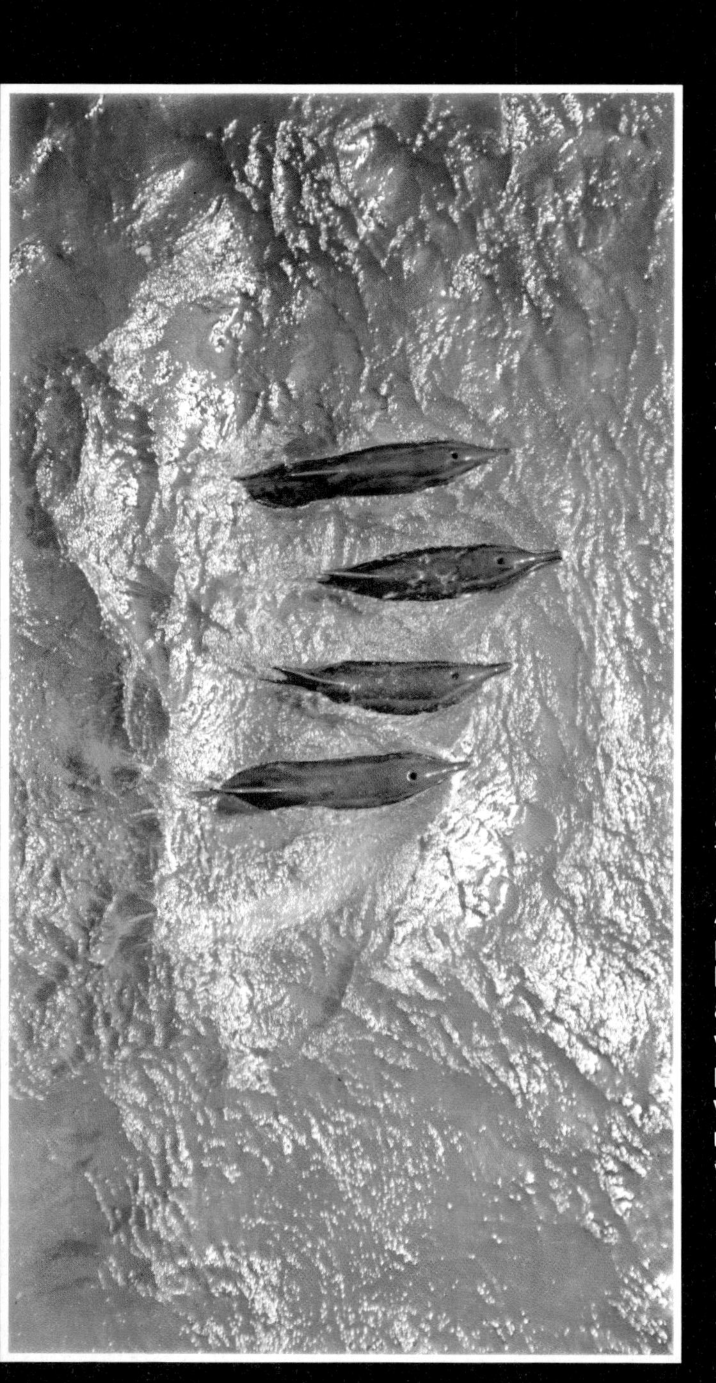

Evolution of animal life

The modern diversity of animal life is a product of over 500 million years of evolutionary development. Through countless generations, natural selection has favoured the individuals best adapted to their way of life by eliminating their inferior competitors. Survival of the fittest means that only they are responsible for producing the next generation, which will inherit at least some of the advantageous features of their parents. Modern animals often seem perfectly suited to their way of life and many also seem extraordinarily complex. Both features have evolved slowly; simple animals in the beginning giving rise to increasingly complex forms with many poorly adapted types arising on the way, doomed to a swift extinction as more successful animals overtook them.

Animal evolution seems to have begun in the sea and progressed in a series of phases, with certain groups becoming dominant for a time. For example, the ability to live on land provided the early amphibia with an enormous new habitat, away from competition with fish. Great numbers of amphibia were produced, only to be eclipsed by hordes of reptiles later on. Reptiles were more independent of water, more able to survive dry weather. Once the reptilian 'design' had evolved, reptiles swiftly took over most land habitats from the less adaptable amphibians, which soon declined in number. Certain evolutionary lines gave rise to birds and mammals which, because of their considerable biological abilities and advantages, in their turn displaced many of the reptiles and now occupy the dominant roles in most of the world's habitats.

Learning from fossils

These processes of continual and progressive changes were much stimulated by climatic changes which have occurred in the past. During certain periods, notably the Triassic 250 million years ago, the world was hotter and dryer and thus a severe disadvantage to amphibia but not to the better adapted reptiles. Some animals seem to have developed a suitable biological design for a particular way of life very early on. As long as their habitat did not change, nor did the animal. Consequently, the living forms are almost identical to their relatives that lived millions of years ago. Scorpions, for example, closely resemble fossils 400 million years old of the animals who were among the earliest to live on land. King Crabs also seem to have changed little in that time. Perhaps the oldest 'living fossils' are found among the molluscs, where generations of two limpet-like species have lived isolated, and unchanged, in the deep canyons of the sea bed for over 500 million years. Their biological design was already 250 million years old before the first Dinosaur hatched from its egg in the Mesozoic Era which extended from 225 to 65 million years ago.

Over half the animals that ever lived are already extinct and our knowledge of them depends largely upon the study of fossils. When animals died, most decomposed and disappeared, just as they would today. However, some happened to be buried in mud, sand, volcanic ash, or in some other way and were preserved from immediate destruction. Over the years their bodies became impregnated with minerals and their surroundings compressed into rock. These preserved specimens are what we know as fossils, and although it was often only the hard tissues (bones, teeth and shells for instance), that remained, a great deal can be learned from close study of them. Chalk and limestone, with their abundant fossil echinoderms (mostly sea urchins) and corals, must once have been the ooze of an ancient sea bed, for these animals are typical of the marine environment. Other animal fossils, lizards for example, are found in rocks that were once desert dust; the rotten vegetation that once grew in warm swamps 350 million years ago is now carbonised and compressed into coal and many fish remains can be discovered in it.

Defining the animal kingdom

The multitude of different animals are classified into natural groups, based on different degrees of similarity of structure, according to a system devised by the 18th century Swedish biologist Carl von Linnaeus. His scheme has been modified to accommodate the greater knowledge now available (and the greater number of species discovered since his time). The Animal Kingdom is now divided into major Phyla, each defined by some particular combination of characters. The Phylum Arthropoda, for instance, includes all those animals with an exoskeleton, segmented bodies and jointed limbs; the Chordata (sometimes called the Phylum Vertebrata) are animals

The 26 Phyla of animals include many minor groups that contain only a handful of species and are of doubtful interest even to the specialist. The major Phyla have to be subdivided into Classes, each of which is further split into Orders. Among the Chordata (one of the 26 Phyla) there are six major Classes: Protochordata (small marine animals), Pisces (fish), Amphibia (frogs, toads, newts), Reptilia (lizards, snakes, tortoises etc.), Aves (birds) and Mammalia. Again these are unequal groupings, the fish species outnumber all the other classes put together, but since most of them are out of sight and unknown to most of us, the importance of the group tends to be overlooked. Mammals, despite the prominence given to them, constitute a mere 4,200 species, only half as many as the birds and less than 0.5% of all animals.

Orders are further subdivided into families; family names are easily recognised because, throughout the Animal Kingdom, they always end in-*idae*. Within the families, individual species are each given a double-barrelled name, the first of which (always italicised and always beginning with a capital letter) is the Genus. The generic name is shared by several closely related species, rather like our own surname. For example birds of the genus *Falco* are all small Falcons, individually distinguished by their specific name (again italicised, but beginning with a small letter); e.g. *Falco tinnunculus* (the Kestrel), *Falco subbuteo* (the Hobby) and so on.

Of course, other general classifications of animals are used. They are grouped as herbivores and carnivores according to what they eat; or as British, American, Australasian and so on, according to their provenance. Such generalisations are often helpful but could be misleading if used as a basis for a formal classification of all animals. One would find such diverse creatures as birds, bats and butterflies lumped together just because they could fly; or silly anomalies like rabbits being classed with badgers because they lived in holes whereas the very similar hares were grouped with sheep because they lived out on open grassland. The natural system of Linnaeus, being based on structural similarities, will tend to group together animals with a similar life and origin, so indicating something of an animal's evolution as well as its present relationships. A logical classification system is an essential framework upon which to base a comprehensive review of the Animal Kingdom. Without it the fascinating diversity of animal life would become a daunting, jumbled nightmare instead of a beauti-

INVERTEBRATES

The First Animals

The Protozoa are usually considered to be the simplest form of animal life since their bodies comprise only a single cell. Presumably the first animals were also single celled forms, some of which evolved into the relatively complex Protozoa living today, whilst others developed into the first precursors of modern multicellular animals. The exact evolutionary pathway followed is the subject of controversy, but it seems likely that multicellular animals originated either as a result of internal subdivision of a large complex Protozoan; or, more likely, from a colonial form of single celled animal.

Today colonial Protozoa are still to be found in rivers and lakes and some of the simplest forms of multicellular animals are biologically little more complex. Some sponges for instance are scarcely more than aggregations of cells, though many of the cells are specialised for particular roles, whereas in a Protozoan colony the units all tend to be the same. Among other living invertebrates we can recognise animals which represent further stages in the evolution of animals. Coelenterates for example (hydras, sea anemones and corals) have their cells in two differentiated layers, separated by non-living jelly, and regionally specialised for catching prey, digestion, reproduction and locomotion.

The flatworms are the simplest animals which are based upon a 3 layered plan, with certain specialised cells in large groups where they all perform one task, e.g. forming a sucker or a reproductive organ. Essentially flatworms are thin 'solid' animals, but the segmented worms (Annelids) have developed further with a fluid filled cavity (the coelom) between the gut and the body wall, providing various advantages to the animal.

Increasing size and complexity go hand in hand; the bigger an animal becomes the more necessary it is to have a skeleton to support it and a nervous system to control its various parts. A blood system is also essential for the transportation of food and oxygen from parts of the body specialised for their collection to other parts needing both to perform some other task. Increased complexity often enables an animal to be more adaptable and capable of overcoming a greater range of environmental difficulties.

Worms and arthropods have segmented bodies, a relatively advanced biological development. They are made of many similar sections, each of which then becomes specialised for a particular role (A worm is like a goods train—lots of identical units strung together with some adapted later to carry coal and others for mail or timber. A highly modified unit forms the locomotive representing the brain and locomotory apparatus.) A segmented body plan is biologically advantageous. It is relatively simple to build up during embryonic development, coordination of its parts can be very efficient and above all, modification of segments for special purposes provides potential for a great variety of adaptations.

In arthropods the segments typically each bear one part of jointed appendages. These have become modified in the head region as mouthparts and on the body as walking legs. The mouthparts are further specialised for biting, sucking and shredding, so adapting their owner to different diets. The walking legs may also develop modifications enabling them to catch food, grip objects, even carry pollen (in bees). The advantages of the two great biological 'inventions', the adaptable segmented body and the strong exoskeleton, are perhaps the main reason why arthropods are so abundant and successful in practically every habitat.

One prominent group of animals which are not segmented, but have still achieved considerable success, constitute the Phylum Mollusca. A shell is their special feature, adapted to perform many functions apart from simple protection. Some molluscs are very abundant, some very large and the cephalopods (octopus, squids, cuttlefish, etc.) are extremely complex and intelligent animals.

Because increasing complexity is often a mark of evolutionary advancement, one should not infer that simple animals are necessarily primitive and inferior. Some have evolved extreme simplicity as part of their way of life. Certain internal parasites for example do not require eyes, legs, brains and other fancy structures for they live bathed in food and do not need to move about. In such animals evolution has produced a very simplified animal rather than a complex, adaptable one.

Many invertebrates are small, obscure, undistinguished things, but the Echinoderms are very distinctive. These are the starfish, sea urchins and sea cucumbers; characterised by having hard calcareous plates in the body surface. They are common, colourful marine animals with a peculiar body organisation. They are not segmented, but have the body arranged in radial sections, usually five in number. They lack complex structures like eyes and a brain, yet have evolved a complicated and completely unique hydraulic system of 'tube feet' for locomotion, controlled by a nerve net. Similarities in early embryonic development suggest that the first vertebrates may have shared a common ancestor with the echinoderms rather than with more complex animals such as the arthropods.

An Invertebrate World

This miscellaneous assortment of beasts constitute the invertebrates. In fact the invertebrates should not be considered as an animal group, but rather as all animals except the vertebrates. The lack of a vertebral column is about the only feature all these diverse creatures have in common. This popular division of animals into vertebrates and invertebrates is convenient but highly lop-sided. Invertebrates account for 95% of all animals: the Animal Kingdom is really an invertebrate world.

The Evolution of Invertebrates

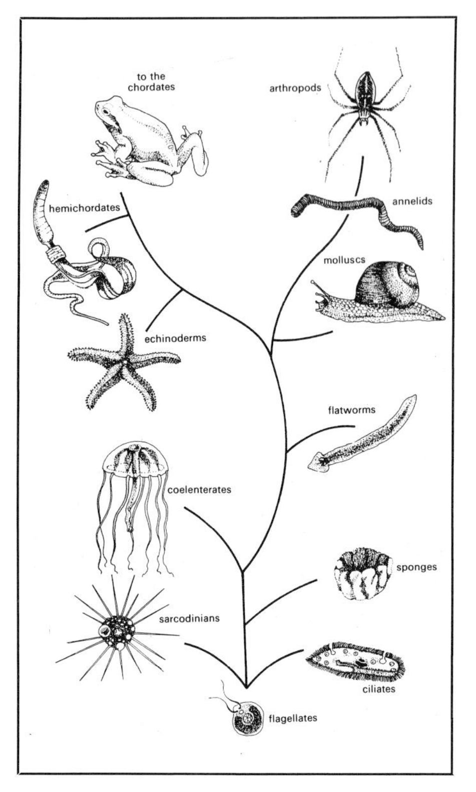

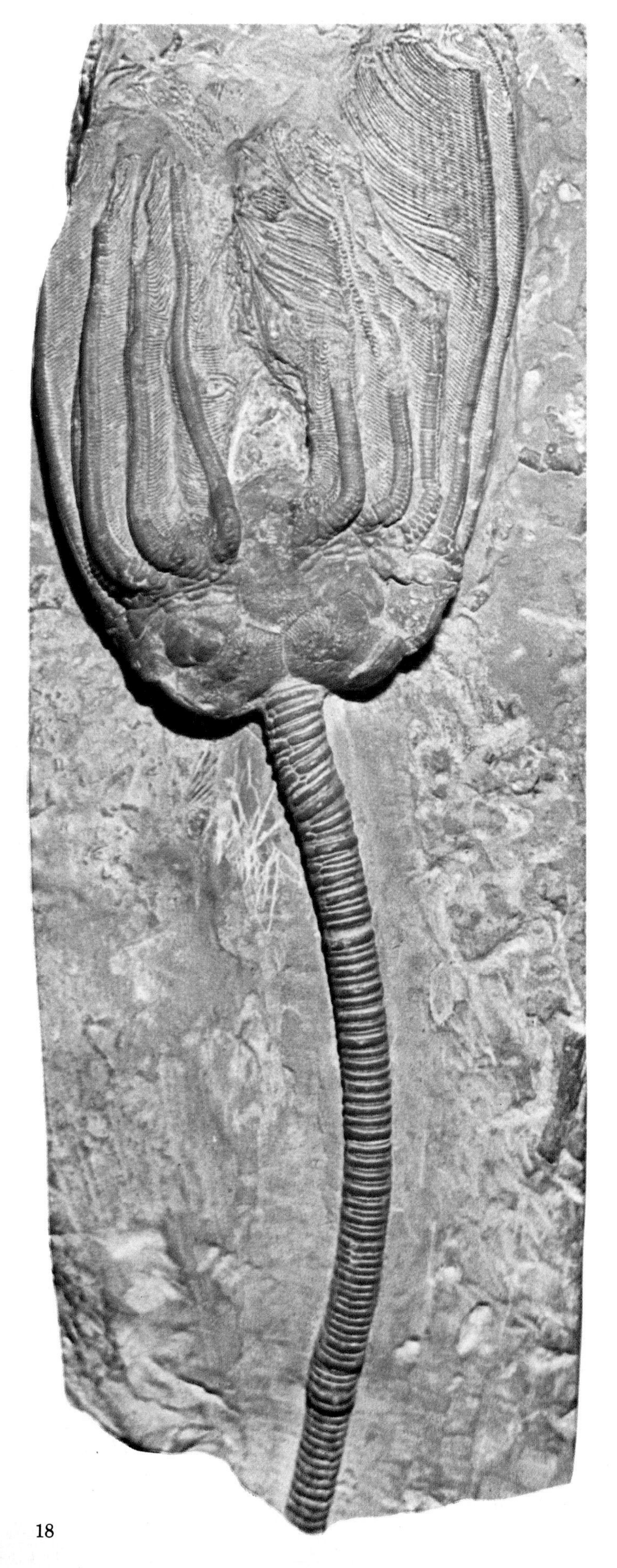

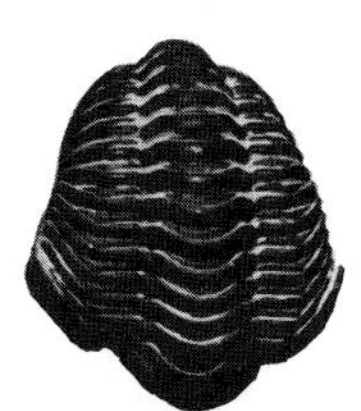

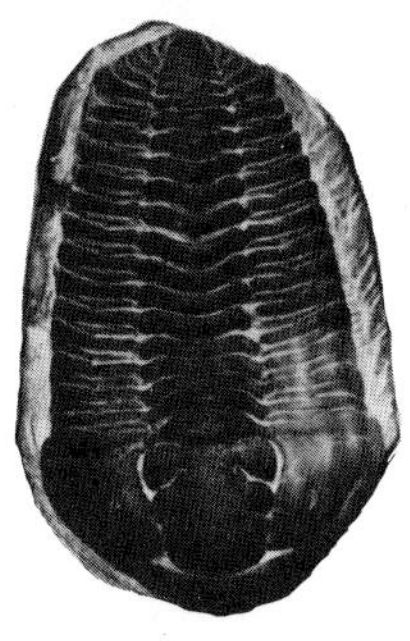

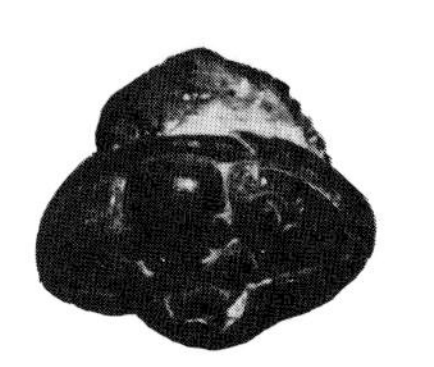

◁ Fossilised Sea Lily, from rocks over 300 million years old, an ancestor of the present forms which have changed very little in that time.
△ Evolution chart showing the possible evolutionary relationships of some of the invertebrate animal groups.
◁ Fossils of Trilobites which became extinct about 225 million years ago.

The Earth came into existence about 5,000 million years ago. Life originated in the salty concentrated oceans in the pre-Cambrian period over 500 million years ago, when the physical and chemical conditions were such that living organisms could arise from non-living matter. From the non-cellular organisms evolved the soft-bodied animals without backbones, the Invertebrates. As the earliest forms of life were soft-bodied and without hard parts, very little fossil evidence exists. Most of the earliest fossils known to us are quite complex in comparison with the lowly organisms which must have existed at the dawn of life. Only a few pre-Cambrian fossils of Jellyfish, Algae, Sponges, Worms, Coelenterates and Echinoderms, have been found.

During the 90 million years after the pre-Cambrian period and the lower-Cambrian period there was a sudden appearance of a wide range of well-fossilised plants and invertebrates. It is quite possible that, even at this time, there existed the major representatives from all the phyla of plants and animals. The Trilobites appear to have been the dominant form of life.

The Trilobites are perhaps one of the most familiar forms of all fossils. They became extinct some 225 million years ago but their group became very successful. These are the Arthropods which contain about three-quarters of all known animal species including Shrimps, Crabs, Lobsters, Barnacles, Spiders, Scorpions, Ticks and Insects.

Another group of Arthropod fossils, originating in the Ordovician period over 450 million years ago, were the Eurypterids. These are the aquatic ancient relatives of the Scorpions. Their usual length was about one metre, although some reached three metres in length. They were most abundant in the Silurian and the Devonian periods and became extinct in the Permian.

The movement of marine animals to the land could take place only when plants had colonised the land to provide food. It is therefore not surprising that the first land animals—probably Arthropods—and plants appear at roughly the same time. The success of the Arthropods in the conquest of land is notable. Certain characteristics in the animal itself were responsible for this. A chitinous shell protected them from drying up and they also possessed appendages which could support their bodies when out of water, and also give them means of locomotion.

△ Ammonites are fossil animals which have been extinct for over 70 million years. They were related to the living cephalopods, similar to the Pearly Nautilus.

▷ Peripatus is an attractive little animal of considerable interest to zoologists because it possesses features that appear to link the segmented worms with the arthropods. It may well be descended from animals that originated at the time when the present annelid worms and the arthropods separated to follow their individual lines of evolution.

Insect explosion

Insects evolved during the Devonian period when legged worm-like creatures, able to breathe air, gradually emerged as the first wingless insects on land. Winged insects, capable of flight, came millions of years later in the Upper Carboniferous period. Giant Dragonflies with wingspans of up to two feet slowly flapped over swampy forests with slender tall trees rising above an undergrowth of fern-like plants, horsetails and creepers. Insects had attained flight 50 million years before the flying reptiles and birds. By the Permian period some 280 million years ago, the insects known today began to emerge. Most insects have survived without much change for over 200 million years. Today, insects are the most successful and numerous group on earth, accounting for three-quarters of the total species of animals.

All living organisms—plants and animals—are made up of fundamental units called cells. Most living systems are composed of cells. In each cell there is enclosed in a microscopic package, all the things necessary to survive in an ever-changing environment. Cells vary in size from a few thousandths of a millimetre in diameter to the huge yolk of an ostrich egg. They also vary in shape; spirals, boxes, spheres, snowflakes and stalked among them. But nearly all the cells have the same basic structure with certain details varying, depending on the cell's function.

Molecular Code

Most cells have a central nuclear region, surrounded by cytoplasm which is enclosed by a cell membrane. The nucleus, the most obvious component of the cell, determines the activity and general properties of the cell. It contains the genetical apparatus that carries the molecular code to be transmitted to the cytoplasm, instructing it what to do. The cytoplasm includes the organelles involved in energy release and the building up of complex molecules. The 'powerhouses' of the cytoplasm are the *mitochondria* since they are involved in the release of energy from carbohydrates and fats.

Fossil formation

Fossils are the hardened remains of plants and animals, or their impressions, and are found in rocks. This is because such animals and plants were covered by mud or sand which later changed into rocks, sometimes over millions of years. Rocks which may contain fossils are limestone, shale, clay and chalk. The animal fossils usually show only the hard parts. The soft parts decay rapidly at death and it is usually the bones that may become fossilised. Very rarely, soft structures such as fins or feathers may be preserved. Sometimes even the bone is dissolved away and just the shape remains as a cavity or mould within the rock type. Other fossils which are formed are footprints of animals made in mud or the remains of the skins of fossil reptiles.

Most of all the living organisms of earlier periods perished without traces. In particular, little is known of the animals with soft bodies. It is only in the finest grained muds, which have not been folded or heated through the ages, that indications of delicate creatures such as jellyfish or worms can be found. Many fossil insects are known from only single specimens. Even so, enough fossils have been found for the study of fossil animals (palaeozoology) and the study of fossil plants (palaeobotany) to be useful in learning about life in the past.

The Simplest Animals

The life of minute single-celled animals which are even capable of reproducing on their own is a fascinating blend of the mundane and the incredible.

All animals which have only one-celled bodies are referred to as Protozoa. They are microscopic in size—between a thousandth of a millimetre to one millimetre in length—yet these animals carry out the process of life within a single cell. There are 30,000 species and all Protozoa live in contact with water—either free water or water which is part of another organism. There are both marine and freshwater species and a large number of parasitic species. Some species live as animals, eating other animals and plants while others live in a similar way to plants, producing their own food by a process of photosynthesis. This is the combining of carbon dioxide and water to form carbohydrates, using energy from the sun. The Protozoa are divided into four classes.

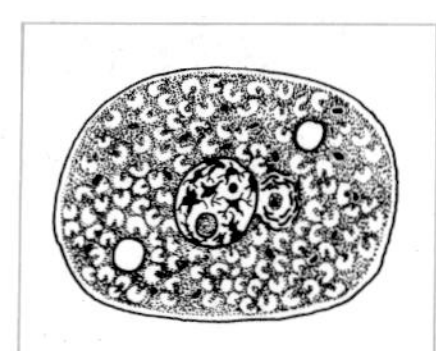

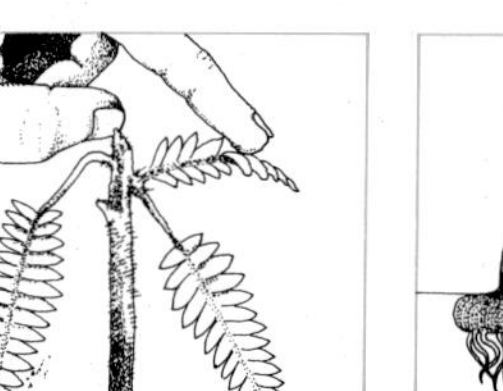

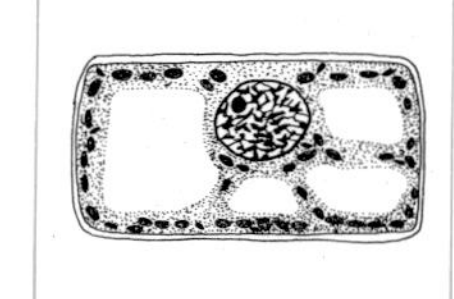

▷ The differences between animals, top, and plants. Most animals have a nervous system which enables them to react quickly when touched. Amongst plants the Mimosa is rare in that it reacts to touch. Most animals move about freely in search of food or partners but most plants rely on chance for fertilisation and can make their own food. Animal cells are enclosed in very thin, delicate membranes and lack supporting structures. The cell walls of plants, made of cellulose, are thick and self-supporting.

△▷ Amoeba, an irregularly shaped protozoan just visible to the naked eye as a slowly moving speck. It crawls slowly over weeds and stones in ponds trapping bacteria by means of pseudopodia (false feet). Amoeba has no natural colour so it has been photographed against a dark background so that the granular nature of the protoplasm can be seen. The tiny plant cells on which Amoeba feeds can be seen in the surrounding water.

△ Radiolaria, tiny marine animals closely related to Amoeba, float in great numbers in the upper layers of the sea water. Each has a beautiful and intricately patterned shell and numerous threads of protoplasm protrude through holes in it, serving to trap food particles in the surrounding water. Huge areas of the ocean floor are covered by Radiolarian ooze which is formed from their skeletons.

The Flagellates

(Mastigophora)

This class of Protozoa possess flagella, thread-like outgrowths which vibrate, enabling the animal to move. Some genera, such as *Euglena* have only one or two flagella and also possess green chloroplasts which allow photosynthesis to take place. However *Euglena viridis*, in putrid water, will absorb food and the chloroplasts lose their colour. Some flagellated green species produce colonies visible to the naked eye. Hundreds are joined to cover the outside of a gelatinous water-filled sphere, which can be up to half a millimetre in diameter.

Ciliophora

This class of Protozoa have cilia—hair-like structures which enable the animal to move and obtain food. Unlike other Protozoa, there are two nuclei—a large vegetative nucleus and a small reproductive nucleus. The ciliates are larger than most other Protozoa and most occur in fresh water and marine water, some being parasitic. Species of *Paramecium* move by vibrating their cilia against the water and at the same time, revolving on their lengthwise axis. They thus travel in a spiral path and can also travel backwards.

Rhizopoda

(Sarcodina)

This class of Protozoa includes the orders **Amoebina, Heliozoa, Foraminifera** and **Radiolaria.** They have no flagella and move by processes called 'pseudopodia'—where the cytoplasm protrudes temporarily to act as a means of locomotion or as a means of feeding.

The food, captured by the cytoplasm flowing around it is ingested into the cell together with a drop of water, to form a food vacuole.

Amoeba occur in ponds and are difficult to see as they are almost transparent. Reproduction is achieved by dividing into two. Heliozoans and Radiolarians have siliceous skeletons with spines radiating outwards from spherical bodies rather like tiny balls with stalks. The animal is inside the latticed skeleton and pushes its pseudopodia through the holes. Reproduction takes place inside the skeleton, half of the protoplasm flowing out through one of the holes, the 'daughter' cell secreting a new skeleton.

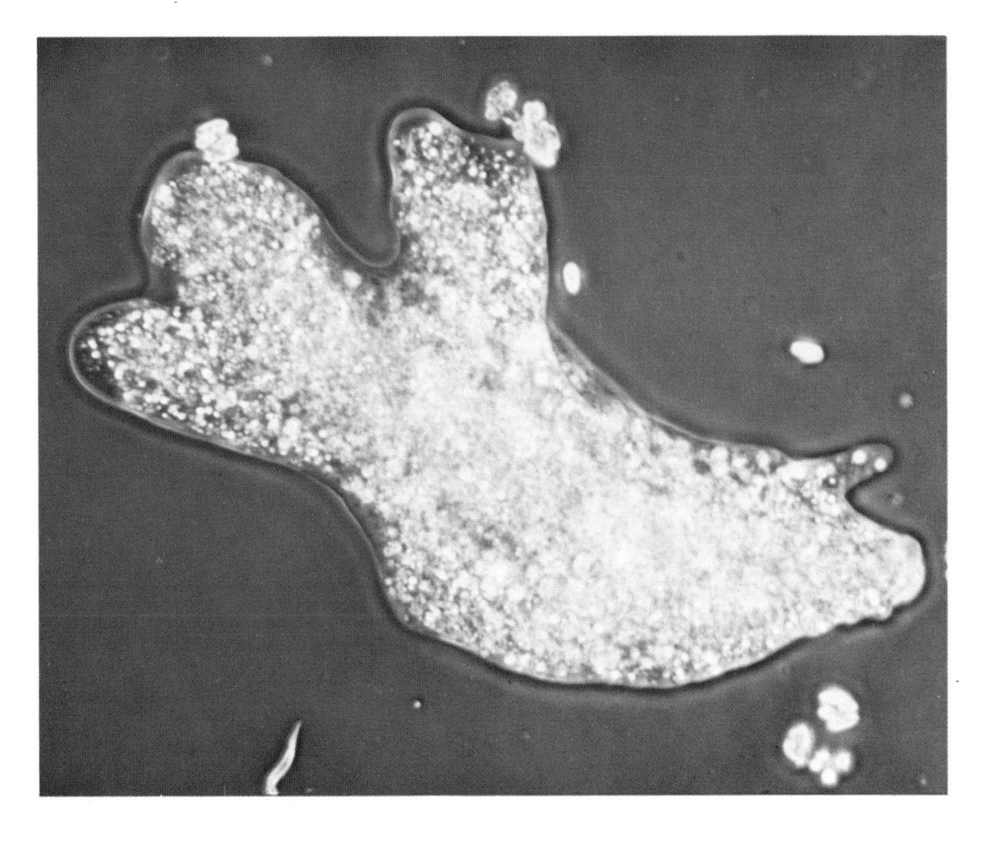

Sporozoa

The Sporozoa are parasitic animals with comparatively complicated life cycles. This sometimes involves asexual and sexual generations with two separate animals acting as hosts. The malarial parasite, *Plasmodium vivax*, is spread by mosquitos and causes malaria in Man, the parasite attacking the red blood cells, with weakness and fever as side effects. Malaria remains a most important microbial disease despite improved controls of the mosquito.

The species *Ghigea anomala*, found in fresh water and which attacks Sticklebacks results in white nodules growing on the fish. These develop into thousands of spores, which in turn infect other Sticklebacks.

Which Kingdom—Plant or Animal?

Plants and animals have many characteristics in common. Both require food to repair damaged tissues, to build up new tissues for growth, and to provide material from which, by chemical change, essential energy for living is released. Both animals and plants have the same chemical compositions, being made up of water, inorganic salts, carbohydrates, fats and proteins. However, a main distinction is that animals lack chlorophyll while plants, using their chlorophyll, are able to make their own food from carbon-dioxide found in the air, water and minerals. In the presence of light they are able to photosynthesize. Animals, however, depend on getting their food in more direct forms, by either eating plants or other animals. This also dictates the form of each, for animals are of a more compact design for moving to find food, while plants need to expose a considerable area of their surface to light. In the lower forms of life distinctions are blurred between the algae, fungi, and bacteria.

An Ocean in a Teaspoon

Microscopic organisms are found in both the animal and plant kingdoms. If a teaspoonful of water were taken from the surface waters of the sea or a pond and a single drop looked at under a microscope, very many animals and plants could be counted. The animals of the **Plankton**—the term given to these organisms—make up the **Zooplankton.** It is made up mainly of Protozoans, the eggs and young stages of fish, the larvae of Molluscs, Starfish, Jellyfish and Crabs. The drifting plant life—the **Phytoplankton**—is made up of single-celled plants such as **Diatoms.** Most of the Zooplankton feeds on the minute plants but the whole of the Plankton is food for many animals, from fishes to Whales. Thus plankton plays a very important role in the food chains of many animals. Indeed, they are the basis for all food in the sea.

Although most of the animals of the Plankton can swim they are entirely at the mercy of the ocean currents, drifters moving where the currents take them. There is a noticeable difference in their positions during the day and the night. At night, most species are close to the surface of the water, migrating vertically away from strong sunlight when day comes.

Aquatic animals without a brain

Stinging thread-cells which can be shot out at prey or used for impressive defence are found only in the coelenterates and are one of the reasons why these slow-moving and primitive creatures have survived and developed so successfully.

Coelenterates, which include the Jellyfish, the Sea Anemones and Corals, are among the most primitive of the multi-cellular animals. Their bodies are apparently a simple assembly of cells, without a brain of any kind yet these creatures are capable of activity. They can sting and capture prey, defend themselves successfully against predators and, in some forms, move away from unsuitable locations.

The word 'Coelenterate' is derived from Greek words meaning 'cavity' and this literally describes the body of the animal. The body of the individual Coelenterate is essentially a hollow sac with a mouth opening at one end. This opening is surrounded by tentacles and is used for both taking in food and discharging waste.

All Coelenterates have radially symmetrical bodies—that is, if the body was divided across any diameter the two halves would be the same lengthwise. (Most other animals are bilaterally symmetrical with a left and a right.) The body consists of two layers of cells with a gelatinous layer between. The inner layer cells perform the function of a gut, both digesting and absorbing the food which the tentacles push in through the mouth. The tentacles of the Coelenterate are covered with stinging thread-cells called *Cnidoblasts* and these are used for both defence and for capturing prey.

Stinging cells

Cnidoblasts are found only in Coelenterates and explain how these slow-moving creatures have managed to survive against predators. Some of the outer surfaces of the tentacles consist of cigar-shaped capsules which contain coiled-up, hollow threads. When the capsule is touched or if potential food is present, the thread is shot out, either impaling the prey or entangling it. Some of the threads have a barbed tip and others eject a drop of poison at the same time as they are shot out and although the individual capsules are very tiny—less than a millimetre in length—their stinging effect is remarkably effective.

Coelenterates are divided into three classes. These are: *Hydrozoans*, made up of Hydroids and Siphonophores; *Anthozoans*, comprising Sea Anemones and Corals and *Scyphozoans*, the Jellyfish.

They fall into two structural types: cylindrical polyps which are anchored to underwater material, and free-swimming discs or bell-shaped 'jellyfish' *(Medusae)*. Medusae reproduce sexually while some polyps reproduce both by budding, forming colonies, and by budding off medusae.

Colonial Hydrozoans

Hydrozoa, of which there are 2,700 species, and having both polyp and medusa forms, is a colony-forming group of Coelenterates. The Hydroid species form colonies by the budding of individual Hydroids, the buds growing long stalks before new polyps are formed. The branching tubes which connect the polyps make up the Coenosarc. *Tubularia indivisa* or Oaten-Pipe forms a mass of small brown stems and can be seen along the edge of the shore at low tide, growing on stones and shells. It can be recognised by its slender brown polyps surrounded by two separate rings of tentacles.

Obelia geniculata forms colonies on seaweed and grows in zig-zag lines. *Obelia* polyps have protective cups and the hydroid is best seen when immersed. It is found all over the world.

Siphonophores are believed to be floating colonies with medusoid individuals forming the floats and the polyps forming the tentacle and digestive units. One such genus is *Velella*, which has a bright blue gas-filled float with a diagonal 'sail' projecting above.

The Life Cycle of the Hydrozoan

The great majority of Hydrozoans alternate between polyp and medusoid stages during their life cycle.

The polyp is the adult animal and is static, either attached by its base to a solid anchorage, or, in Siphonophores, forming part of a floating colony.

Compound Hydroids branch by producing buds which do not separate. Instead, they develop into colonies, with each branch possessing a terminal polyp.

Below the mouth tentacles is a second ring of tentacles and above these are special buds. Cells develop within these buds which eventually open, releasing free-swimming medusae to join the plankton of the sea. The medusae, some of which are male and some female produce eggs and spermatozoa, by whose union, ciliated larvae are formed. The larvae drift to the sea bed and fix themselves to rocks to begin new Hydroid colonies.

Siphonophore colonies are formed by budding. The original polyp develops a float which produces budding zones. From these zones the various colony members arise. Sometimes medusae are formed but they release the germ cells whilst still attached, and never become free.

◁ Sea Anemones are usually sedentary, fastening themselves to fixed objects such as shells and stones. Here three or four are attached to a Whelk shell, inhabited by a Hermit Crab. Sea Anemones make excellent protective camouflage for the crab, at the same time sharing the crab's food, a relationship called 'symbiosis'.

△ This hydroid colony is a species of Clava, fixed to a piece of seaweed. Each polyp has its tentacles spread to catch prey—microscopic planktonic organisms.

The Hydra

The **Hydra** is a fresh-water Hydroid of the class *Hydrozoa*. It takes the form of a solitary polyp which is temporarily attached to water plants.

The body is a simple, elongated sac with mouth and tentacles at one end and the colour of the body varies from green (*Chlorohydra viridissima*) to grey-brown *(Hydra fusca)* or red-brown *(Hydragrisea)*. The Hydra progresses by looping, attaching itself first by the base and then by the tentacles.

Hydra reproduces in two ways. If the surrounding conditions are favourable, buds form on the main body of the parent Hydra. The buds ripen and a mouth and tentacles develop. The mouth leads directly to the cavity of the main body until the tentacles of the new, small Hydra begin to gather food. At this point the Hydra separates from the parent Hydra. Up to eight buds can be produced at one time.

Hydras, which are hermaphrodite, also reproduce sexually. Male sperms are expelled into the water from thickenings which form at the mouth end of the animal. Eggs are formed in thickened areas near the base and, when ripe, they are fertilised by the sperm from the surrounding water. The eggs develop into a two-layered ball of cells, with a hard outer covering which protects them from cold or from dehydration. The cysts are released into the water and are carried away by currents. After some weeks, the development starts again inside each cyst. The protective covering breaks open and a small hydra emerges.

Portuguese Man-of-War

Siphonophores can be divided into two sub-groups. There are the *Calycophora*, with swimming bells, beneath which the main body with its tentacular, digestive, and reproductive organs is suspended; and *Physophorida*, which have either a gas-filled float or a combination of float and swimming bells.

Physalia, the **Portuguese Man-of-War,** is probably the best known, or perhaps the most notorious, of these creatures. It prefers warm tropical seas, but sometimes finds its way into the Gulf Stream and British waters. It is a surface-dwelling Siphonophore, although reputedly capable of reducing the buoyancy of its large, blue-crested float so that the entire animal submerges.

Beneath the float are suspended a great variety of polyps, some with branches dangling below the main body like a multiple snare. All of these tentacles are heavily charged with stinging cells and may extend 6 ft (2 m) below the float.

Although the sting of this animal is very painful, it is rarely fatal to humans.

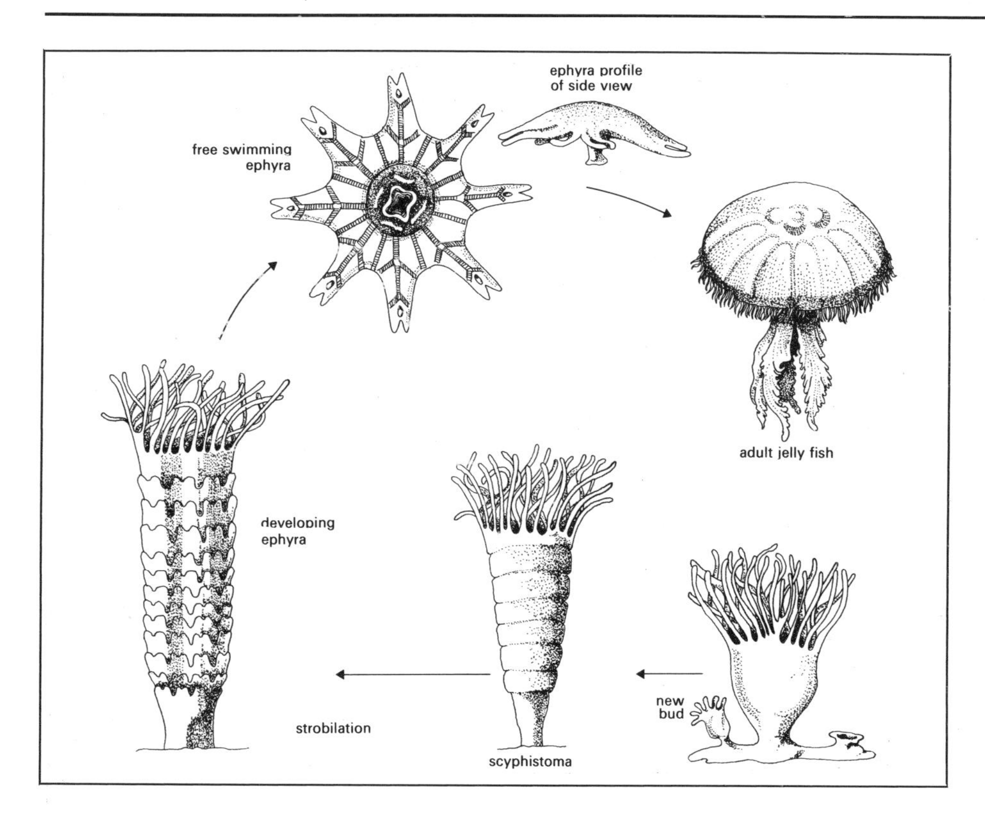

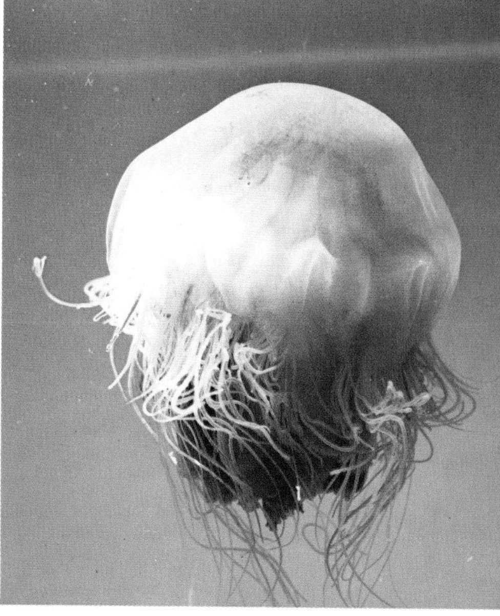

◁ The Jellyfish life cycle diagram gives the stages of development from when the fertilised egg has become a sedentary hydroid. The budding of many separate ephyrae or miniature Jellyfishes from one strobiloid scyphistoma is a multiplication phase, increasing the numbers many times for each successful fertilisation.

▽ The crown of stinging tentacles surround the gullet of a Sea Anemone into which prey is fed. The microscopic stinging cells or cnidoblasts by which the prey is captured are among the most complex of all cells and are unique to the coelenterates.

△ Lion's Mane Jellyfish: the largest jellyfish, which consists of only two layers of cells including the stinging cells hanging below.

Jellyfish

The class *Scyphozoa* includes more than two hundred species of true **Jellyfish.** All are free-swimming except one, the **Stalked Jellyfish,** *Haliclystus*, which has a stem, branching out into tentacled heads, joined by tissue.

Although Jellyfish are often found washed up on beaches and rocks they are really dwellers in the open sea. Some, the **'Crown Jellyfish'**, are deep-sea forms, found well below 3,000 ft, whilst others swim almost on the surface.

Jellyfish range in size from a few millimetres in diameter to the largest, *Cyanea capillata*, which, in Arctic waters, can be over 6 ft (2 m) across.

Jellyfish swim by pulsations of the swimming bell and one group, the *Cubomedusae*, can reach speeds of two knots. This sometimes deadly group usually live in tropical seas, but even in British waters, two species *Cyanea capillata* and its smaller cousin, *Cyanea lamarki*, can raise painful weals on the skin.

One species, *Chrysaora hyoscella*, is used by young Whiting as protection. Although the tentacles are venomous, the fish shelter beneath the animal, out of reach of other predators.

One Mouthed, Many Mouthed

Most **Jellyfish** are predators. They feed as they move through the water, spreading their tentacles over the widest possible area. Tiny fish and planktonic animals are stunned and trapped by the stinging cells and passed along to the mouth by the mobile tentacles.

The **Disc-Jellyfish,** which includes *Cyanea* and *Aurelia,* and the **Moon-Jelly,** have a single mouth, the corners of which form a frill or fringe of tentacles. *Cyanea* is one of those which captures its food by using stinging cells, but *Aurelia* secretes a fine mucus on the top of its swimming bell and food adheres to this. When the plankton is caught it is transferred to the edge of the bell by ciliated cells and then transferred to the mouth by tentacles.

Other groups of *Scyphozoa*, including *Rhizostoma* and *Cassiopea* are known as many-mouthed Jellyfish. In these creatures the oral tentacles are fused and pierced by fine canals, each leading to a tiny mouth. These Jellyfish live on microscopic planktonic organisms which are sucked up through the tubes into the mouths.

Sea Anemones

There are over one thousand species of **Sea Anemones,** found in water ranging from inter-tidal levels to the deepest parts of the ocean. Small *Anthozoans* have been found on rocks dredged up from the Philippine Trench, a depth of 35,000 ft.

These animals reproduce in a variety of ways. Some produce eggs which develop into polyps; others split bodily or separate off parts of the base which grow into new polyps.

The Sea Anemones common to British waters are solitary animals although one species, *Corynactis viridis*, has the separate creatures growing so close together that they can be mistaken for a colony.

Most Sea Anemones use a basal disc to anchor themselves to a rock or other hard surface, but some live commensally with a Hermit Crab, attaching themselves to the crab's shell. The crab receives protection from the stinging cells of the Sea Anemone which, in turn obtains food from the host. When the crab moves to a larger home he takes the Sea Anemone with him.

The Soft Corals

In southwest British waters, during low spring tides, a species of coral with the rather unpleasant name of **'Dead Man's Fingers'** is found.

This is *Alcyonium digitatum*, a soft coral, so called because it does not have a horny or limestone skeleton. Instead it has an external skeleton made of a tough jelly-like material, secreted by special cells, and having within it countless tiny calcareous spicules.

Through the mass of the skeleton run canals, linking a great number of polyps. These expand over the entire surface of the colony when it is immersed, causing it to blossom into a complex of white or orange coloured tentacles.

It is quite widespread and is present in great quantity in the inter-tidal waters of Indian and Pacific oceanic coral reefs, where it covers large areas in clumps, rather like fleshy seaweed.

Horny Corals

One large group of 'flower-animals' is the *Octocorallia*, in which the polyps usually have eight tentacles. Of this genus the most diverse belong to the order *Gorgonaria*, the **Horny Corals,** of which there are approximately 1,200 species found in marine waters.

These animals are of attractive appearance, supported on an inner core made of calcarous spicules embedded in a network of horny material. They are almost plant-like in their growth, coloured in vivid yellows, reds, oranges and purples, and with some having a fine display of luminescence.

One kind of Horny Coral, the **Sea-Fan,** is found in the English Channel, although it flourishes more strongly in warmer waters.

The Sea-Fan has an axial skeleton which grows from a securely mounted base. The polyps cover this branching skeleton, each being borne in a tiny pit. The entire colony forms a fan-like shape commonly presented to the prevailing currents so that there is the maximum area available for catching food.

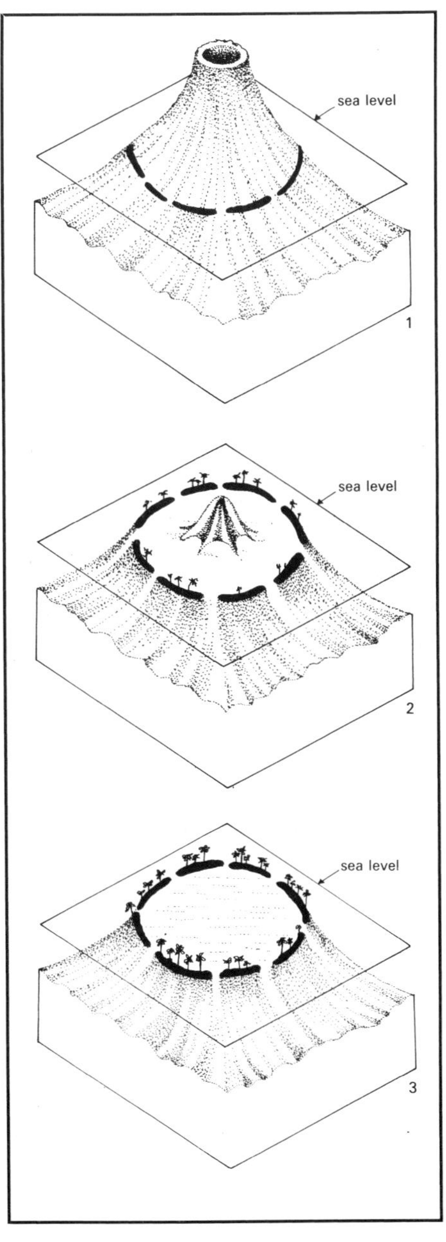

△ Coral atolls are formed by coral forming in the sea around an active volcano. Eruptions cause the land to sink, leaving the coral to grow.

▽ Structure of a single coral polyp. Coral reefs are formed from myriads of these budding continuously but remaining cemented together in a limy skeleton.

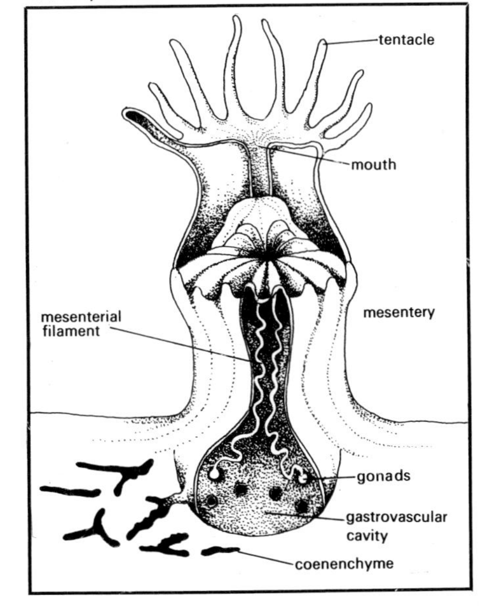

The Formation of a Coral Reef

The **Stony Corals, Madreporarians,** form enormous colonies. There are more than 1,000 species, living in marine water of up to 9,000 ft deep.

They become more prolific and diverse, however, in clear warm seas at depths between 15 and 150 ft.

The stony skeleton of the Coral is secreted by the polyps and is almost pure calcium carbonate in the form of crystallised aragonite. It becomes coloured yellowish brown or green by *zooxanthellae*, organisms which live in symbiosis with the Corals. These organisms receive carbon dioxide and other nutrients from the Coral. The algae can photosynthesise and so make food. It is believed by many authorities that the zooxanthellae are essential to the growth of a coral reef.

Each day's growth is clearly marked by rings in the coral, and, due to seasonal changes, so is the annual enlargement. The rings of Coral known to have lived 300 million years ago, in the Devonian Age, indicate that then, there were nearly four hundred days in the seasonal year.

'Crown of Thorns' versus Coral

Anthozoans are carnivorous animals, armed with stinging cells. Even so, they have their enemies and are food for other sea dwellers. **File Fish** and **Butterfly Fish** browse on the Coral but are relatively harmless. *Acanthaster planci*, the **'Crown of Thorns' Starfish** is, however, a very different foe.

At present, this animal appears to be experiencing a population explosion, has already destroyed a quarter of the Great Barrier Reef, and has spread to the Indo-Pacific atolls.

A single Starfish destroys about twenty square inches of Coral in one day, and there are areas of reef where the entire surface is one interlocking mass of Starfishes.

Control experiments have been made using **Triton** molluscs, the only predator known to eat the 'Crown of Thorns', but their rate of feeding is too slow to have any real impact.

It seems probable that the only real chance of control is to discover the form and habitat of the larval stage of the Starfish and mount an attack at that point.

It may be that the onslaught of the 'Crown of Thorns' is not as sensational as it appears. Some authorities believe that this is only one in a continuous series of attacks, but that others in the past have gone unnoticed because scientific interest was not aroused.

The Worms

Man loftily refers to the worm as a lowly creature because of its simplicity, forgetting that worms have adapted so well that they have no need to be any more biologically advanced than they are.

The Worms include the Platyhelminthes, or flatworms, which have flattened and unsegmented bodies. Amongst them are classified the planarians, free-living, mostly acquatic and less than an inch in length; the flukes, which are parasitic; and the tapeworms, which are also parasitic (Man is numbered amongst their hosts) and can reach a length of 40 feet.

The Nematoda, the round worms, include the Ascarids, Hookworms and Eelworms.

The Nemertea, the ribbonworms, are usually found in shallow seas or coastal waters and some species can be 30 feet in length.

The Annelids, advanced enough to have brains, include the marine bristleworms, the earthworms, some of which can grow to over 10 feet in length, and the parasitic leech.

Flatworms, Flukes and Tapeworms

(Platyhelminthes)

Flatworms include free-living **Flatworms** *(Turbellaria)*, parasitic **Flukes** *(Trematoda)*, and **Tapeworms** *(Cestoda)*. Free-living Flatworms live under stones and rocks, among seaweed and in freshwater ponds and lakes. Some species can live on land but are confined to very humid areas, hiding beneath logs and leaf mould during the day, emerging only at night to feed. These are the giants of the group, some reaching over 2 ft (60 cm) in length.

Many Flatworms are parasitic and are known as Flukes *(Trematoda)*. They usually measure no longer than a few centimetres. Many Flatworms infest the gut of vertebrates—birds, fishes and mammals, including man. The **Liver Fluke** has a typical life cycle, the adults developing in a host, such as man, where eggs are passed out in faeces. These eaten by a second host, such as a Snail, hatch into larvae which eventually infect the first host (man), again by eating its way back through his skin and into the bloodstream where it is carried to the liver.

Tapeworms vary greatly in size, that of Man reaching 20 yards in length. They live in the intestines amidst digested food. Most of them are ribbon-like.

Segmented Worms

(Annelida)

True Worms have segmented bodies and they include the familiar **Earthworms** and **Leeches.** The Annelids attain the largest size of any worm-like invertebrate; the great **Australian Earthworm,** for instance, reaches a length of over 9 ft (3 m).

There are three classes of worms: the **Polychaete Worms** *(Polychaeta)*, the **Freshwater** and **Land Worms** *(Oligochaeta)*, and the **Leeches** *(Hirudinea)*. Many Polychaetes are strikingly beautiful and are coloured red, pink or green and some are iridescent. Polychaetes have a well-developed head which bears eyes, antennae and a pair of feeding palps. The **Fanworms** burrow in sand but the head is covered with feathery antennae, which trap minute particles of food from the surrounding water.

The fresh water and land worm species contain the familiar **Earthworms.** Some fresh water species burrow in the bottom mud and silt; others live among submerged plants. All of these worms are covered with *setae* (bristles), which help in movement. Worms are hermaphrodite, possessing both male and female organs but a worm copulates with another to exchange sperm.

The **Leeches** consist of 300 species of marine, fresh water and terrestial worms. Although they are popularly considered bloodsuckers, a large number of them are not parasitic. The Leeches are the most specialised Annelids, the parasitic species having adapted to suck blood from their hosts. Most Leeches will suck blood from a wide variety of hosts; for example, mammals, including man, are the preferred host of *Hirudo* but this leech will also attack amphibians, Snakes and Turtles. Bloodsuckers feed infrequently but, when they do, they can consume an enormous quantity of blood. One species, *Haemadipsa*, will ingest ten times its own weight.

Roundworms and Ribbonworms

(Nematoda; Nemertea)

The **Roundworm** group contains some of the most widespread and numerous species of all multicelled animals. Many are found in the sea, in fresh water, and in the soil. They occur from the polar regions to the tropics and in all types of environments, including deserts, hot springs, high mountains and in deep marine water.

Sometimes, the species occurs in very large numbers; one square metre of seabed mud off the Dutch coast for instance, contained 4,420,000 Roundworms. Some Roundworms are parasitic, attacking virtually all groups of animals

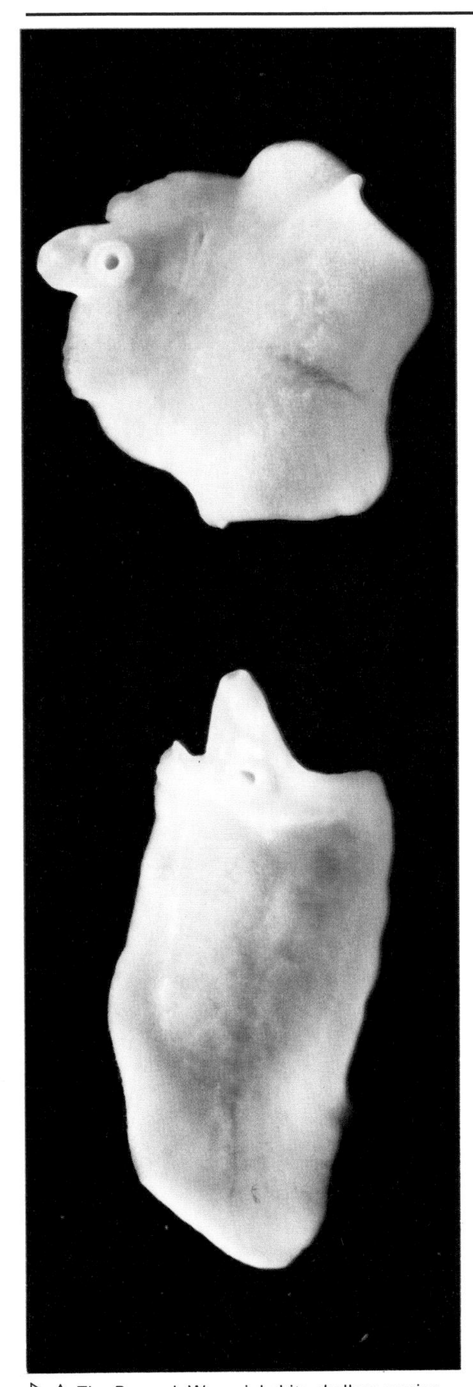

▷ △ The Peacock Worm inhabits shallow marine water and lives in a tube built in the sand. The worm sticks its 'head' out of the tube and collects small organisms on the feathery tentacles which are brightly coloured and when they are extended, a group of the worms can look rather like a cluster of flowers, quickly disappearing back into the tubes if danger threatens.
◁ △ Most Lugworms live in sand or muddy sand and their burrows can be recognised by coiled castings.
△ Liver Fluke: at one stage, the Fluke is in an invertebrate host. The adult is 1½ in long and ½ in wide.

and plants. Many infest foodcrops and domesticated animals and cause great losses to farmers.

The **Ribbonworms** appear to be an offshoot from the free-living Flatworms but are more highly organised. The group has approximately 570 species, most of which are marine and seabed dwellers. Many of the species are difficult to see with the naked eye as they are pale in colour but a few are brightly coloured, with patterns of yellow, red, orange and green.

A Parasitic Way of Life

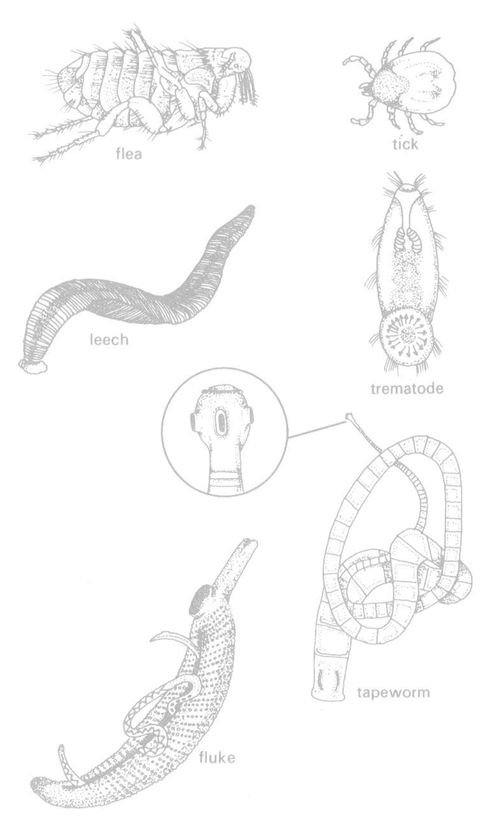

Parasites have evolved various body modifications; ectoparasites, which live on the outside of the host have hooks or suckers. Endoparasites have no gut and breathe anaerobically.

When an animal has evolved a parasitic way of life it usually means that there are modifications in its body structure to adapt it to its parasitic environment. Those parasites which live on the outside of a host *(Ectoparasites)* are usually less modified than those which live inside the host's body *(Endoparasites)*.

Ectoparasites are invariably provided with organs which enable them to cling fast to their host. *Polystoma*, a freshwater **Fluke,** has a number of suckers and hooks so that it can attach itself to the gill of a tadpole.

Some endoparasites which live in the intestine are constantly in danger of being dislodged, so they have clinging mouth parts. **Tapeworms** have hooks on their heads. Large tapeworms avoid being swept from a host's body by muscular movements.

When the endoparasite is living inside the host, it is surrounded by a rich food supply so that a gut is not necessary to the creature.

There is no oxygen supply inside an intestine so the parasitic animal respires anaerobically. This is the release of energy from food material by a process of chemical breakdown which does not require oxygen. Blood and tissue parasites, on the other hand, receive a plentiful supply of oxygen.

A parasite must ensure that its eggs reach another host so that the species will survive. As a very high percentage of eggs never survive to reach a new host, the parasites produce eggs in very large quantities.

Although various kinds of Molluscs such as Snails, Mussels and Octopuses appear so different, all have the same basic structure. Each has a muscular but soft 'foot', locomotory in some; a mantle, a sheet of tissue enveloping most of the internal parts; and in the general case, an external, protective shell secreted by the mantle. In some the shell may be rudimentary or absent.

Water-adapted molluscs possess feathery 'gills' or ctenidia. Land-adapted forms breathe air. Within the mouth, most kinds possess a spiny 'tongue' for rasping food.

An outstanding feature of the group is their diversity to suit widely different environments. There are six classes, each with specialized adaptations. In the primitive *Amphineura*, the marine Chitons, the body is comparatively simple and protected by an articulated shell. The *Scaphopoda*, a small marine group of Tusk Shells, are worm-like and each is enclosed in an open-ended circular shell. The *Pelecypoda* or bivalves such as Oysters and Clams, has representatives in both fresh and sea water. The *Gastropoda*, Snails and Slugs, includes members well adapted to life on land, with others at home in water. The marine *Cephalopoda*, Squids, Cuttlefishes and Octopuses, are different from all the others in the special form of the tentacled foot. The *Monoplacophora*, a very primitive class and now represented by a single species, *Neopilina galatheae*, was recently found in deep water in the Pacific Ocean. Its only known relatives are fossils up to 500 million years old. The body, although looking like a Limpet, shows segmentation similar to that of the segmented Worms. The trochophore larval stages of some molluscs resemble larvae of the Annelids.

Molluscs—on Land, in Marine and Fresh Water

◁ Sea Slug moving over a hydrozoan coelenterate colony. Despite its name, the animal is most colourful and attractive when extended. It lacks a shell and has no true gills.
▷ These snails are in a state of complete rest. In many warm countries, at the onset of the dry season, they climb some support and seal themselves against the drying atmosphere. They regain normal activity when rain falls once more.
▷ Part of the margin of a Scallop, valves apart. Near the upper rim are tiny pallial eyes or ocelli. The margin of the mantle is seen to be fringed. Further in are the broad plates of the gills, enlarged for collecting food.

Snails and Slugs

(Gastropoda)

These are perhaps the most familiar Molluscs. Their class name gives a clue to the body form since they are 'gut-footed' in the sense that the body, markedly lop-sided, has its food canal pushed out into a visceral hump on the upper side and ending near the mouth on an otherwise symmetrical, ventral foot. The hump is often spirally twisted and encased in a one-piece shell within which the whole body can be withdrawn and sometimes sealed. Some Slugs have no shell visible at all, others may have vestiges. The head bears the mouth, tentacles and eyes, and in the mouth is a rasping radula which scrapes particles

from vegetable food material. The mantle forms a dorsal covering sheet enclosing the mantle cavity. In water forms this is water-filled and houses, usually, a single feathery gill. In land Gastropods the cavity is air-filled and forms a respiratory surface by possessing many small blood vessels in its lining.

Many Gastropods are unisexual, such as the Whelk, but some like the land Snails and sea Slugs are hermaphrodite.

Sea Butterflies

(Opisthobranchia)

The **Sea Butterflies** are more or less symmetrical bilaterally, that is, have similar right and left sides. Along each side of the foot is developed a lateral, fin-like lobe, that when flapped produces a swimming movement—hence their name. The genus, *Clione*, includes delicate pink and blue creatures sometimes occurring in large numbers. They may form part of the food of Whales.

Limpets and Whelks

(Prosobranchia)

The **Common Limpet** *(Patella vulgaris)* is a very common shore animal. It moves slowly by creeping when the tide covers it and returns to the same spot where it clings tightly by its foot before the tide ebbs. When immersed it breathes by a ring of delicate gills on the edge of its mantle but when exposed it retires under its simple conical shell, completely protected.

The **Common Whelk** *(Buccinum undatum)* is a snail-like gastropod and carries a shell-closing horny sheet. It is carnivorous. Its mantle at the respiratory aperture is prolonged into a tube-like growth. Its powerful radula is capable of rasping through the covering of a shellfish. Empty Whelk eggcases, are found on beaches as groups of hundreds of whitish, horny capsules.

The **Abalone, Ormer** or **Ear-Shell** *(Haliotis)* is a relative of the Limpet but not found in the inter-tidal zone. Its very beautiful shell is a useful source of mother-of-pearl.

Common Snails

(Pulmonata)

The name Snail is applied to two groups of prosobranch gastropods, one aquatic and the other terrestrial. **Water Snails** of some kinds, such as *Paludina* species, the fresh water **Winkles**, are equipped with gills and thus breathe under water. Others, species of *Limnaea*, the common **Pond Snails**, possess air-filled mantle cavities and must surface to breathe. **Land Snails**, such as *Helix* species, are all air breathers, as are their close relatives, the **Slugs**.

The common **Pond Snail** *(Limnaea stagnalis)*, very frequent in fresh water, is recognisable by its two triangular, non-retractile tentacles, each with an eye at the inner front side. It has no operculum like the fresh water **Winkles.**

The commonest land Snail is the **Garden Snail**, *(Helix aspersa)*, with a shell stronger than that of its aquatic relatives. It lacks an operculum, but when hibernating, seals the opening with dried mucus. The head bears two pairs of retractile tentacles, the longer, upper pair each having an eye at its tip. Slime manufactured by the foot facilitates movement over dry surfaces, hence snail trails. The Garden Snail is hermaphrodite but pairs cross-fertilise and each lays eggs in the soil. *Helix pomatia* is the larger **Roman Snail**, for some a great table delicacy. The largest of all land Snails are the **Achatinas** of tropical Africa.

The Clam Family

(Pelecypoda)

As the name indicates, these are the 'hatchet-footed' Molluscs. The mantle is two lobed and secretes a shell in two hinged halves, either equal as in **Mussels** or unequal as in **Oysters**. From the two-sided shell, the group is sometimes called the Bivalves and because of the gill structure they are also known as the *Lamellibranchiata*. They are all aquatic with both marine and fresh water forms. Most are generally inactive slow movers and spend much time partially buried in mud or sand, often with the valves closed. The **Sea Mussels**, *(Mytilus)*, common sea shore inhabitants, often attach themselves to surfaces by means of strong threads, forming byssus attachment organs, and so become sedentary. In **Pelecypods**, the mantle becomes the respiratory surface while the gills are food filters, collecting particles as water washes over them from a special tube or siphon. Among the bivalves are the fresh water **Swan Mussels** *(Anodonta)* and **Orb Shells**, *(Sphaerium)*. Marine forms include **Scallops, Cockles, Razor Shells, File Shells** and **Clams**.

The Oyster

(Ostrea)

The edible **Oyster**, *Ostrea edulis*, when adult, is fixed to the sea bottom by its larger left-hand valve. The foot is rudimentary and the animal feeds on tiny pieces of organic matter delivered on water currents. The animal is hermaphrodite but one Oyster is never male and female at the same time. It is first male, liberating sperm and later female, becoming 'white-sick' with eggs and 'blue-sick' as the fertilised eggs develop into larvae, eventually to be set free by the million. Most of the larvae perish, the survivors settling as 'spat'. The American species, *Ostrea virginica*, is unisexual.

Oyster shells are composed of layers, conchiolin outside, then crystalline calcium carbonate like arragonite, and next to the mantle, nacre. If a foreign body, particularly a fish gut parasite, gets between the mantle and the nacre, there is secreted more nacre to surround and isolate it. Thus is born a pearl. The pearl-forming process can be initiated by man artificially and results in what is known as a 'cultured' pearl. Species of *Meleagrina*, **Pearl Oysters**, are found in Australian waters. Oyster shells also yield valuable mother-of-pearl.

Burrowing Clams

This description is applied to those bivalves that burrow for long periods into the substratum, exposing only their comparatively long respiratory tubes. Some will surface at intervals and move freely.

Among the more interesting is the **Razor Shell**, *Solen ensis*, an elongated form that burrows vertically and when immersed by the tide, exposes the two siphons at the sand surface, appearing like a keyhole. If disturbed, it ejects water from the siphon and contracts deeply into its burrow under the power of its muscular foot.

Another quite remarkable burrower is the rock boring **Piddock**, *Pholas dactylus*, that cuts into soft rock by means of horny teeth on its shell forming a round cavity into which it passes. With increase in size, the deeper cuts become greater in diameter than those at the surface. The Piddock thus becomes a prisoner in a cell of its own making.

The Cephalopods

(Cephalopoda)

These are sometimes called the 'head-footed' molluscs, the foot being modified into a ring of sucker-bearing tentacles around the head. There are eight in the **Octopus** and **Argonaut**, ten in **Squids, Cuttlefishes** and *Spirula*, and many in the **Pearly Nautilus.** All are marine. Other features are the unprotected body exposure of some and the fusion of the mantle with the body, except ventrally where a cavity houses the gills and allows body wastes to escape. Leading out from this cavity is a stout tubular aperture or funnel. In Cephalopods, except the Nautilus and some Octopuses, water discharged from the funnel can carry black dye (sepia) produced in a gland of the gut which opens near it. Members of the group can creep on the sea bottom or swim actively by expelling water very forcibly from the funnel. This jet propulsion results in the animal moving head hindmost. There are two main sub-groups, *Belemnoidea* (dibranchiates or two-gilled), including the Octopods and Decapods and the *Nautiloidea* (tetra-branchiates or four-gilled) including only the Pearly Nautilus.

Nautilus

(Nautiloidea)

The **Pearly Nautilus** *(Nautilus pompilius)* is the only living representative of a unique form of cephalopod, recorded in fossils for over 500 million years. It has an external shell and many small, retractile, suckerless tentacles. There is no ink-sac. The shell, pearly and beautifully coloured, is spirally formed and consists of a series of chambers separated by curved cross plates. The last formed chamber houses the animal and is largest. Another is added as the previous is outgrown. The chambers communicate by small holes in the septa and a body extension, the siphuncle, passes through these around the spiral coil. It is possible that gas produced by this varies the bouyancy so that the animal can rise or fall in the water. The head, when extruded, possesses two large but simple eyes, laterally placed. Dorsally it is covered by a fleshy hood, mottled and warty. This seals in the retracted body. The Nautilus is a bottom feeder on Crustaceans, usually in deep water, but can come to the surface.

Cuttlefish

(Belemnoidea: Decapoda)

These are ten-armed forms; the suckers are stalked, with horny rims. Two central tentacles are longer than the others with suckers at the tips only. They are used in catching prey and can be retracted from their bases. There is an internal calcareous shell, the 'cuttlebone'.

The common **Cuttlefish** *(Sepia officinalis)* lurks on the bottom waiting for prey that when caught is transferred to the shorter tentacles and thence to the beak-like jaws. When swimming, a Cuttlefish uses either the beating of its lateral flesh folds or 'fins' to move slowly, or a jet propulsion mechanism from the funnel to dart rapidly backwards. To escape enemies, it discharges ink. The head bears a pair of complex eyes, comparable with vertebrates. The male forms sperm enclosed in tubular spermatophores and these are passed to the female by one of the long arms, said to be hectocotylised. After fertilisation, the female lays groups of blackish eggs resembling bunches of grapes. Cuttlefish species range in size from a few centimetres to over 6 ft (2 m) long.

Squids

(Belemnoidea: Decapoda)

These are also ten-armed, similar to Cuttlefishes in general form and habits. The internal shell is horny, not calcareous, and known as the 'pen'. **Squids** are generally more active swimmers and smaller species may move in groups, darting and swerving as they go. Some are very large and *Architeuthis princeps* is reputed to be the largest of all invertebrate animals, more than 48 ft (16 m) long. The more common small Squids of British waters are species of *Sepiola* and *Loligo*.

The eggs of these Squids are produced in narrow cases, each containing as many as 200 eggs. They may be up to 4 in (10 cm) long, gelatinous and sticking together in mop-like bunches.

A Squid relative, *Spirula,* is unique in possessing a spirally coiled shell, separated into chambers as in *Nautilus*, but almost completely hidden by folds of the mantle.

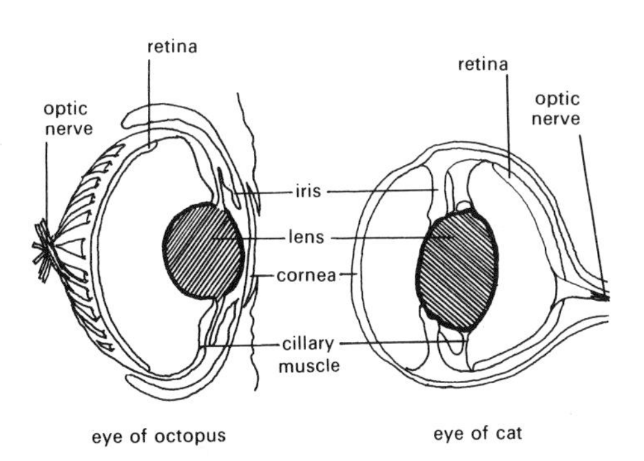

◁ Nautilus: section through the shell of a Pearly Nautilus, clearly showing the successively enlarged chambers which house the growing animal.
△ Diagram comparing the eyes of the invertebrate Octopus and a vertebrate Cat. Although the eyes are formed differently and the animals are so dissimilar in all other structural features, the eyes are remarkably alike.
▷ Common Octopus: The Octopus inhabits holes or crevices where it awaits prey, then darting out for it. The animal has camouflage colouration, its body colour blending with surrounding rocks.

The Octopus

(Belemnoidea: Octopoda)

Octopuses are shell-less eight-armed cephalopods, the arms of equal length and with unstalked suckers. Otherwise, the body is basically similar to its other class members. Octopuses are less active swimmers than the Decapods but can move rapidly by the same jet propulsion mechanism.

Species of Octopus and allies such as *Eledone* and *Cirroteuthis* vary in size from a few centimetres across the tentacles to as much as 18 ft (6 m) in *Octopoda apollyon* of western America. All are capable of rapid colour changes in the skin pigment cells. Some are reputed to subdue prey by venom secretion and certainly in one case the poison is fatal to man.

In some Octopuses, during mating, the male inserts a tentacle into the female's mantle cavity to deposit his spermatophores. It may then be broken off to remain there. This event gave rise to the term 'hectocotylus' since the detached arm was once thought to be a separate animal parasite of the Octopus and called *Hectocotylus.*

The mother Octopus lays many hundreds of eggs in long strings which she fixes to the inside of her den. She shows maternal care by watching over them and aerating them with jets of water. From each hatches a tiny Octopus but few survive to maturity.

The **Argonaut** or **Paper Nautilus** *(Argonauta argo)* is a relative of the Octopus. The female forms a beautiful external shell up to 6 in (15 cm) across but it is not comparable with the shells of other molluscs. It is formed only to cradle the eggs. The male Argonaut is much like an Octopus in shape but only a centimetre or so long.

A good deal of research has been carried out on the nature of nervous activity in Octopuses and their behaviour has also been studied in some detail. The Octopus is often credited with some degree of 'intelligence' but this is clearly not comparable with that of human beings.

Arthropods–The Successful Millions

In their variety of species and in actual numbers, there are more Arthropods in the animal kingdom than all the other animals together. Although there are still many species yet to be discovered, there are over three-quarters of a million described Arthropods.

The name Arthropod means 'jointed limbs' and the most characteristic feature of Arthropods is that their limbs and bodies are made up of segments. In the worms, which also have segmented bodies, the body wall is soft and flexible, but in the Arthropods the outer skin contains a substance called chitin and is hard and waterproof. In some species, Lobsters for instance, the outer skin also has lime salts in it and this makes the covering even harder. This rigid 'skin' forms the skeleton of the Arthropod, (exoskeleton). This is exactly opposite to the vertebrates where the skeleton is inside the body and the soft tissue on the outside. All Arthropods shed their hard skins periodically to allow them to grow and develop. The skin also restricts water loss from the body which explains why these animals are among the most successful colonisers of land.

The limbs, which are attached to the body segments in pairs, also have the chitinous covering except for the areas where the limbs actually join the body and between the various joints. Here, the skin is thinner and more flexible and allows movement to occur. The limbs are extremely variable in both structure and function: those on the head are often specialised into sensory organs or antennae. The pairs behind the antennae often have the function of grasping and handling food and are usually referred to as the 'jaws', while the trunk limbs are used for walking or swimming, or wings for flying.

Small aquatic Arthropods, such as the Water Fleas and Shrimps, may breathe through their body wall, while others develop special gills. The land forms, such as some Spiders have lung books which are really modified gills—or they may have a system of internal tubules (tracheae) which have openings to the outside of the animal, as in the insects.

Arthropod adult males and females often differ in structure and mating usually results in the production of yolky eggs.

Arthropods are generally divided into five distinct classes. *Onychophora*, a very small primitive group including Peripatus: *Crustacea*, which are mainly aquatic forms and include the small Water Fleas, Barnacles, Woodlice, Shrimps and Prawns, Lobsters and Crabs; *Myriapoda*, including Millipedes and Centipedes and *Arachnida*, including the Spiders, Scorpions, Harvestmen, Mites and Ticks, the Sea Spiders and King Crabs. The largest group, which includes all the insects, is *Insecta*.

▽ Goose Barnacle: Barnacles are marine arthropods and are very different from other crustaceans. They are completely sedentary, being attached to the sea bed by a long stalk and are found in deep waters. The cirri or feeding limbs sweep the water for planktonic food, which are then transferred to the mouth. Goose Barnacles are sometimes found washed up on the shore.

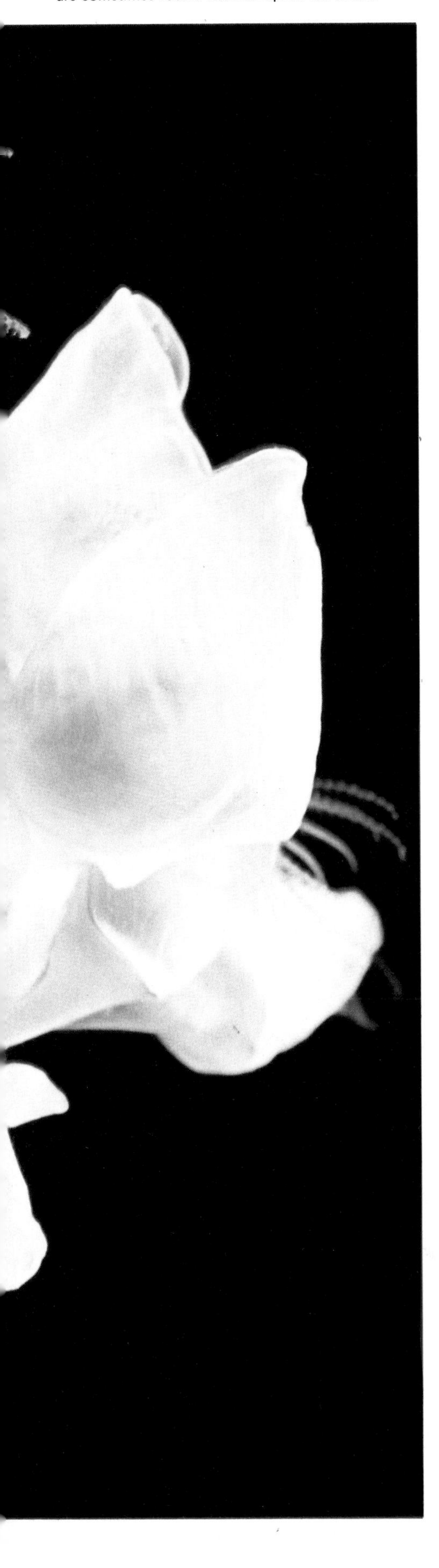

Peripatus

(Onychophora)

Peripatus is sometimes considered as a 'connecting link' between the annelid worms and the arthropods, and has some features common to both of these groups. It is usually recognised as a very primitive arthropod and may well be a descendent of animals that originated near the time when the present annelid worms and the arthropods separated to follow their individual lines of evolution.

Species of Peripatus are rare and found only in the tropical forests of the world. They are always confined to moist situations and by day hide under rotten bark or in crevices in rock. They are active only at night.

Peripatus is small, 2 to 3 in (5 to 8 cm), soft bodied, has a velvety appearance and is without obvious segmentation. It is commonly brightly coloured —blue, green, orange or even black forms are frequent. There are many pairs of small unjointed legs, each leg ending in a pair of claws for gripping. Peripatus feeds at night on small insects and other animals. It explores by means of its sensory antennae and the prey is caught by slime which is ejected from small structures on the head. This slime hardens to form a rubbery solid in which the prey is entangled. These glands are also thought to have a protective role. The mouth is surrounded by a circular lip and this adheres to the prey whilst two pairs of jaws tear the food.

Peripatus breathes by a very simple system of tracheae. The openings to these tubes, unlike other arthropods, cannot be closed and are thus the major sites for water loss. In addition small bladder-like structures on the legs may be everted and act as a gill when the air is moist.

An unusual type of mating occurs in some species of Peripatus in that the male injects the sperm into the skin of the female. The sperm then pass through the body of the female to the ovary where fertilisation occurs. Some species lay eggs in moist conditions and in others the young develop internally and a primitive type of placenta may be formed. The period of gestation is about twelve months and from 30 to 40 young are produced at one time. After birth there is no evidence of parental care.

Crustaceans

These are primarily aquatic animals and most live in the sea, there are some freshwater forms and a few are terrestrial **(Woodlice)**. Generally the crustaceans are not well adapted to life on land; it is probable that this habit may still be evolving.

There are about 26,000 known species of crustacea and these may be conveniently divided into two groups *(Entomostraca* and *Malocostraca)*. The *Entomostraca* are generally small species such as **Barnacles, Fairy Shrimps** and the numerous microscopic forms including the **Water Fleas** and **Copepods**. Many of these minute organisms are important bases in food chains. The *Malocostraca* are larger forms such as Crabs, Lobsters, Shrimps and Woodlice.

Water Fleas and Small Crustacea

Many of these are microscopic animals that occur in vast numbers. Their bodies are often transparent. The **Fresh Water Flea** *(Daphnia)* has its body protected by a bivalve shell or carapace and, as its name suggests, it moves through the water by a series of hops or jumps. The antennae are the main swimming organs and these also create a current which brings food particles to the animal. **Copepods** *(Cyclops)* are often associated with water fleas.

The primitive **Fairy Shrimp** *(Chirocephalus)* often occurs in large numbers in small pools that dry up in summer. It swims upside down with graceful undulations of its body and limbs. **Brine Shrimps** *(Artemia)* live in salt pools and marshes but are now extinct in Britain.

Barnacles

Barnacles are exclusively marine and differ from other crustaceans in that they are completely sedentary for most of their lives. A few Barnacles live parasitically on other animals.

The arrangement of their bodies is different from other crustaceans. Each animal is enclosed in a fold of skin that produces a series of calcified plates which gives a hard protective covering to the body. This shell is only open when the animal is feeding. During this activity, delicate, feathery limbs are extended through the opening. These limbs rhythmically grasp the water and trap minute food particles which are suspended in the sea water. The limbs are then withdrawn into the shell and food is passed to the mouth. Each individual has both male and female reproductive organs although cross-fertilisation usually occurs.

There are two main types of Barnacle; the **Stalkless Acorn** or **Rock Barnacles** *(Balanus)* and the **Stalked Goose Barnacles** *(Lepas)*. The former can be found in vast numbers encrusting rock surfaces in the inter-tidal zone. Each Barnacle consists of a conical shell with its broad base cemented firmly to the substrate. Bàrnacles can tolerate exposure at low-tide but are only able to feed when submerged.

The Goose Barnacles are deep sea forms but can occasionally be found washed up on the shore. These are attached to the substrate by a stalk of variable length.

Barnacles are not always sedentary. In their young larval stages they can swim freely and disperse to new habitats, where they settle once more.

Woodlice or Slaters

(Isopoda)

The woodlice are the only large group of crustaceans that are truly terrestrial. Some members of this group *(Isopoda)* are aquatic; the **Hog Slater** *(Asellus)* is common in freshwater ponds, whilst the **Shore Slater** *(Ligia)* occurs just above the high-tide mark. *Limnoria* burrows into wood and causes extensive damage to piers and pilings.

The body of all woodlice is flattened. They feed mainly on decaying plant and animal matter. Some breathe by means of primitive tracheae which are developed on the abdominal limbs. These limbs may also serve to form a brood pouch for the developing young.

The terrestrial forms are always confined to moist situations beneath stones or bark, and develop various devices to conserve the water content of their bodies.

All woodlice are nocturnal and some, **(Pill Bugs)** are able to roll their bodies into a ball, which also serves as a protection against predators.

Fresh Water Shrimps and Sand Hoppers

(Amphipoda)

These are characterised by a side to side (lateral) compression of the body. The limbs are specialised and enable the animal to swim jerkily, to crawl on its side or to jump or hop. They breathe by gills which are continuously ventilated by a water current maintained by the posterior limbs. Most are scavengers and feed on detritus which is raked up by the antennae.

The fresh water **Shrimps** are commonly found in ponds and streams. The larger male often carries his mate with his front legs holding her in front of him, a habit adopted by their marine relatives as well. The **Sand Hoppers** or **Beach Fleas** *(Talitrus)* are common high up on the shore or in the strand line, but are confined to moist conditions since they still rely on gills for their respiration. They are nocturnal and found under moist weed.

Mantis Shrimps

The **Mantis Shrimps** *(Squilla)*, so named because of their superficial resemblance to a Praying Mantis insect, are a small group of mostly tropical species. They differ from other Shrimps in that the carapace is reduced and the abdomen is very long. Mantis Shrimps live in burrows and capture their prey by a rapid extension of a pair of large raptorial legs. These limbs are spiny and the end joint folds back on the inner part, like the blade of a pocket knife.

Shrimps and Prawns

These are swimming decapods. The body is laterally compressed with a long 'tail' or abdomen. The swimming legs are well developed and the end of the body is formed into a tail fan which is used as an organ of propulsion. When this is swept vigorously under the body a rapid backward movement is effected.

Shrimps and **Prawns** are encountered mostly in shoals along coastal waters and some are luminescent. Shrimps *(Crangon)* are smaller than Prawns *(Palaemon)* and only have a small beak or rostrum between the eyes.

Most Shrimps and Prawns capture their prey with the chelipeds although one deep sea-form, *(Sergestes)* has armed antennae which it lashes in the water to catch prey. The Pistol or Snapping Shrimps generally live in burrows and have one cheliped much enlarged; by a special mechanism this snaps shut and makes a loud snapping noise which frightens away potential predators and may also capture small prey.

Crab Courtship Behaviour and Life Cycle

Courtship in Crabs varies, but generally a male spends some time actually wooing his intended mate. In the **Fiddler Crab** the brightly coloured cheliped is waved and used to attract a mate, whilst other crabs make sounds by rubbing parts of their bodies together. Mating occurs by the under surfaces of the two crabs coming together and the sperm are passed to the receptive female in packets. After mating the sperm may be stored in the female's sperm storage organs until the eggs are laid. The eggs are fertilised immediately and attached to special limbs on the abdomen. This is normally tightly flexed against the rest of the body, but is lifted for brooding. The eggs are bright orange and the entire mass often resembles a sponge.

The eggs actually hatch at a fairly late stage in development and give rise to a small, free swimming zoea larva. By this means the primitive nauplius larva is by-passed. The zoea larva is recognised by a long spine on the carapace, and after a series of moults develops into a megalopa larva which resembles a minute lobster. This larva is eventually transformed into a small crab by the abdomen flexing under the body. Crab and other Crustacean larvae form a very high proportion of the marine plankton.

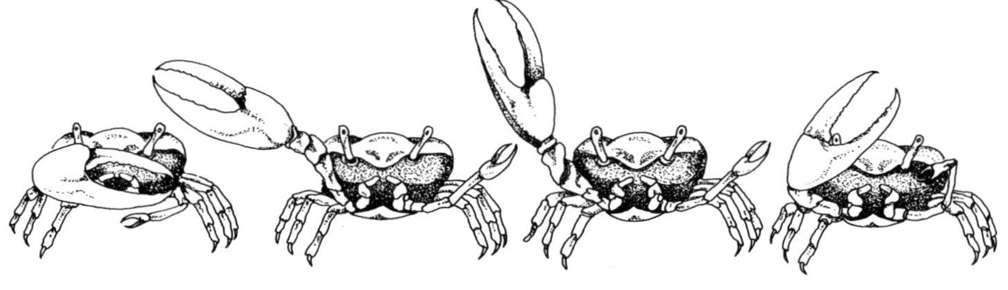

Shrimps, Lobsters and Crabs

(Decapoda)

These are the most highly specialised crustaceans, and form the largest group containing about one third of all known species. They are mostly marine, though some have invaded fresh water and a few Crabs are amphibious or even terrestrial. The name *Decapoda* is derived from the possession of five pairs of walking legs.

The front of the body is covered by a continuous calcified shield or carapace whilst the hinder part or 'tail' is composed of separate plates joined with a softer material allowing flexibility. The limbs are specialised to perform various functions. The front limbs handle the food; one pair forms large pincers or chelipeds which may be asymmetrical in shape and size. The chelipeds capture the prey. The more posterior limbs are locomotory and adapted for either walking or swimming. Some of the limbs have gills associated with them. In the males one pair of limbs is concerned with the transference of sperm to the eggs as the female deposits them.

△ Swimming Crab: not all species of crab can swim and many of those that can have developed a slight widening of the hind legs, as flat round paddles, which assist in swimming.

△ ◁ The 'attractive' waving technique of the Fiddler Crab's display.

◁ Fiddler Crabs have one claw larger than the other. This is usually brightly coloured and is used to attract the attention of females and in the defence of territory.

▷ Crayfish live in temperate fresh waters and generally feed on small animal life and plant food, although many species are scavengers. There is a considerable difference in size between the various species, the largest, found in Tasmania growing to 9 lb in weight while the smallest, a North American species is only 9 mm long when fully mature. The Common European Crayfish breeds at the end of November and the young crayfish are born 6 months later. After three to four years, Crayfish are mature and it is recorded that they can live for up to twenty years.

Lobsters and Crayfish

These closely resemble Shrimps and Prawns in appearance, although instead of swimming they generally move by walking or crawling. The **Crayfish** are fresh water forms and live in holes in river banks. Some Lobsters dig burrows or retreats. The female Lobster produces eggs only in alternate years. The eggs are fertilised externally and become attached to the small swimming limbs where they develop. The reddish eggs are easily seen and the female is said to be 'in berry'. It is an offence to take them in this state.

Crabs

These are distinguished from other decapods in that the carapace is enlarged and expanded laterally. The abdomen or 'tail' is permanently recurved under the rest of the body and cannot be seen from above.

Most Crabs move by a typical sideways walking **(Edible Crab, Common Shore Crab)**, while others can swim by having the last pair of legs flattened into hairy paddles. The **Ghost Crabs** *(Ocypode)*, are very fast runners and may live many miles from the sea, although they migrate back to the sea to breed.

Hermit Crabs *(Eupagurus)* are specialised to live in the empty shells of gastropod molluscs. Their body is completely reorganised; the abdomen is soft, asymmetrical and spirally coiled to fit within the spiral chamber of the shell. As the crab grows it becomes too large for its shell and looks for a larger home. Young **Coconut Crabs** *(Birgus)* live in shells, but as adults they cannot find any large enough and consequently redevelop protective plates on the abdomen.

Crabs are particularly vulnerable to attack by predators when they are moulting and usually seek protection in dark crevices or under stones. A few are expert burrowers. The **Fiddler Crabs** *(Uca)* live in burrows in the inter-tidal zone. From such burrows the males wave their single, enlarged and brightly coloured pincer to disturb intruders. **Spider Crabs** utilise other animals and sea weeds, which grow on their rough carapace as a means of protection. The **Sponge Crab** *(Dromia)* cuts pieces of sponge which it places on its carapace to give the amusing appearance of the crab wearing a hat. The ultimate stage in seeking protection is to live inside the body of another animal (commensalism). The **Pea Crabs** *(Pinnotheres)* live in the mantle cavities of Cockles and obtain both protection and food from the Cockle.

Most Crabs are carnivores capturing their prey with the large pincers. Some develop a mechanism whereby they can sift sand and strain off microscopic food particles.

△ Soldier Crabs: these are burrowing crabs and when the tide comes in they dig themselves air-tight and water-tight chambers, remaining there until the tides goes out again. They have stalked eyes.

△ ▷ Millipedes (top), in spite of their name, have only 200 legs. They have a segmented, worm-like body and on the head is a pair of antennae, jaws, and a group of simplified eyes.

▷ Centipedes are predators upon insects, worms and slugs and have a poisonous bite. They are found all over the world in temperate and tropical regions, and can be as small as a few millimetres long or as large as 20 cm in length.

King Crabs

There are only five living species of **King Crabs** *(Limulus)*, although the fossil record shows that they were once a much larger group. They are marine animals occurring in the shallow coastal waters of North East America and parts of the Pacific.

The body is divided into a horse-shoe - shaped cephalothorax, (these animals are also known as **Horse-Shoe Crabs**), and abdomen which terminates in a long moveable spine. The spine is used in burrowing and to restore the animal to its right position.

The Crabs feed mainly on worms and molluscs, which they catch with chelate appendages. The food is macerated by the spiny bases to the limbs before it is passed to the mouth.

During the breeding season the males and females congregate in the inter-tidal zone and the males climb onto the backs of the females. The eggs are laid in batches in the sand.

Centipedes and Millipedes

(Myriapods)

These are rather secretive terrestrial arthropods that are usually found in the soil or beneath stones or bark. They are widely distributed in both temperate and tropical regions.

The body is composed simply of a head and a many segmented trunk of similar leg-bearing segments. Although Centipedes and Millipedes are often grouped together in the Myriapoda, many of their features are quite distinct.

The Millipedes or thousand-leggers *(Diplopoda)* generally have cylindrical bodies, with each segment bearing two pairs of legs. They are not very agile animals and move by slowly burrowing through soil or rotten wood. Various protective devices have evolved in connection with their slow movement. Most Millipedes have a tough integument and some, the Pill Millipedes, can roll into a ball. A few Millipedes have glands which secrete a fluid distasteful to

potential predators. One specialised type of soft Millipedes *(Polyxenus)*, have small tufts of hairs like small pin cushions, these are easily detached and cause severe irritation. Millipedes feed on decaying vegetation.

Mating is a complex process in which the male twists his body around the female. The eggs are laid in nests which may be guarded by the female.

The Centipedes *(Chilopoda)* have flattened bodies with only one pair of legs on each segment. They are active creatures and most (*Lithobius*, for example) move by running. A tropical species *(Scutigera)* has very long legs and can run extremely fast. Other soil dwelling species *(Geophilus)* burrow through soil or humus. All Centipedes are active carnivores and have a pair of large poison jaws to capture and kill their prey.

Reproduction usually occurs by the male spinning a web and placing in it a packet of sperm (spermatophore) which the female picks up. Some Centipedes lay eggs in cavities in decaying wood and then brood them. Other species lay their eggs in the soil and show no parental care.

◁ Bird-eating Spiders do not eat many birds and the term is somewhat imprecise. They use their webs as a kind of burrow and leap out on their prey.
△ Black Widow Spiders have various names in different parts of the world. In Australia for instance, they are called Fire Spiders. The bite is venomous and causes a burning sensation, cramps and sometimes paralysis but human victims usually recover quite quickly.
▷ Trapdoor Spider: these spiders build tunnels which have hinged lids.

Spiders and their Relatives

(Arachnids)

The *Arachnida* is a large class of arthropods which includes the **Spiders, Mites** and **Ticks, Scorpions, Harvestmen** and other small groups. They differ from the insects in having generally four pairs of walking legs and never developing wings. The body is divided into two main parts, the cephalothorax and the abdomen. The form of these varies in the different groups: Spiders always have a 'waist' and the whole body is relatively short; Mites, Ticks and Harvestmen lack a 'waist' and usually have a somewhat rounded appearance; Scorpions also lack a 'waist' but the abdomen is extended into a long whip-like tail with a sting end.

Spiders

The Spiders are carnivorous, either hunting their prey, as in the case of the **Wolf Spiders** *(Lycosidae)* and the **Jumping Spiders** *(Salticidae)* which leap upon their victims, or devising sometimes elaborate techniques for capturing their prey; the latter method usually involves some type of silk web, the silk being produced by spinning organs or spinnarets under the abdomen. The **Crab Spiders** *(Thomisidae)* do not use a web to ambush their prey; as their name suggests, they are flattened and crab-like and they lie in wait in flowers for unsuspecting insects, often assuming the colour of their background (cryptic colouration).

Once caught, the prey is held by the paired fangs or chelicerae, (adjacent to the mouth), and poison is injected from glands within the chelicerae; this paralyses the prey. Spiders do not chew up food but inject it with digestive juices and then suck in the softened tissues. There is a pair of sensory feelers or pedipalps in addition to the chelicerae, and in male Spiders these also have a function in mating. The last joint of the pedipalps in males contains the palpal organ in which sperm is stored before mating; the sperm is then transferred directly to the female from the pedipalp.

Mating, however, is always preceded by courtship and this is particularly important in the spiders because the male has to overcome the natural killer instinct of the female. In order to avoid being attacked, males make themselves recognised by a variety of techniques. Those species with good eyesight use visual signals made with pedipalps and chelicerae in a way reminiscent of some crabs. In species in which the female must be approached across her own web, the male must vibrate the web in a recognisable fashion. However, even when courtship is successfully completed and mating has taken place there is every chance that eventually the male will be consumed by his partner.

Spiders often lay their eggs within silk cocoons and guard them until the young hatch; they may even carry them around on their body. Once hatched, many young spiders are dispersed by allowing themselves to be blown along in air currents attached to a gossamer thread of silk produced from their spinnarets; this method of dispersal is called ballooning.

Some spiders attain quite large

proportions. The famous so-called **Tarantulas** *(Theraphosidae)* can be up to 3 inches (8 cm) long with a 10-inch (25 cm) leg span. They live in silk burrows and dart out rapidly to catch their prey, injecting it with a powerful poison which can be very troublesome to humans though unlikely to be fatal. The **Black Widow** of North America is another notorious large spider with a strong poison; it lives in sheltered places and is often found under houses. Again the poison is rarely fatal but children in particular are at risk.

△ The distinctive colouration of the Orange Kneed Tarantula may be a courtship display. Tarantula is a group name for hairy spiders and this species is found in Mexico.

Scorpions

(Scorpionidea)

True **Scorpions** are restricted to the warmer regions of the world. They are secretive, nocturnal animals that hide under stones or in burrows by day.

The body of a Scorpion consists of a cephalothorax covered by a shield or carapace and a longer abdomen which has a broad fore region and a narrow 'tail' arched over the back, terminating in a sting. The possession of a poisonous sting makes these animals among the most feared arthropods.

Scorpions are entirely carnivorous and feed mostly on insects. The prey is captured by a pair of large pincers and either killed or paralysed by the sting.

Mating is preceded by an elaborate courtship display in which the two Scorpions raise their abdomens and move around in circles. The male then grasps the female's pincers with his own and together they walk backwards and forwards. This 'dance' may last for several hours after which the male deposits a package of sperm on the ground and guides the female to it. The eggs develop in the body of the female and the young are born alive. The young Scorpions immediately climb onto the back of the female and she transports them. After their first moult they leave their mother and become independent.

The **False Scorpions** *(Pseudoscorpionidea)* and **Whip Scorpions** are not closely related to the true Scorpions. The False Scorpions are very small animals, (less than 1 cm), which often live in leaf litter under bark or stones. They are commonly quite numerous, but are rarely seen. Their abdomen does not terminate in a 'tail' region or possess a sting. They are carnivorous and the female lays her eggs in a silken cocoon which she guards.

The **Whip Scorpions** are tropical species with long sensory fore legs and an abdomen ending in a long whip-like flagellum. They are nocturnal animals which feed on other insects.

Harvestmen and Sun Spiders

(Opiliones, Solpugida)

Harvestmen closely resemble spiders which have very long flexible legs. However, they do not have the distinctive waist between the two main regions of the body, and never spin webs.

Harvestmen live in both temperate and tropical regions and prefer humid conditions. Many species are nocturnal and omnivorous, feeding on small invertebrates and dead animal and vegetable matter. After mating, the female Harvestman lays her eggs deep in the soil by means of a special egg-laying tube.

Sun spiders *(Galeodes)* are tropical arachnids which are distinguished from the true spiders by the enormous fangs used to kill other animals including small vertebrates.

Spiders webs

Web making is characteristic of many species of spiders, the webs being constructed of silk for the purpose of capturing prey. The silk is produced from spinneret organs under the abdomen. The shape of the web ranges from the irregular tangle webs produced by the *Pholcidae* (the cobwebs in houses), through the horizontal sheet webs of the **Money Spiders** *(Linyphiidae)* and the characteristic webs of the **Funnel Web Spiders** *(Agelenidae)* to the geometric orb web of the *Argiopidae* including the **Garden Spiders.**

The orb web may consist of up to four kinds of silk thread and the method of construction is slightly different for each species. The first stage is the positioning of strong boundary lines within which the rest is placed. Then the 'spokes' are spun, radiating from the centre at which a platform is constructed. A strengthening spiral connects the 'spokes' near the centre and finally, after some manoeuvring, a viscid spiral is left on the outer part of the web to form the effective part of the snare. On completion of the web the spider waits either at the centre of the web or in a nearby retreat holding a signal thread of silk which will vibrate when the web is moved by prey.

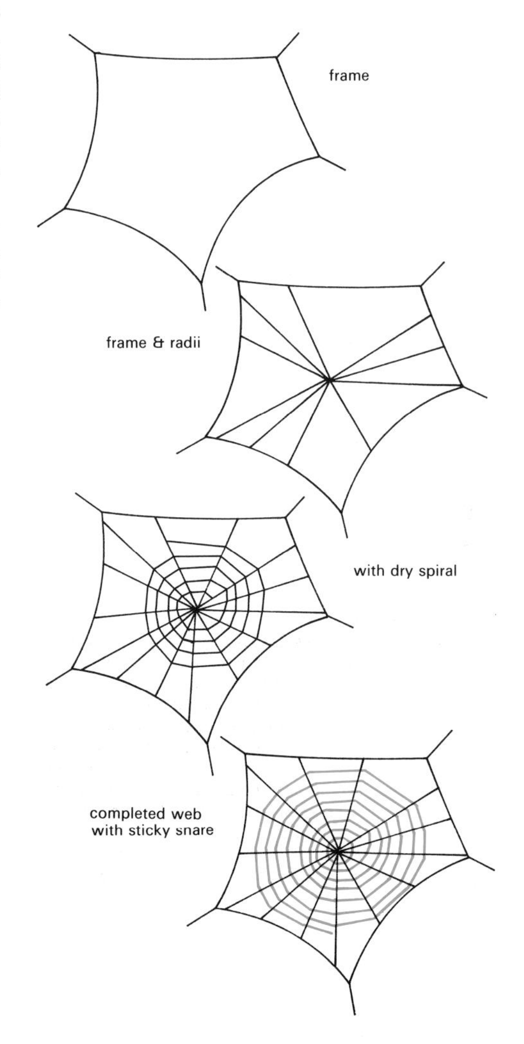

Mites and Ticks

(Acarina)

These arachnids have small bodies which are not obviously divided into regions, four pairs of legs and mouth-parts adapted for piercing and sucking.

The **Mites** differ from the **Ticks** in having relatively soft and smaller bodies and although a large number are parasitic on animals and plants many others are free living and can be found almost anywhere on land and even in fresh and salt water *(Hydrachna)*. Mites are an important constituent of the soil fauna. Man has long had problems with Mites in his own home. Some attack his food, such as the **Cheese Mite** *(Tyroglyphus)*. The **House Mite** infests the mattress of his bed and the upholstery of his furniture and the dead bodies form much of the dust which causes allergies in some people. The parasitic forms are found on both plants and animals. Some of the galls seen on plants are the result of mite infestation; 'red pimple' of sycamore is an example. The **Red Spiders** *(Tetranychus)* that infest plants and cause damage in greenhouses are really mites. Others *(Sarcoptes)* are specially adapted for burrowing under animal skin, causing 'itch' in humans, 'mange' in dogs and 'scaly leg' in poultry.

Ticks *(Ixodes)* are blood suckers and parasitise animals including humans, cattle, Sheep and poultry, often spreading disease. However all Ticks leave the host in order to deposit their eggs which are laid on the ground in vast numbers; to complete the life cycle the young must climb up vegetation in the hope of being picked up by a passing host. Sucking is achieved through a probe that carries recurved teeth so that it is not easily dislodged; for this very reason it is almost impossible to remove a Tick once it has become established. When gorged with blood some species distend to quite large proportions.

△ Larval Water Mites on a bug. Over 15,000 species of Mites and Ticks are known and in a very wide variety of habitats. Both land and water living are known and are parasitic on animal and plant life.
△△ Scorpion female and young: as soon as the young Scorpions are born, they climb on to the mother's back and may remain there for up to two weeks.
▷ Sea Spider; the body area is very small.
◁ Diagrams show the web of an Orb Spider in the making. From the top: the positioning of the boundary lines, the placing of the 'spokes', the strengthening spiral, and finally the 'sticky' viscid thread.

Arthropod poison

Scorpions, spiders and centipedes produce poison in specialised glands and inject it into their victims.

Scorpions have a sting on the tip of the tail which is curved under so that it can only be used when the tail is curved over the back. The poison normally kills the prey, but its effect on humans varies according to the species; it may have only a local irritating effect equivalent to a hornet sting, or an effect on the nervous system equivalent to some snake venoms and even fatal. *Androctonus*, a Scorpion which occurs in the Sahara Desert, can kill a dog in seven minutes or a man in six hours.

The poison glands of spiders occur in the chelicerae, near their mouths, and the venom is injected into the prey through the tip of the fangs. Normally the venom is used to paralyse the prey by poisoning the nervous system. Most spider poisons do not harm man but a few, such as the Black Widow of the warmer parts of North America, can produce severe symptoms and even death, particularly in young children. By contrast some **Wolf Spiders** *(Lycosa)* inject a poison which decomposes the tissues of the prey.

The poison 'fangs' of Centipedes are really modified front legs, and are used to paralyse and kill prey. Some large tropical species can produce symptoms in man equivalent to a bad hornet sting and *Scolopendra gigantea* has been known to cause a human death.

Sea Spiders

These are exclusively marine animals occurring at various depths. Most species crawl slowly over seaweeds although some are able to swim.

Sea Spiders are often considered to be doubtful relatives of the true arachnids, having four pairs of walking legs but a very reduced abdomen. They are carnivorous and browse on small sedentary animals. The male is responsible for carrying the eggs until they hatch. Special limbs, or ovigers, are used for this purpose and clusters of eggs are cemented to them.

The Insects: a Million Known Species

The Insects make up the largest of all the groups of Arthropods. There are more kinds of insects than there are of all other animals put together. No one knows exactly how many kinds of insects there are, but almost one million species are known and many hundreds of new ones are discovered every year.

Insects share the usual arthropod features of jointed limbs and segmented bodies, but it is quite easy to distinguish an adult insect from other types of arthropods. The insect body has three main regions: a head, a thorax, and an abdomen. The head carries one pair of antennae or feelers, which the insect uses to smell and to feel its way about, and the thorax bears three pairs of legs. The legs give the insects their alternative name of *Hexapoda*, which means 'six feet'. In most insects the thorax also carries one or two pairs of wings. The presence of wings is the surest way of distinguishing insects from other arthropods because no other arthropod—no other invertebrate, in fact—has any kind of wing. But this does not mean that all insects do have wings; many primitive insects, such as the Silverfish, have never had wings, and many others, including the Lice and Fleas, have lost their wings during their evolution. The abdomen of the adult insect never bears limbs, but it may have 'tails' or pincers at the end. These are very often concerned with mating and egg-laying. Young insects are less easy to distinguish from the other arthropods, and even from some worms. They never have fully developed wings, and they often have more than three pairs of legs or, sometimes, none at all. Young insects nearly all have to undergo a complex series of changes called metamorphosis before they become mature.

The insects owe their huge numbers partly to their small size, which enables them to make their homes in nooks and crannies unavailable to larger animals, and partly to their extraordinary adaptability. Few insects live in the sea, but they have invaded every other habitat and there are very few materials, either plant or animal, which are not eaten by some insect or other. Blood, nectar, wood, dung, wool, leaves, and fungi are just some of the materials which insects use for food. Such diets obviously call for very different feeding methods, and we find a correspondingly wide range of mouths among the insects. To mention just a few examples, there are the powerful biting jaws of Beetles and Locusts, the 'hypodermic needles' of the Mosquito and the Aphids, the delicate 'drinking straws' of the Butterflies, and the spongy 'mops' of the House Flies.

The insect world is divided into about thirty orders. Most orders have a name which refers to some feature of the wings and usually ends in -ptera, which is the Greek word for wings. We thus get Diptera (= two wings), Lepidoptera (= scale wings), and so on.

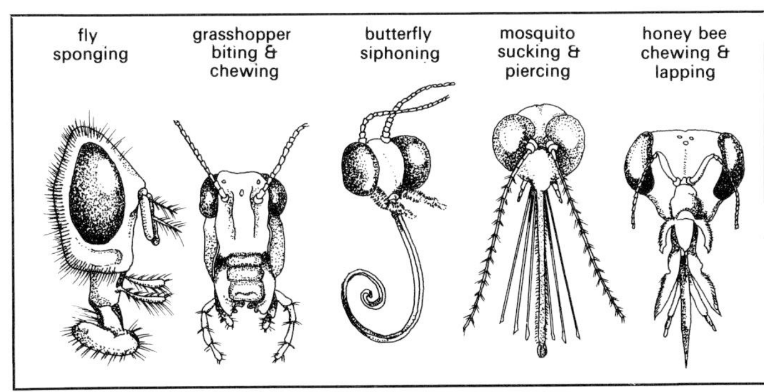

◁ Common blues mating, the blue male above, the brownish female below.
△△ Insects mouth parts, adaptations to different feeding modes.
△ A Damselfly alighted, diaphanous wings together.

The Primitive Wingless Insects

(Apterygota)

The 2,500 or so insects in this group never have any trace of wings and it is thought that they never have had any during their entire history. There are four orders in the group, but they are not closely related. The only really familiar member of the group is the **Silverfish.** It belongs to the order *Thysanura*, whose members have a carrot-shaped body and three slender filaments at the hind end. The body is covered with silvery scales, which come away when the insect is touched. Silverfish live indoors and feed on starchy materials, including the glue used in food cartons.

The order *Diplura* contains a number of small soil-dwelling insects, many of which resemble Silverfish except that they have only two tail filaments. The order *Protura* contains even smaller soil-dwellers, some less than 2 mm long and lacking antennae.

Springtails of the order *Collembola* are small creatures almost all of which live in the soil or in leaf litter. They get their name from the forked 'spring' at the hind end. When disturbed, a springtail releases this spring and it is catapulted away from danger. There are about 1,500 known species of Springtails.

Mayflies and Stoneflies

(Ephemeroptera and Plecoptera)

The **Mayflies** and **Stoneflies** spend their early lives in ponds and streams, and the adults are rarely found far from water. Mayflies are very delicate insects with spindly legs and two or three long filaments at the hind end. The one or two pairs of thin wings are held vertically above the body when the insects are at rest. Mayflies are most often seen in the evening, when large swarms 'dance' over the water. The young insects *(nymphs)*, have three tail filaments and they eat various plant and animal matter in the water. Some nymphs take two years to grow up, especially in the cooler regions. When fully grown, they come to the surface and split their skins. The winged insects emerge and fly away but, unlike other insects, which never moult after getting their wings, the Mayflies have to change their skins once more before they are fully mature. Adult Mayflies live only for a few days and they never feed. There are about 1,300 species.

Stoneflies are generally rather more sturdy than Mayflies and they spend most of their time settled on stones or waterside plants with their wings folded flat over their bodies. The nymphs look similar to those of Mayflies, but have only two tail filaments. Most of the 3,000 or so species prefer stony streams.

Dragonflies and Damselflies

(Odonata)

The **Dragonflies** are the most agile of insects, making incredibly fast turns as they dart after the small insects on which they feed, hovering and even flying backwards. They are long-bodied insects with four gauzy wings which rustle as they fly. Their heads are large and composed mainly of the huge eyes which are necessary for picking out their prey in flight. The spiky legs are held under the head and used rather like a net to scoop their victims from the air.

Young Dragonflies, known as nymphs, live in water and are fiercely carnivorous. They stalk about on the bed of the pond or stream and catch worms or other small animals by shooting out the lower 'lip' and impaling them on its spines. Their wings develop gradually on their backs as they grow up.

Damselflies belong to the same order as the Dragonflies and have the same kind of life history, but they are much more delicate insects. They fly rather weakly and spend much of their time sitting on the waterside vegetation. Dragonflies and Damselflies together account for nearly 5,000 species. They are found nearly all over the world, but they are most numerous in tropical regions.

Crickets and Bush-Crickets

(Orthoptera)

Crickets and **Bush-crickets** are easily distinguished from **Grasshoppers** by their very long, thread-like antennae. The true crickets appear rather more flattened than the other insects in the group and, although they have long back legs, tend to run more than jump. Crickets generally live on the ground and feed on almost anything. The **House Cricket** is a native of North Africa, but it has become a pest in buildings all over the world.

Bush-crickets are less flattened in appearance than the true crickets and the fully winged species are distinctly wedge-shaped when at rest. They live among the vegetation and feed mainly on smaller insects. The females can be distinguished by the broad, sabre-like ovipositor or egg-layer at the hind end. This is quite different from the needle-like ovipositor of the true crickets. Bush-crickets are sometimes (incorrectly) called **Long-Horned Grasshoppers** because of their long antennae, and some American species are known as **Katydids** because their 'songs' seem to repeat this word.

Male crickets and bush-crickets 'sing' by rubbing the bases of their wings together and their calls are generally of a much higher pitch than those of the grasshoppers. Most of them are active at night. There are about 5,000 species of bush-crickets and 1,000 true crickets and, like the grasshoppers, they are found mainly in the tropical regions.

Grasshoppers and Locusts

(Orthoptera)

Grasshoppers are plant-eating insects which generally live close to the ground and blend in well with the vegatation. Their antennae are usually much

shorter than the body, and they use their very long back legs for jumping. There are normally two pairs of wings, which lie neatly along the sides of the insect when it is at rest and give it a distinctly wedge-shaped appearance. Few grasshoppers fly very far at any one time, and many have such small wings that they cannot fly at all. Mountain-dwelling grasshoppers are nearly all flightless. Most of the 6,000 or so grasshopper species live in tropical areas, but grasshoppers of one kind or another are found nearly all over the world.

Male grasshoppers usually attract the females by chirping. They produce this buzzing sound by rubbing the hind legs over the wings. Tiny studs on the legs pluck hard veins on the wings and make them vibrate. The insects are active only while the sun is shining.

Locusts are large grasshoppers living in the warmer regions. They have distinct breeding centres where, from time to time, they build up huge populations. Vast swarms then spread out from these centres and destroy crops wherever they land.

Cockroaches

(Dictyoptera)

Cockroaches are rather flattened insects with a broad shield concealing the head from above. They have long, spiky legs and run very fast. Some species also fly well, but many are wingless. They are omnivorous creatures, eating almost any kind of plant or animal matter. Female cockroaches lay their eggs in little purse-shaped containers, which some species then carry around protruding from their abdomens until the eggs hatch.

Most of the 3,500 or so species of cockroaches, including all the large ones, come from the warm parts of the world. Some of these warm-region species have now become world-wide pests, however, and are especially common in kitchens and warehouses where temperatures remain fairly high.

Praying Mantis

(Dictyoptera)

Almost all the 2,000 or so species of the **Praying Mantis** live in the tropics and, although belonging to the same order as the **Cockroaches,** they are highly specialised for feeding on living insects. The 'neck' is long and mobile and it is surmounted by a head which can swivel in almost any direction. Large eyes enable the Mantis to pick out any small insect moving near it. The front legs are very large and spiny, and when the insect is at rest they are folded up in front of the face. It is this attitude that has given the insects their name, for they resemble people in prayer. When another insect is spotted, however, the legs are shot out and the spiny sections snap shut around the prey like the blades of a penknife. The powerful jaws then tear the prey to pieces. Some tropical species are so brightly coloured that other insects mistake them for flowers and actually land on them to search for nectar. They meet their death very quickly.

Mating is a perilous business for the male mantis. He is somewhat smaller than his mate and he usually ends up as her next meal. After mating, the female lays her eggs in a mass of frothy liquid which hardens to form a cocoon.

Termites

(Isoptera)

Termites are social insects, living in large colonies which are similar in many ways to those of the ants. The two groups are quite unrelated, however. Most of the 2,000 or so known Termite species live in the tropical or subtropical regions and nearly all of these feed on wood doing tremendous damage to buildings. Some of the species actually make their nests inside tree trunks and wooden beams, but most nests are built under the ground or in huge mounds of soil called *termitaria.* The termites then construct tunnels linking the nests with the feeding grounds.

Most Termites are small insects, rarely more than about half an inch long. They are usually very pale, hence their alternative name of white ants, and, apart from their horny heads and jaws, their bodies are soft. Unlike an ant colony, which is founded by a queen, a Termite colony is started by a male and female together. These are known as the king and queen and although they have wings to start with, they lose them before actually beginning the nest. The offspring of the royal couple form the work force of the community. They are always immature insects and are generally wingless. Some of the workers have huge jaws and are known as soldiers. Their job is to defend the colony. At certain times of the year some of the juveniles carry on developing past the usual stage and acquire wings. These are the new kings and queens and they fly out to mate and form new colonies.

◁◁ Long Horned Grasshopper, with long delicate antennae and tapering pronotum. The powerful hind legs give considerable jumping ability.

◁ The American Cockroach is a native of South America, but has been carried world-wide by ships.

▽ Termite hills are made from soil glued with saliva. Inside are elaborate chambers and runways. The shape may influence temperature control.

Stick Insects

(Phasmida)

Stick Insects, often called **Walking Sticks** in the United States of America, are leaf-eating insects which have evolved a truly amazing similarity to the twigs among which they live. Their slender bodies, more than a foot long in some species, often carry warts and spines resembling the buds and prickles on the twigs and they are very difficult to see when resting in their natural habitat. Many of the species are wingless, but some have large hind wings and fly well. The front wings are usually no more than small flaps.

Males are absent or very rare in many stick insect species and the females lay fertile eggs without mating —an example of virgin birth or *parthenogenesis.* The eggs, which often look just like seeds, are scattered freely on to the ground below. Most stick insects live in the tropics, but several species occur in the United States of America and southern Europe, as well as in Australia and New Zealand. The tropical leaf insects, whose bodies are extremely flat and leaf-like, are closely related to the stick insects. Together, they account for about 2,000 species.

Earwigs

(Dermaptera)

Earwigs are a small group of insects, with only about 1,000 species, and most of them can be recognised by the prominent pincers at the hind end which are used in defence. Approximately an inch long, Earwigs are generally brownish in colour with or without wings. The front wings, when present, are very short and squared off at the rear, leaving most of the abdomen exposed. The hind wings are large, but they are very thin and elaborately folded when packed away under the front ones. Many Earwigs fly well, but they are generally reluctant to spread their wings. They are generally active at night, feeding on a variety of plant and animal matter. By day, they hide away in dark crevices. Most of the species live in the warmer parts of the world, and a few from the tropical regions live as parasites on bats and various rodents.

Some of the Earwigs actually care for their eggs when they have laid them, and the young often stay with their mothers until they are nearly grown up.

Booklice and Thrips

(Psocoptera) (Thysanoptera)

The order *Psocoptera* contains nearly 2,000 species of small, winged or wingless insects variously known as **Barklice, Booklice, Dustlice,** and **Psocids.** The winged species, with their wings held roof-like over the body, could be mistaken for aphids, but they have larger heads and much longer antennae than these. They crawl about on tree trunks and other plants, feeding on pollen grains and microscopic algae and fungi. Many wingless species live indoors, often among old books and papers where they feed on traces of mould. Most of the indoor species can be recognised by their broad back legs.

Thrips are very tiny insects with narrow black or brown bodies. Some of the 3,000 or so species are wingless, but many have minute feathery wings. They are most often seen nestling inside flowers but, true to their alternative name of **Thunderflies,** they take to the air on thundery summer days and cause much annoyance. Several species cause damage to crops by piercing the leaves and sucking out the sap.

Lice

(Mallophaga, Anoplura)

Lice are very small parasitic insects that live and feed on birds and mammals. They are flattened and wingless and they have strong claws with which

they cling to their hosts. Most Lice are very particular about the animals upon which they live and each kind of Louse normally keeps to either just one kind of host animal or to a group of closely related animals. The eggs are normally glued to the hair or feathers of the host, and the emerging young lice are very similar to the adults. Young

△ Stink Bugs live in most parts of Europe and show a red and black colouration as a warning to predators. The Bug has a most appalling taste when eaten and birds avoid eating it.
◁◁ Stick insect here has its wings open but when folded the whole body closely resembles twigs.
▷ Golden-eyed Green Lacewing fly. Eggs are laid on a slender stalk and the larvae feed on aphids.
▽ Cicada with its sheathed stylets penetrating a plant stem is sucking its juices.

birds and mammals generally acquire Lice from their parents in the nest.

About 3,000 kinds of Lice are known at present, but there are probably many more yet to be discovered. Two distinct orders are recognised. The *Mallophaga* contains the **Biting Lice,** also called **Bird Lice** because most of them live on birds. They have biting mouths and they chew the skin and feathers of their hosts, often causing much irritation. The order *Anoplura* contains the **Sucking Lice,** which have piercing mouths and which feed on blood. All the Sucking Lice live on mammals.

The True Bugs

(Heteroptera)

The **True Bugs** are a very variable group of insects, but they all possess piercing mouths adapted for sucking the juices of plants or other animals. The jaws have evolved into needle-like structures which fit together to form two slender canals. These are plunged into a plant or animal and the juices are sucked up through one canal while the bug's saliva goes down through the other. The antennae never have more than five segments. Wings are usually present and folded flat over the body when at rest. The front wing is very characteristic, having a horny basal portion and a membranous tip.

Many of the larger terrestrial bugs are known as **Stink Bugs** because they exude an unpleasant smell, and some of them exhibit brilliant warning colours. Most of the other terrestrial species have sombre colours which blend in with their surroundings. The majority of insects in this large and widely distributed order feed on plants, but there are many predatory species, such as the **Assassin Bugs** which attack other insects and the **Bed Bugs** and their relatives which suck blood from birds and mammals. The water-living bugs are nearly all carnivorous. They include surface-dwellers, such as the **Pond Skaters,** and truly aquatic species such as the **Backswimmers** and **Water Boatmen.** All must come to the surface to breathe, however.

Cicadas and Aphids

(Homoptera)

The order *Homoptera* contains a large number of sap-feeding insects. Their mouths are very similar to those of the *heteropteran* bugs, but their wings, when present, are usually held roof-wise over the body instead of being folded flat. The **Cicadas** are the largest members of this group, some of them being two inches long. Their bodies are vaguely frog-like and most of them live in the trees when adult. They are difficult to see as they sit on the bark, but they advertise their presence by the very loud and shrill calls of the males. The sound is produced by the very rapid vibration of a little membrane on each side of the body. Cicadas are also famed for the very long life histories of some species; the nymphs of one North American cicada for instance, spend 17 years feeding under the ground before becoming adult. Most of the 1,500 or so species of Cicada live in the warmer parts of the world.

Many of the smaller members of this order can jump well and they are

known as **Leaf Hoppers** or **Plant Hoppers.** They are abundant in leafy habitats and they sometimes damage crops. One group contains the familiar **Frog Hoppers** or **Spittle Bugs,** whose nymphs live on various plants and surround themselves with froth.

The **Aphids** are rather pear-shaped insects, rarely more than about 3 mm long. The best known are the **Greenfly** and **Blackfly** which plague gardeners all over the world. As well as damaging the plants with their sharp beaks, the insects carry numerous plant diseases. Most of the Aphids seen in summer are females, and all reproduce very rapidly by bringing forth one or more offspring every day. Enormous numbers of young build up in this way. Most are wingless, but there are also many winged specimens which fly off to attack fresh plants.

Lacewings and Antlions

(Neuroptera)

The **Lacewings** and their relatives are carnivorous insects with four delicate and densely netted wings. There are about 4,500 species. Lacewings are generally green or brown and they feed on Aphids and other small insects. Many of the green species have beautiful golden eyes. The green Lacewings lay their eggs at the tips of slender stalks, which can often be seen on the undersides of leaves during the summer. The rather spiky, shuttle-shaped larvae also feed on Aphids and often camouflage themselves with the empty skins of their prey.

Adult **Antlions** are quite large insects, looking rather like Dragonflies except that they have much larger antennae and fly much more lazily. The larvae live in sandy ground and make little pit-fall traps to catch ants and other insects. The prey fall into the pits, right into the large jaws of the Antlions. Most Antlions live in the tropical regions. The **Ascalaphids** are closely related to the Antlions, but they are swift fliers and they have very long antennae. They catch food in the air like Dragonflies, and their larvae hunt on the ground.

Caddis flies

(Trichoptera)

Caddis Flies are rather hairy, moth-like insects which, with very few exceptions, spend their early lives in the water. The adults have four wings which are clothed with fine hairs and are usually a rather dull brown color. They are active mainly by night, and can be found resting on the waterside vegetation by day with their wings folded roof-wise over the body and their antennae held out in front of the head. There are about 3,500 species.

Caddis larvae live in water and many of them build tubular cases around themselves. The cases are made of silk, to which the larvae glue small pieces of leaf, twig, or gravel. Each species makes its own characteristic type of case and remains in it throughout its larval life. Some larvae, however, live freely on the stream bed, and some make little nets on the water plants. Small plants and animals are swept into these nets by the current and the caddis larva feeds on them.

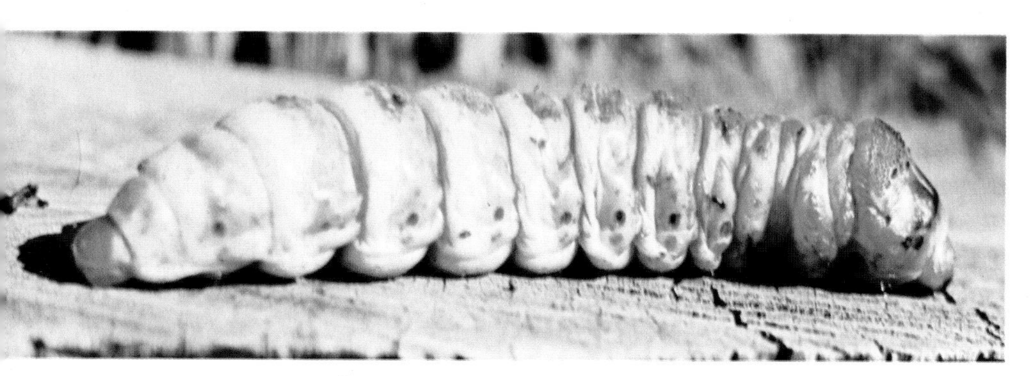

△ The Witchety grub of Australia is large, but is typical of most beetle grubs.

Scorpion flies

(Mecoptera)

The *Mecoptera* is a small order with only about 400 members. All of them have the head drawn out into a prominent 'beak' with the jaws at the end. The most common members of the order are the **Scorpion Flies,** so called because the tail end of the male is turned back over the body like the tail of a scorpion. The wings of most Scorpion Flies are mottled with brown spots. The insects fly rather weakly and spend most of their time crawling on bushes and other plants. They are quite harmless and feed on an assortment of materials, including dead insects, ripe fruit, and even bird droppings. Young Scorpion Flies are like the caterpillars of butterflies and moths and they live as scavengers in the soil.

The related **Hanging Flies** look rather like Crane-flies, but they are easily distinguished by their four wings and the long 'beak'. They live mainly in the warmer parts of the world and they spend most of their time hanging from twigs and catching small flies with their long back legs.

Beetles

(Coleoptera)

The **Beetles** make up the largest of all the insect orders, with more than 250,000 species. They are found all over the world, in every possible habitat, and they include the bulkiest insects as well as some of the smallest. The **African Goliath Beetle** can be as large as a man's fist and it weighs more than four ounces, and at the other end of the scale, there are minute beetles less than one millimetre long. Some beetles are wingless, but the majority have four wings. The front ones, when present, are always in the form of horny cases called *elytra.* They meet in a straight line in the middle of the body, usually covering the abdomen and giving the insects good protection. Although most beetles can fly, the majority rarely do.

The hard elytra are partly responsible for the enormous success of the beetles because they provide protection; but the ability to feed on a wide variety of foods has also been important. The beetles have strong, biting jaws and they can tackle almost any solid food material. Some also use their jaws to suck up liquids.

Water Beetles

(Coleoptera)

Many beetles, belonging to several different families, live in ponds and streams. The young stages or larvae possess gills and can get their oxygen direct from the water just as fishes can. The adults however, are air-breathing insects and normally have to come to the surface to renew their air supplies. They hang just below the surface and draw air into the space between the body and the elytra. The body's breathing pores open into this space and they can take in air just as they would if living on land. Some very small beetles can absorb enough oxygen from the water in this way without having to come to the surface.

Some water beetles, including the large **Diving Beetles,** are fierce carnivores. They use their broad, hair-fringed legs to swim rapidly through the water in search of small fish and other animals. Other species are plant-feeders and they are more often found crawling on the water plants. The **Whirligig Beetles** live on the water surface during the summer months, skimming round in circles and looking for small insects to eat. These shiny black oval beetles have two parts to each eye, one looking down into the water and one looking out over the surface.

Glow-worms

(Coleoptera)

The **Glow-worms** are small beetles which give out a greenish light from their hind ends. They belong to a family called the *Lampyridae,* which contains more than 1,000 species. Many of them are better known as **Fireflies** and most of them live in the warmer parts of the world. Both sexes normally give out light, but the female often glows more brightly. She is very often wingless and she rests on the ground shining her yellow-green light to attract a mate. Young glow-worms, which look rather like woodlice, generally feed on slugs and snails.

Ladybirds

(Coleoptera)

Ladybirds are predatory beetles belonging to the family *Coccinellidae.* This group contains about 5,000 species, distributed all over the world. They are nearly all domed or rounded beetles and

△ Different coloured, olive-green male and brown female Diving Beetles. The male has specially formed front feet with which to hold the female when the beetles are mating.

△ ▷ The Stag Beetle may be as much as 5 cm in length. Its grubs feed on decaying wood.
▽ A Long-horned Beetle is recognised by the antennae which far exceeds its body length.

their bright colours warn birds that they have unpleasant tastes. Many are red or yellow with black spots. Ladybirds feed on Aphids and other small insects throughout their lives. The shuttle-shaped larvae are often blue with various coloured spots.

Stag Beetles and Scarabs

(Coleoptera)

The **Stag Beetles** and **Scarabs** belong to a large group of beetles whose antennae are expanded at the tips to form a number of small flaps. Their

larvae are fat and white and permanently curved into the shape of a letter C. Stag Beetles belong to the family *Lucanidae,* of which there are about 1,000 members. They get their name because the males of many species have enormous jaws resembling the antlers of a stag. These jaws cannot bite, however, because their muscles are not strong enough. The males use them much as stags use their antlers to battle over the females. Larval stag beetles live in rotting wood.

The Scarabs belong to the family *Scarabaeidae,* which has about 19,000 members and includes the giant **Hercules** and **Goliath Beetles.** Like the Stag beetles, the Scarabs and their relatives live mainly in the tropical regions. These insects can draw their antennal flaps together to form a club and the males of many species carry huge horns on the head or thorax. The jaws, however, are of normal size and are usually concealed from above by a broad head shield. The Scarabs themselves are dung feeders and the best known are the **Tumblebugs** whieh trundle balls of animal dung around until they find suitable places in which to bury it. Holes are dug with the aid of the broad, spiky legs and eggs are then laid in the buried dung. The larvae feed on the dung until they are fully grown.

A Wanderer butterfly emerging from the chrysalis. The process is the last stage in a complete metamorphosis through preceding egg and then caterpillar stages. Under favourable conditions, the wings, which are expanded by blood being forced into them, dry out and then harden off in about 2 hours. The butterfly is then ready to use them for flight.

Moths and Butterflies

(Lepidoptera)

With about 100,000 species, the **Moths** and **Butterflies** form one of the largest of the insect orders. Some females are wingless, but the insects normally have four wings which are clothed with minute scales and it is these that provide the striking colours and infinite patterns of the wings.

Most moths and butterflies have a tubular mouth *(proboscis)*, which is coiled up under the head when the insect is at rest, but unrolled and used rather like a drinking straw when the insect is drinking nectar or other juices. Some adult moths have no proboscis, however, and do not feed. A few very primitive moths have biting jaws and feed on pollen.

The *Lepidoptera* is divided into about twenty super-families. The butterflies belong to just one of these and the moths are divided up among the rest. The moths are far more variable than the butterflies, ranging in wingspan from only 5 mm to the giant **Silk Moths** with spans of 250 mm (10 in). It is difficult to point out differences between butterflies and moths. Butterflies are day-flying insects, but so are many moths, so it cannot be stated that

butterflies fly by day and moths fly by night. One of the best distinctions is the appearance of the antennae: those of the butterflies are always clubbed while moth antennae are very varied—long, short, thread-like, or hairy, but rarely clubbed. Moths which do have clubbed antennae, such as the brightly coloured **Burnets,** can be distinguished from butterflies by their wings. The hind wings always have a bristle springing from the base and projecting forwards under the front wing. Known as the *frenulum*, this bristle helps to hold the wings together.

The young of butterflies and moths—caterpillars—are basically worm-like in appearance, but they have three pairs of legs near the front and generally five pairs of stumpy legs further back. Many species are hairy. Most caterpillars feed on leaves, but some eat wood and the **Clothes Moth** caterpillars eat wool. When fully grown, the caterpillars turn into *pupae*. Many moth caterpillars spin silken cocoons around themselves before pupating. Many butterfly caterpillars hang themselves upside down from a silken pad before turning into pupae.

The Swallowtails

(Papilionidae)

The **Swallowtails** are nearly all large and very colourful butterflies which get their name from the tail-like extensions of the hind wings. They belong to the family *Papilionidae,* which contains about 700 species of butterflies found mainly in the tropics. One of the strangest looking species is the **Racquet-tailed Swallowtail,** whose narrow and rather lacy wings bear tails shaped like tennis racquets. Only one Swallowtail lives in the British Isles, where it is confined to the Norfolk Broads, but several live in Europe and North America. The family also contains the mountain-dwelling **Apollo Butterflies,** whose thinly scaled wings lack tails and are basically white with black and red spots. The most majestic of all butterflies, however, are the **Birdwings** of south east Asia and the New Guinea region. Green, yellow, and velvety black are their predominant colours and they generally have very long and narrow front wings. The hind wings are relatively small and they do not usually bear tails. **Queen Alexandra's Birdwing** from New Guinea has a wingspan of more than ten inches in the female and it is the largest known butterfly.

The Hawk-moths

(Sphingidae)

Perhaps the most distinctive of all moths are those belonging to the family *Sphingidae* generally known as **Hawk-moths** or **Sphinx moths**. These moths are found throughout the world and are diverse in shape and size. But as a family they possess many unique attributes.

The **Death's head Hawk,** *(Acherontia atropos),* of Africa and southern Europe is one of the larger moths, with a wing-span of 5 to 6 in (14 cm). Every summer, these insects migrate long distances northwards over the Alps and sometimes reach Scotland. The uncanny image of the human skull on the thorax has given the moth its name and made it well-known throughout the world.

Many of the Hawk Moths can hover over the flowers upon which they feed, beating their wings at incredible speed and probing their long tongues deep into the blossoms to suck in the nectar. **The Humming-bird Hawk** *(Macroglossum stellatarum)* is so named because it so closely resembles those tiny, wonderful birds. Notable amongst these hovering Hawk Moths is the **Convolvulus Hawk** *(Herse convolvuli),* which possesses a tongue nearly 5 in (12 cm) long, over twice the length of the insect itself. The caterpillars of the Hawk Moths are often large and distinctive with bold stripes on the body. Nearly all of them have a large pointed horn near their tails.

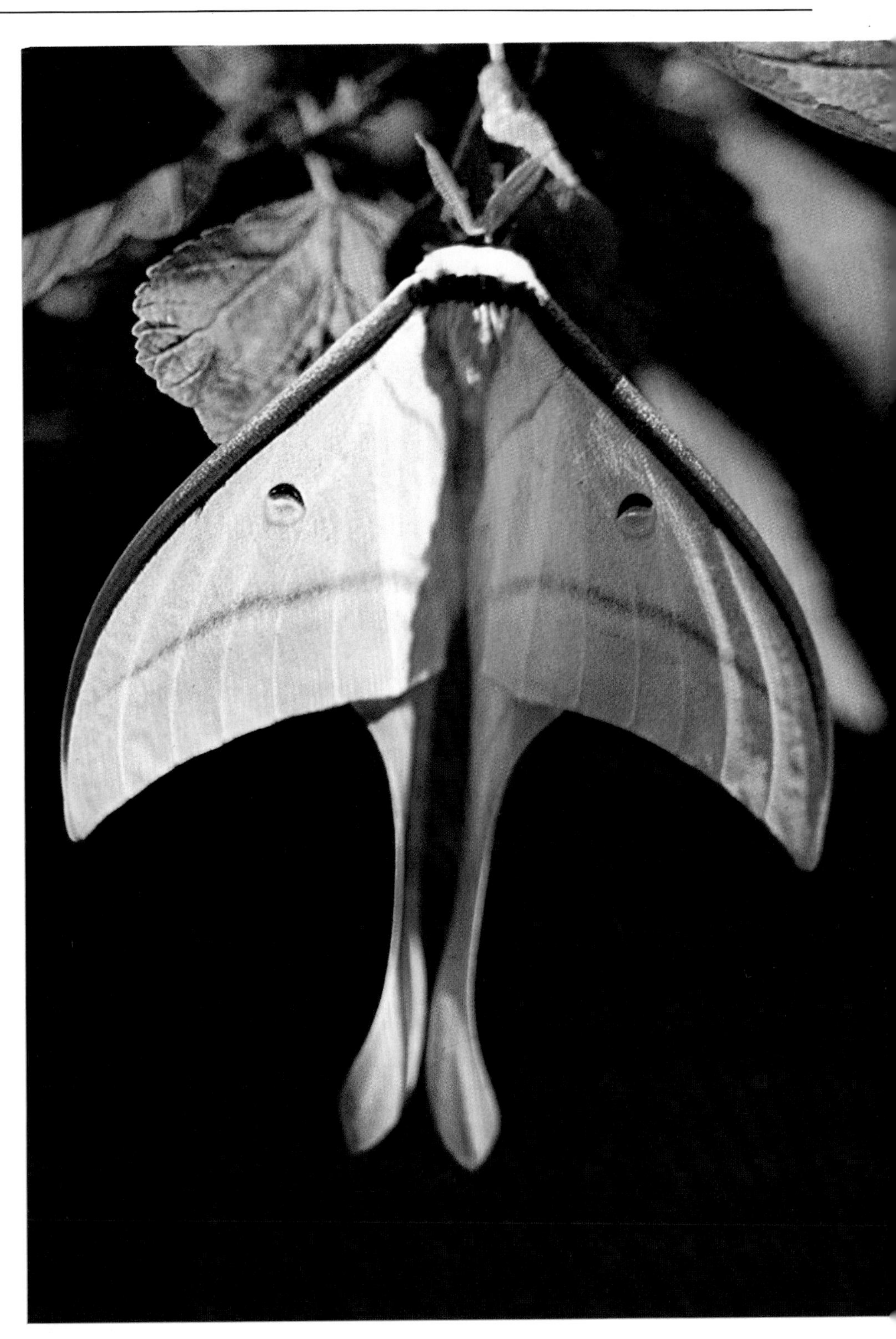

△◁ Swallowtail butterfly. The projecting 'tail' on the hindwings is characteristic of Swallowtails but is not present in all species.

△▷ The Moon Moth has elongated hindwings forming tails.

△ Assam Silk Moth caterpillars will feed on a variety of leaves—oak, apple, hawthorn or willow.

◁ Privet Hawk Moths pairing. The female emits a scent from an organ near the tail; the male upon picking up the scent, instinctively flies upwind until he finds the female. Hawk Moths will often stay together for as much as 24 hours.

The Giant Silkmoths

(Saturnidae)

The giant **Silkmoths** include some of the largest moths in the world. The **Atlas Moth** from India, in common with some other species, may have a wingspan of more than ten inches. These moths belong to the family *Saturnidae,* which includes the **European Emperor Moth** and the **North American Cecropia Moth** but which is best represented in the tropics. They are nearly all heavy-bodied moths and their wings usually carry translucent eye-spots. The hind wings are often prolonged into 'tails', as in the beautiful green **Luna Moth** of North America. The antennae are feathery, especially in the male, and are used to pick up the scent of a female from more than a mile away.

The caterpillars of this family are stout and often brightly coloured and very attractive. Many are covered with spiky outgrowths. When fully grown, the caterpillars spin dense cocoons and many species, especially some of those from India, yield commercially valuable silk. Most of the world's silk, however, comes from the cocoons of the cultivated **Silk Moth,** a much smaller moth which belongs to the family *Bombycidae.*

True Flies

(Diptera)

The **True Flies** can be distinguished from most other winged insects because they have only one pair of wings. The hind wings have been modified into a pair of pin-like organs called *halteres.* They act as balancing organs while the insects are in flight. The flies include about 75,000 species, with an enormous range of shapes and sizes. There are tiny **Fruit-flies** and **Midges,** long and slender **Crane-flies,** heavily-built **Horse-flies,** and furry **Bee-flies.** All adults feed on liquids, but their mouths are very varied. **Horse-flies, Mosquitoes,** and many Midges feed on the blood of various vertebrates and they have needle-like mouth-parts which pierce the skin and suck out the blood. **House-flies** and **Bluebottles** feed on decaying matter, soaking up the fluids with a spongy 'mop'. The brightly coloured **Hover-flies** obtain nectar from flowers in the same way.

Young flies are even more varied than the adults, although all of them are legless. Some, including the **Leather-jackets** (young Crane-flies), eat plants and others live as parasites inside other animals. The majority, however, eat decaying matter. The common **House-fly** is a typical example. The adult lays its eggs in rotting matter, such as animal dung or household refuse, and the little white maggots soon hatch out. They are carrot-shaped, with the narrow end at the front and no real head. Each maggot sucks up the putrid liquid around it and grows so quickly that it has to change its skin twice in a week. Its third skin then hardens into a little brown barrel-shaped object, inside which the larva pupates and turns into an adult fly.

◁ A Crane-fly develops from a soil inhabiting grub, the destructive 'Leather Jacket'.
△ A fully developed hive of Bees can contain as many as 60,000 bees: half that number will swarm.
▷ The Bulldog Ant is notorious for the strength of its biting jaws and holding power.

Mosquitoes

(Diptera)

The **Mosquitoes** are a family of slender flies whose females feed on blood. There are nearly 2,000 species, distributed all over the world. The wing veins of the mosquitoes are covered with scales and the antennae of the males are densely feathered. Both sexes have a long beak or proboscis, but only the females can pierce animal skins and obtain blood. The males normally feed on nectar and other plant juices.

Mosquitoes all breed in water. The eggs float on the surface and hatch into worm-like larvae which are very swollen at the head end. The larvae spend much of their time hanging just below the surface, wafting microscopic organisms into their mouths with their numerous bristles. When disturbed, however, the larvae swim away with a rather violent wriggling motion. When fully grown, they turn into comma-shaped pupae which do not feed but which are able to swim about very rapidly. The adults emerge from the pupae at the water surface.

Female Mosquitoes are involved in the transmission of several serious human and animal diseases, including malaria and yellow fever. The Mosquitoes pick up the germs when feeding from infected people and pass them on next time they take a meal.

Fleas

(Siphonaptera)

Fleas are small insects which, in the adult state, suck blood from birds and mammals. Like the Lice, they have lost their wings during their evolution. Wings are of no use to insects that spend most of their lives crawling through fur or feathers. The adult Fleas are generally brownish or yellowish creatures, greatly flattened from side to side and never more than about 8 mm long. They have long back legs and are very good jumpers. Fleas lay tiny, pearly white eggs, but they do not glue them to the host. Many of the eggs fall off into the nest or sleeping quarters of the host and hatch there into little maggots. These are not parasitic and they feed on the debris in the nest, including the droppings of their parents. The maggots pupate when fully grown and the new adult Fleas emerge to climb on to the host when it next returns to the nest. About 95 per cent of the 1,600 or more Flea species are parasites of mammals, and the others attack birds.

Ants

(Hymenoptera)

Distantly related to the Bees and Wasps, the 10,000 or so species of **Ants** all belong to the family *Formicidae.* The majority live in the tropical regions and they are all social insects, living in large or small colonies. Some species are an inch or more long, but the majority are much smaller. They can all be recognised by the sharply bent antennae and the narrow waist which bears one or two vertical lobes. Fully developed males and females bear four wings to start with, but the majority of ants in a colony are wingless. All ants have strong, biting jaws and many are equipped with stings.

The ant colony is normally founded by a single mated female who is known as a queen. Although she has wings at first, she breaks them off before beginning her nest. After she has reared her first few young ants she does nothing but lay eggs. Nearly all of her young grow up to be wingless females called workers. These ants build the nest and look after the queen and the younger ants. A large colony may have more than 100,000 ants in it, all working smoothly together for the good of the community. The ants are nearly all omnivorous creatures, but some eat only animal material and others eat only plants. Many like sweet substances and collect Aphids honeydew.

At certain times of the year the ants rear fully developed males and females. These leave the nests in swarms and fly off to mate. The females then look for places in which to make new nests, but few of them ever manage to raise a family.

Wasps

(Hymenoptera)

The name **Wasp** is used for many different groups of the *Hymenoptera,* but the true wasps, also known as **Yellowjackets** in the United States of America, all belong to the family *Vespidae.* These have black and yellow or brown and yellow bodies and can be distinguished from similar insects in other families by the way in which their wings are folded lengthwise and laid along the edges of the body when at rest. The eyes are also characteristic in that they are deeply notched on the inner side and often almost crescent shaped. Adult wasps enjoy nectar and other sweet substances, but the young are fed on other insects. Wasp females possess powerful stings.

The True Wasps include both solitary and social species. The solitary ones lead lives like those of most other insects except that they make nests for their eggs. Having made a nest and laid her eggs, however, the female has no more contact with her offspring. The social wasps, like other social insects, form colonies in which mother and offspring live side by side and in which all the members work for the good of the community. Wasps living in tropical regions often form permanent colonies, but colonies in temperate regions are usually only of an annual nature. A mated female or queen hibernates during the winter and starts her nest in the spring. She rears the first of her young herself and these all become workers—winged females, but smaller than the queen. These workers take over the building of the nest and they feed their younger sisters. The queen goes on laying eggs through the summer and some of the later grubs eventually grow up into male wasps and fully developed females. These fly off and mate and the females look for places to hibernate. The rest of the wasps die in the autumn.

Bees

(Hymenoptera)

Many of the **Bees** are very like the Wasps in appearance and behaviour, but there is one major difference; Bees feed their young on nectar and pollen, whereas wasps rear their young on other insects. In association with their pollen-collecting habits, bees are also much more hairy than the wasps. Their hairs are minutely branched and well suited to picking up pollen from the flowers. Most bees also have broader hind legs than the wasps, and they do not fold their wings longitudinally when at rest.

As with wasps, there are both solitary and social species. They belong to several families, but most of the social species belong to the family *Apidae.* The hairy **Bumble Bees** form annual colonies which behave rather like the wasp colonies, being started by a mated queen in the spring and rearing new queens in the late summer. The **Honey Bees** form permanent colonies, ruled over by a queen who is so specialised for egg-laying that she can do no other work. She can start a new colony only by taking a swarm of workers with her to do the building and feed the young. Secretions from the queen, known as 'queen substance', somehow control the workers and ensure that everything runs smoothly in the nest. The Honey Bee workers have evolved wonderful methods of telling each other where to find food and the insects make the very best use of any source of nectar.

Ichneumon Flies and other Parasites

(Hymenoptera)

Distantly related to the Wasps, the **Ichneumon Flies** are slender insects which spend their early lives feeding in or on caterpillars and other young insects. They are usually brown or black and they range from a few millimetres to more than an inch in length. The females of some species have extremely long egg-layers or ovipositors which are longer than the rest of the body. All of the species have relatively long antennae, which the females use to seek out suitable hosts. Having found a victim, she lays her eggs inside it or else attaches them to the outside. When the Ichneumon grubs hatch they begin to feed on their host. Sometimes there is only one Ichneumon grub to a host, but there may be many in some species. The result is always the same, however: the grubs gradually eat away the host until it is merely an empty skin. By this time the Ichneumon grubs are fully grown and they turn into pupae either inside or outside the host's skin. Adult Ichneumons emerge later. Ichneumon flies control the numbers of many insects in this way, and they share this task with several other groups of *Hymenoptera.* One group of very small insects called **Fairy Flies** grow up inside the eggs of other insects.

Insect Communication

The vast majority of insects lead solitary lives and, except when looking for a mate, they have no need to communicate with each other. Insects can attract their mates by sound, as among the **Grasshoppers** and **Cicadas,** or by scent or visual display. Many female moths, including the **Emperors** and **Silk moths,** emit scents which can attract males from far away. Female **Glow-worms** use visual signals by flashing their lights on and off. Colour patterns and unusual flight behaviour also help to bring males and females together. Male **Mayflies,** for example, swarm together and 'dance' over the water. The females are attracted to these swarms and are taken off by the males.

The greatest development of communication occurs among the social insects, which live in colonies and work together for the common good. Chemical signals are of great importance here. They are used for laying trails to and from the nest, so that other members of the colony can find food, and the insects also exchange information by rubbing their antennae together and passing on the scents that they have picked up. The **Honey Bee** workers have a very elaborate system for telling each other where to find food. A bee returning from a successful trip dances on the combs in the nest and the other bees follow her movements very carefully. The speed and direction of the dance tells the other bees the distance and direction of food.

▽ Grasshopper stridulating: the inside back legs have 'pegs' which, rubbed over a vein, make sound.

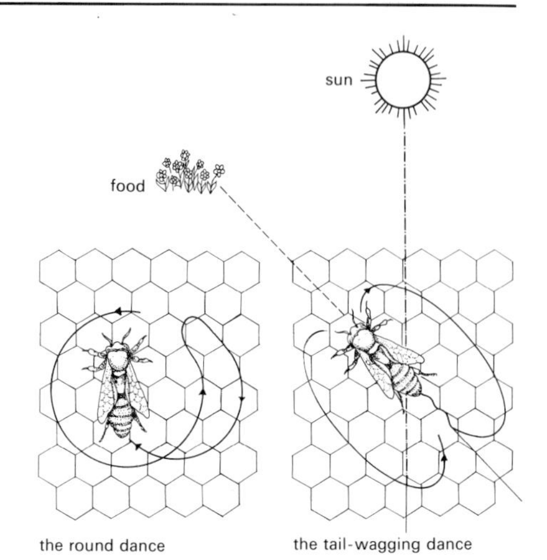

△ Bee dances: a round dance means food near. An eight, tail wagged, means distant food, orientated by the direction of the sun, indicated by the angle of the eight on the comb.

Insect Metamorphosis

Some **Aphids** and a few other insects give birth to active young, but the majority of insects start life as eggs. The young creature that emerges from the egg is usually very unlike the adult, however, and it has to undergo considerable changes as it grows. These changes are collectively called *metamorphosis* and they follow one of two main pathways.

Insects such as the **Earwigs, Grasshoppers,** and **Dragonflies** undergo what is known as a *partial metamorphosis.* The young insect looks vaguely like the adult, but it is much smaller and it has no wings. It is called a *nymph.* Like all young arthropods, it has to change its skin several times as it grows, and at each of these moults it gets more like the adult. Wings gradually develop on the outside of the body and get larger at each moult until the adult stage is reached. The primitive wingless insects and the **Lice** have no wings to grow, and their metamorphosis consists of just a size increase.

Butterflies and **Moths, Flies, Bees, Beetles,** and **Lacewings** are among the insects which undergo a *complete metamorphosis.* The insects which hatch from the eggs are very unlike the parents and they are called *larvae.* They change their skins several times, but do not become any more like the adults. Butterfly caterpillars, for example, merely become bigger caterpillars at each moult. But then they reach full size and they moult again to become a resting stage called a *pupa* or *chrysalis.* The larva's body is here re-built to form the adult or *imago,* and it is not until this stage that wings appear. Insects often remain in the pupal stage for several months.

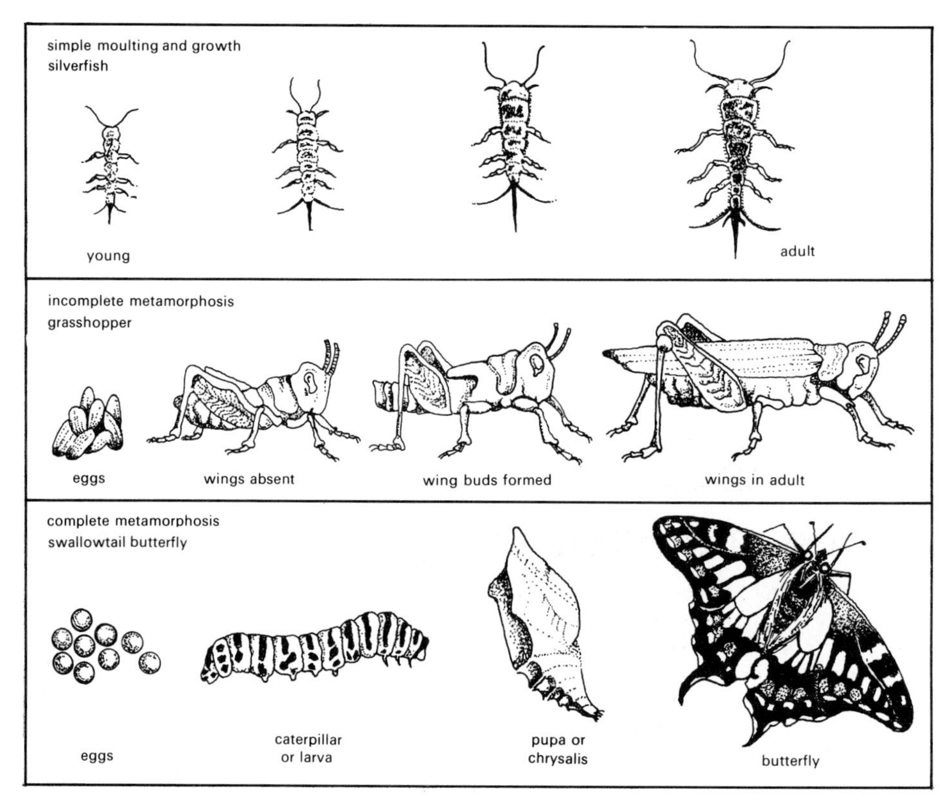

Warning Colours and Mimicry

Wasps and other insects with stings or distasteful and poisonous bodies do not usually go in for camouflage. They generally have bold colours and patterns and they expose themselves freely to the eyes of birds and other predators. But they do not get eaten, and it is believed that the bold or bright colours act as a warning to their enemies. Young and inexperienced predators try the insects, but they soon learn to associate the bold colours with unpleasantness and they then leave the insects alone. Bright red, red and black, or yellow and black are the most common warning colours.

◁ The harmless Hoverfly mimics the Wasp.

Not all brightly or boldly marked insects are distasteful, however. Many insects which are good to eat have evolved similar patterns. Any insect which looks vaguely like a wasp is likely to be avoided, and so insects with such patterns derive great benefit. These harmless and edible insects which resemble unpleasant ones are called mimics. **Hover-flies** are among the commonest mimics of the wasps and bees.

Insect Camouflage

Insects have a great many enemies among the birds, lizards, and small mammals and they have evolved many methods of concealing themselves or deceiving these enemies. The simplest method of camouflage is to merge in with the background, and many insects do this merely by having the same general coloration as their surroundings. **Grasshoppers,** for example, are mainly green with brown stripes which resemble the grass stems. Many caterpillars are green, too, with pale stripes which resemble the veins of the leaves. **Moths** that rest on tree trunks by day are generally brown or grey, with stripes that merge with the bark crevices and break up the outline of the wings. The latter are pressed right down against the bark to get rid of tell-tale shadows. **Stick Insects** and many caterpillars take things even further and look just like twigs. Some **Plant Hoppers** look like thorns as they rest on the twigs, and some moths look like bird droppings. Camouflage of this kind is very effective because the predators do not give the insects a second look.

▽ The Geometrid Moth caterpillar blends perfectly with its beech twig background. Camouflage colouring such as this protects the caterpillar from being spotted by potential predators.

Several insects employ a different form of deception and pretend to be larger and fiercer than they really are. The **Eyed Hawkmoth** and several other species have large eye-spots on their hind wings. The spots are normally covered, but they are exposed when one of these insects is disturbed. Suddenly confronted with a large pair of 'eyes', the attacker turns round and flees.

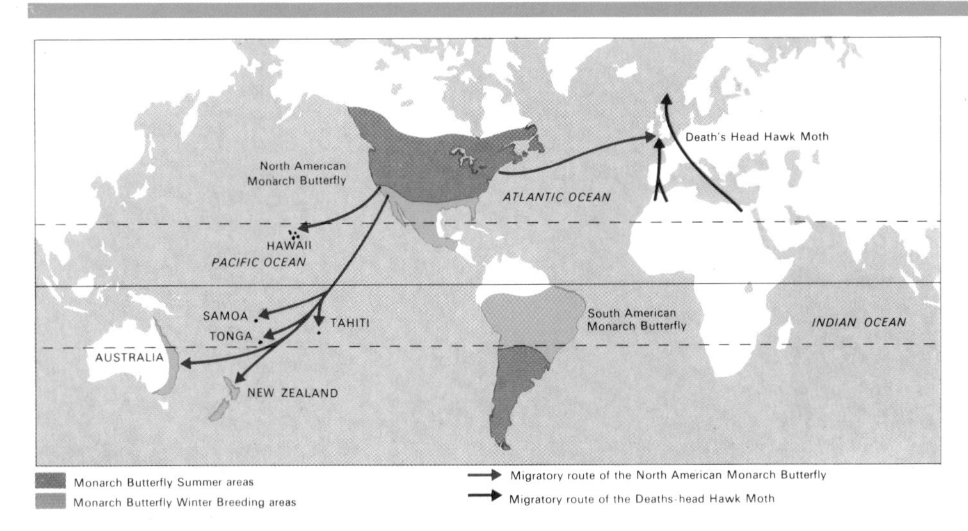

◁ Seasonal migratory routes of the Monarch butterfly and the Death's Head Hawk Moth. The Monarchs travel distances of more than 3,000 miles, sometimes flying as much as 80 miles a day, non-stop.

Insect Migration

Migration is the regular movement of an animal species between two areas or habitats. It is perhaps best known among the birds, many of which spend the summer in the northern hemisphere and then, at the approach of winter, fly south to the tropics. Few insects make such spectacular journeys, but there are some notable travellers among the butterflies and moths. One of the best known is the **Monarch** or **Milkweed,** a native of the Americas, although it is now widely distributed across the Pacific to Australia and New Zealand. This striking orange and black butterfly is common over much of Canada and the United States during the summer, but the insects move southwards in the autumn and meet up with each other to form huge swarms. When they reach the southern United States and Mexico they settle in the trees and spend the winter in a state of partial hibernation. The butterflies begin the northward journey in the spring, but they travel rather leisurely and do not get very far before they lay their eggs and die. Their caterpillars soon grow up into new adults, and these move northwards again. Another generation is reared on the way, and the butterflies that finally arrive in Canada are the grandchildren of the ones that left the previous autumn. Some individuals fly on towards Europe and Australasia. A similar migration takes place among the Monarchs of South America, moving backwards and forwards between southern Tierra del Fuego and the Brazilian tropics. Entomologists have very little idea of how the insects find their way and keep flying in the right direction.

Denizens of the Sea Bed

The most easily recognised of the Echinoderms are the Starfish and these can be found in seas all over the world. Sea Cucumbers of tropical and warm seas are not as familiar as the Sea Urchins, which live on the rocky floor of the sea. The graceful Feather Stars are the most prolific of the surviving modern Crinoids.

Echinoderms or 'spiny-skinned' animals are found in all the world's oceans from the shores to very deep waters; some species have been taken from depths of up to 6,000 metres.

Echinoderms live only in marine water and the majority of them are 'bottom dwellers', moving slowly over or lying on the sea bed.

Over 5,000 species of Echinoderms have been identified and these are classified into five groups: Starfish or Sea Stars (Asteroidea); Brittle Stars (Ophiuroidea); Sea Urchins (Echinoidea); Sea Cucumbers (Holothuroidea) and the ancient and primitive Crinoids which include Sea Lilies and Feather Stars.

In general, Echinoderms are radially symmetrical, the body being divided into five parts around a central axis and protected by a hard exoskeleton.

The greatest number of Brittle Stars, Sea Urchins, Sea Cucumbers and Crinoids are found in the warmer waters of the Indian and Pacific Oceans, particularly off the shores of Borneo, the Philippine Islands and New Guinea. The cooler Atlantic Ocean does not support numbers to the same extent. The greatest density of Starfish is found in the north west Pacific Ocean.

When adult, Echinoderms move very slowly but, at the larval stage, they are free-swimming creatures forming part of the plankton life of the sea. These floating larvae are dispersed by oceanic currents and in this way over-population of any particular area of sea bed is prevented. The oxygen requirements of slow-moving bottom dwellers, such as Sea Lilies, Sea Cucumbers or Urchins, are minimal. Many of these creatures are therefore able to live successfully in water of considerable depth, where very low oxygen concentrations prevail.

Sea Lilies

(Crinoidea)

The first living **Crinoids** seen by Man were dredged from the sea late in the nineteenth century. Fossil Crinoids were already known but it was assumed that the class were extinct. In fact, more than 600 species survive today and are descendents from Palaeozic ancestors. The majority of ancient Crinoids possessed stalks and were attached to material on the sea bed and at least one species trailed an attachment 60 ft long! There are now only 80 species with stalks in existence and none of these have stalks longer than half a metre.

These Crinoids, called **Sea Lilies,** are found in deep water, between 750 ft and 15,000 ft below the surface of the sea.

Held on top of the stalk, the cup-like body of the Sea Lily has five long arms which may divide further into branches. Detritus material is caught on each arm and becomes entangled with slime in an open groove. A single groove leads down each arm so that each of the five converge on the upturned mouth. Food particles are swept down grooves by the action of microscopic hairs or cilia.

Simple sex organs are found on tiny projections from each arm. Sperms and eggs are released into the sea where fertilisation takes place. Tiny larval forms develop which are dispersed by sea currents.

Feather Stars

(Crinoidea)

Feather Stars are the most prolific of the surviving modern Crinoids. Named for the graceful, trailing arms which radiate from the central body, many are beautifully coloured or patterned, particularly those from shallow, tropical waters. Specimens of these delicate creatures have been taken from considerable depths—some from as deep as 4,500 ft below the surface of the sea. Altogether about 550 species of Feather Stars have been identified.

Feather Stars begin life as a fertilised egg which develops into a stalked larva, similar to that of the Sea Lilies. But at this point the Feather Star larva breaks free from the stalk and becomes a free-swimming adult organism. Swimming is accomplished by waving the long, slender trailing arms. When needed, extra anchorage is provided by small outgrowths from the underside of the body. These are termed 'cirri' and are short and thick in the species which live on a rocky sea bottom, and thinner almost root-like in those that live on a soft sea bottom.

Feeding is essentially the same as in Sea Lilies except that Feather Stars from shallow waters are able to trap living plankton of both plant and animal origin as well as detritus material. Like Sea Lilies, Feather Stars have no special sex organs and many release sperms and eggs into the water. However, in some, *Antedon* for instance, eggs are fixed to the outer surface of certain areas of the arms.

Sea Cucumbers

(Holothuroidea)

Sea Cucumbers are an unusual and fascinating class of Echinoderms. About 500 species have been described, most of which display the elongated cucumber shape and they vary in length from a few centimetres to nearly a metre. The largest species of Sea Cucumber is *Stichopus* which is found off the Philippine Islands. North American and European species measure up to 12 in (30 cm). The greatest numbers of Sea Cucumbers are found in the world's tropical or warmer waters, some of them at considerable depths. Most are dull in colour but certain tropical species, found in shallow waters, have vivid, striped patterns.

Sea Cucumbers are sand or mud dwellers. They lie on one side and move sluggishly on three rows of tiny, tube-like feet. The burrowing species have no tube-feet but are endowed with sets of muscles which enable the body to be squeezed and contracted through mud. In the non-burrowing Cucumbers, the mouth is encircled by a crown of slimy, food-trapping tentacles. These sweep the water for plankton or sift sand and mud for tiny creatures or detritus. The giant Cucumber, *Stichopus*, is said to fill and empty itself of sand at least three times a day.

During spawning, eggs are released into the sea. But 30 brooding species, which live in the cold waters of the Antarctic, catch their eggs with their tentacles and move them to tiny brood pouches on their body surface, where they develop and hatch into the larval form.

Starfish

(Asteroidea)

Starfish are the most easily recognised Echinoderms and are found in seas all over the world but, like the other classes, the greatest numbers live in the Indian and the Pacific Oceans. More than a thousand species are known. Some are very brightly coloured but most are a drab yellow. In size, they average between 4 and 8 in (10 and 20 cm) in diameter, the largest measuring up to 3 ft. Projecting in rows from the underside of each arm are tiny, fluid-filled tube feet. Minute sets of muscles contract to permit fluid to pump in and out and a simple nervous system controls the overall movement. In those Starfish which crawl over the sea bed, the walking feet are suckered and secrete an adhesive substance. Other species of Starfish have more pointed tube feet, which help the creatures to burrow into sand.

The mouth of the Starfish is on the underside of the central body disc—a necessary adaptation to a bottom-dwelling carnivore that feeds on molluscs, small crustaceans and marine worms. Many Starfish crawl over their prey and the more advanced kinds, *Asterias* for instance, capture their food with their long arms.

A remarkable feature of Starfish is their power of regeneration. If an arm or part of an arm is lost or damaged the creature is able to grow a new part. In certain circumstances it can even renew portions of the central disc.

There is normally one breeding season in the annual cycle of Starfish and when ripe, the sex organs almost fill the arms. On spawning, eggs and sperms are released into the sea for subsequent fertilisation. Here they develop into the larval form.

◁ This brilliant Sunstar illustrates the radial symmetry of the echinoderms but the five-rayed form is more usual.
▽ Sea Cucumber, the mouth can be seen encircled by a crown of food-trapping tentacles.

The Forceps-Carrying Sea Stars

(Asteroidea)

A good example of this advanced group of Starfish is the world-wide genus, *Asterias*. The arms, normally but not necessarily, five in number, radiate from a central disc which contains the mouth, gut and other essential body parts. A series of spine-bearing calcareous plates project from the flesh and provide a considerable physical protection. Further protection is provided by numerous small pincers or *pedicellariae* which cluster between the spines. These spines have toothed jaws which capture small creatures crawling over the body.

A system of hundreds of fluid-filled, suckered tube feet propel the Sea Star over the substratum but speed of movement is not essential as much of their prey consists of rock-bound Mussels or Clams. The predatory Sea Star crawls over the shell and, in a humped position, fixes its tube feet to the surface. The arms start to pull on the shell but this causes the Clam to shut tight instantly. The attack may be prolonged for some hours before the victim gives way and the shell is breached. The Sea Star now turns part of its stomach inside out and pushes it through its mouth so that it flows over and engulfs the open Clam. The stomach secretes digestive juices which pour over the fleshy victim. Eventually, a digested, mushy fluid is sucked in by the Sea Star as it pulls the stomach back inside its body.

Sea Urchins

(Echinoidea)

Sometimes known as **'Sea Hedgehogs', Sea Urchins** are closely related to **Heart Urchins** and **Sand Dollars.** Their bodies consist of a hard, spherical skeleton with long protective spines. They vary in colour from dull brown to green, white or red, whilst a few are attractively banded. Most Sea Urchins measure from 2 to 4 in (5 to 10 cm) in diameter while a number of Indo-Pacific Urchins measure nearly 12 in (30 cm) across. In some, the spines may be tipped with poison which inflict irritating wounds on predators.

Amongst the spines are found small, pincer-like *pedicellariae*. Some of these secrete a poison that paralyses small animals and may be used in defence.

These tough little creatures generally live on rocky floors and move on articulated spines that act as stilts, and on long protruding tube-feet. Some Urchins burrow, often to escape the effects of wave action at the surface. A European species, *Paracentrotus Lividus,* tunnels through rock. Sea Urchins are scavengers and consume a variety of material including seaweeds.

Spawning involves the shedding of eggs into the sea water where fertilisation takes place. Brooding takes place in a few instances, largely amongst cold water species.

▽ The perfection of form of a Sea Urchin's chewing equipment, called 'Aristotle's Lantern'. The gut of the Urchin here has been partially dissected.
▷ The larger Sea Urchins are sometimes used for food, the edible parts being the soft, reproductive organs suspended from inside the upper shell.
▽▽ The five-rayed basic plan of most echinoderms is clearly displayed by the Brittle Stars. If an arm is detached another will soon regenerate to replace it. One species, Ophiactis savigni is self-dividing.

Aristotle's Lantern

(Echinoidea)

Sea Urchins feed on a variety of plant and animal material and to accomplish this they require an efficient mechanism. With the exception of **Heart Urchins,** they have a remarkable piece of equipment called 'Aristotle's Lantern'. This is an arrangement of 'teeth' and does in fact resemble a lantern in appearance. Five hard plates, the 'teeth', are arranged in circular fashion around the mouth. The ends of the plates are pointed and sharp and each tooth is bound to its neighbour by a set of muscle fibres. Another set of muscles moves the lantern a little way out and back into the mouth. The teeth muscles

allow each tooth to move separately so that the entire machinery works as a food shredder. Despite this, an Urchin may take weeks to consume a small clump of seaweed.

Cake Urchins and Sand Dollars

(Echinoidea)

Cake Urchins and **Sand Dollars** are relatives of Sea Urchins and are of a flattened appearance. They live mostly in tropical or warm seas. The body is made of a hard, fused test with a great number of very small spines which enable the creature to burrow into sand. All Cake Urchins and Sand Dollars are burrowers and some of them cover themselves completely. Others project above the surface, usually at an oblique angle. A number have been observed underwater and have been seen to throw up a pile of sand, completing their burrows in 15 to 20 minutes. **Keyhole Sand Dollars** can bury themselves rapidly in the space of 3 to 4 minutes. They do this by rotating the forward edge of the body and literally slice their way through sand.

The food of these curious creatures comprises minute organisms that live in sand. Sand Dollars possess a form of Aristotle's Lantern.

Reproduction is similar to that of Sea Urchins and only one brooding species is known.

Brittle Stars

(Ophiuroidea)

Brittle Stars, also known as **Basket** or **Serpent Stars,** are the most beautiful of the Echinoderms. Five long, slender arms trail from a small, central disc and colours are usually mottled or banded. The discs are often less than $1\frac{1}{2}$ in (3 cm) across although some species of Basket Star measure up to 4 in (10 cm) across. There are 1,600 species in existence and are generally found in the warmer waters of the world. Most crawl over the sea bottom, some swim rather feebly and others burrow into mud. The body is armoured with skeletal plates, and spines on side plates give grip when the animal crawls or burrows. Brittle Stars move by flexing the arms and feed on detritus and tiny marine animals which may be raked through the mouth by the arms. The mouth is in the middle of the underside of the central disc and is framed by hard chewing plates.

Like Starfish, Brittle Stars possess strong regenerative powers and can re-grow severed arms. Very few reproduce by division. Brooding takes place among certain arctic and antarctic species.

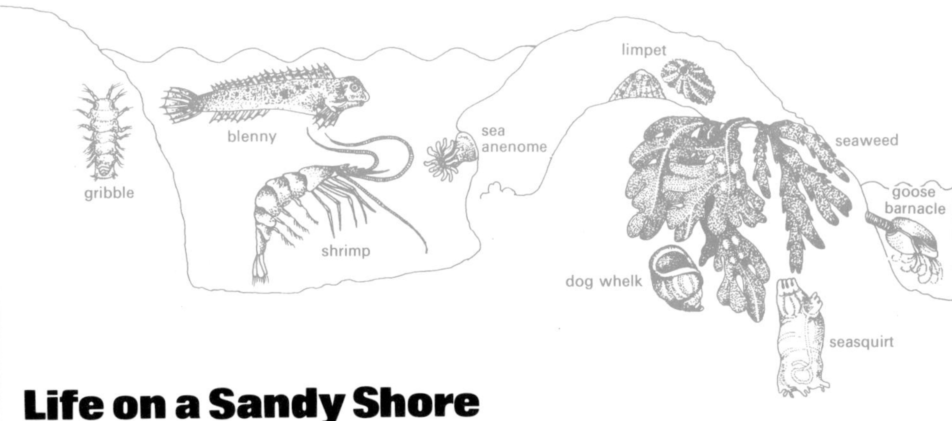

Life on a Sandy Shore

Rocky shores present a marine wonderland of plant and animal life. A firm anchorage is provided for seaweeds and these serve to protect rock from excessive erosion by wave action. Thus, seaweeds are found on all rocky shores apart from those that are open and exposed to heavy seas or gross industrial pollution. A variety of animals shelter in rock pools or beneath the fronds of seaweed. Many of these actually feed on seaweeds or scavenge on their broken or decaying remnants. Others, **Mussels** and **Barnacles** for example, sift rich sources of plankton from coastal waters. Every day a shoreline is subjected to the rise and fall of the tide and much of the plant and animal life is forced to withstand alternate periods of complete submergence and total exposure. Some animals, such as high or mid-shore **Winkles, Whelks, Topshells** or **Limpets,** move and feed when under water but remain fixed when the tide recedes; this ensures that their fleshy body parts are not exposed to the drying effects of the atmosphere. Although rocky pools represent a watery haven for these creatures, **Sea Anemones, Prawns, Crabs,** and small fish such as **Blennies** and **Lumpsuckers,** live in the pools too. These animals suffer from the fluctuations of seasonal or day and night temperatures and also from varying oxygen and salinity levels. A heavy shower may lower the salinity of a pool rapidly whereas a period of hot sun may cause evaporation so that a rise in salinity results.

Life on the Rocky Shore

At first glance, a flat expanse of sand may appear to be barren and devoid of living things. There is no firm anchorage for seaweeds nor are there sheltered rock pools. Sand-dwelling creatures are rather special and in order to survive they must burrow away from the crashing waves. By so doing, they do not expose themselves to the drying effects of the atmosphere and are protected from fluctuations of seasonal temperatures. At only 6 in (15 cm) below the surface, a constant temperature is maintained all year round.

Sand dwellers include certain **Marine Worms, Crustaceans, Molluscs, Echinoderms,** even fishes. Between them they adopt a variety of ways of obtaining food. Some **Tube Worms** put out collecting tentacles at high tide which collect surface detritus material and tiny particles of decomposing seaweed. As the **Lugworm** excavates its burrow it swallows sand, digesting the detritus material caught amongst the grains, and leaves coiled sand casts on the surface. *Nepthys,* the **White Worm** attacks and devours fellow sand inhabitants using its powerful jaws. The crustacean **Sand Hopper** feeds on detritus at the surface of wet sand. Its relative, the **Common Shrimp,** is surprisingly voracious and will consume anything up to the size of a **Ragworm.** In tropical sands, burrowing **Crabs** and molluscan burrowing **Snails** are very common. The best-adapted feeders include **Cockles** and **Razor Shells,** which at high tide project part of the shell above the sand, sucking in plankton-rich sea water.

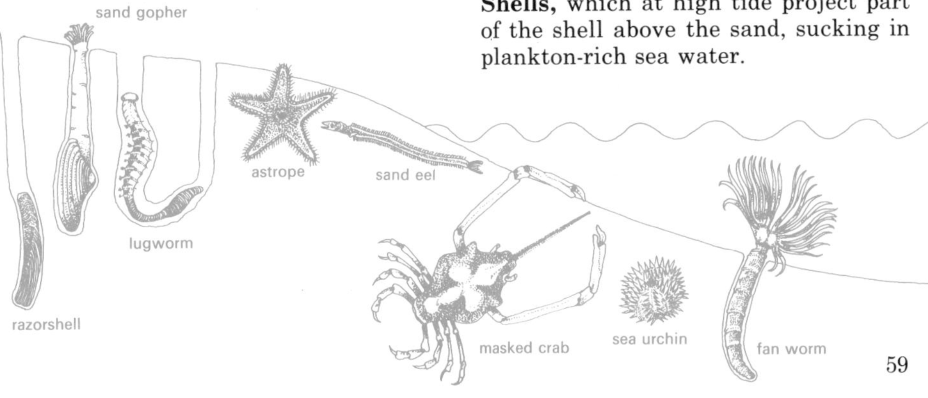

Of the many varied forms of life which exist in water, few are better adapted for a liquid environment than fishes. In order to move through the water efficiently fishes have developed forms which are closely related to their way of life.

All fish, both marine and fresh water varieties, belong to one of two distinct groups. Bony fish *(Osteichthyes).* which are the largest group, have as their name implies, a bony skeleton. Swordfish, Mackerel, Herring and Barracuda are all kinds of bony fish. The second group are the Cartilaginous fish *(Chondrichthyes)*, and these have skeletons of cartilage. There are three classes within the group, each with a different shape and different systems of locomotion.

Sharks, Rays and Chimaera are the three classes of the Cartilaginous fish. There is a third group of fish-like creatures which is not usually included in the general fish classification. These are the *Cyclostomes* of which there are comparatively few species. Lampreys and Hagfish are examples.

Hundreds of species of fishes in waters throughout the world have evolved into numerous and varied shapes to suit their environment. The streamlined bodies of some species allow the fish to swim quickly away from enemies while other species, such as Globe fish or Puffers, with their shortened, rounded bodies, achieve their protection by having projecting spines. Puffers are also able to inflate

themselves and project these spines—a very efficient weapon against enemies. Other species of fish have poisonous spines—the Weever fish for instance, has poison spines on its fins and gill covers and this defence, combined with the habit of burrowing in sand when danger threatens, is ample protection.

The power to produce electric discharges has been developed in four families of bony fish as well as in Torpedoes and Rays. These electric organs can develop voltages of more than 600 volts in the water. Fish that live on the ocean bottom are generally slower moving and would have very little chance of escaping from a predator if it were not for their ability to camouflage themselves. Turbots, and some species of Plaice are among those species which can change the colour of their skin to match the background colouring. Colour-change is also used by some species as a warning to predators that they are unpleasant to eat—Dragonfish and Triggerfish, for instance, give striking warning colour patterns. Sea Horses and Pipefish often appear to resemble sea weed—even developing protuberances which look like 'leaves' and spots and stripes are found on fish which live against a variegated background or among coral.

Fish have radiated into a wide and varied class of vertebrates, living in most available niches of the world's waters. Here Stripey Fish and Surgeon fish live around coral reefs.

FISH

The Evolution of Fish

The Devonian period, which occurred approximately 320 million years ago, was the golden age of fishes. Four classes of fishes had evolved and have been identified. These were the *Agnatha*, creatures without jaws, the *Placoderms*, which had bony plates in the skin, the *Chondrichthyes* (the cartilaginous fish) and the *Osteichthyes* (bony fish).

The most primitive of the chordate animals *(Chordata)* are the Hemichordates. These are simple, organised, burrowing marine worms which include *Balanoglossus* and *Dolichoglossus*.

In external appearance they resemble invertebrates rather than vertebrates. The identifying chordate structures are a dorsal nerve cord and a shorter ventral one, gill slits and a notochord. The larvae are of extreme importance. Small, bell-shaped organisms drifting with the ocean currents, these tornaria larvae look very similar to the larvae of echinoderms. It is possible that both groups evolved from a common ancestor.

The few animals which look more like invertebrates than vertebrates are known as the *Urochordata*. They include the Sea Squirts, *Ciondia* and *Oikopleura*. They are chordates because they possess, at some stage in their lives, a stiffening rod (the notochord), gill pouches and a dorsal tubular nerve. The Sea Squirt's larva, free living and tadpole shaped, has the notochord and nerve chord which is lost in the adult. Although these modern chordates have undoubtedly changed from the original unspecialised forms, they are not very different from the early chordates, which provide a vital link between the invertebrates and vertebrates.

There are no fossil traces to link the earliest chordates with the earliest fishes as both were probably soft-bodied. The first fish fossils date from the Ordovician period, over 400 million years ago. The Devonian period saw a great expansion in the development of fish and many fossils were found.

A typical fish of the jawless class is *Cephalaspis*. Flattened in shape for a bottom-dwelling life, this was an armoured fish with a bony head shield and a heavy, scaled body.

The Sharks, Rays and Chimaeras probably evolved from the same ancestors, becoming adapted to all kinds of existences by the time of the Upper Palaeozoic era. Many species became extinct but 600 survive and among them are one of the world's most feared fish—the man-eating sharks.

The skeleton of these fishes is unique, because it is not made of bone, but of a softer substance called cartilage. The hard parts of the fish are the teeth and the spines of the fins.

The bony fishes *(Osteichthyes)* are the most abundant, diverse and complex. They evolved in the mid-Devonian era and, due to their efficient mode of swimming and feeding, became the most successful class. Two distinct types of bony fishes existed in the beginning—the ray-finned fishes (the *Actinopterygii)* which can be sub-divided into three orders, and the *Chondrosteans*, which have almost died out except for the Birchirs *(Polypterus)* found in African fresh waters.

In the Permian era, Holosteans largely replaced the Chondrosteans. The only remaining survivors, the Bowfin and the Gar-pike, are found in the lakes and streams of North America. These survive, in spite of competition from higher bony fish.

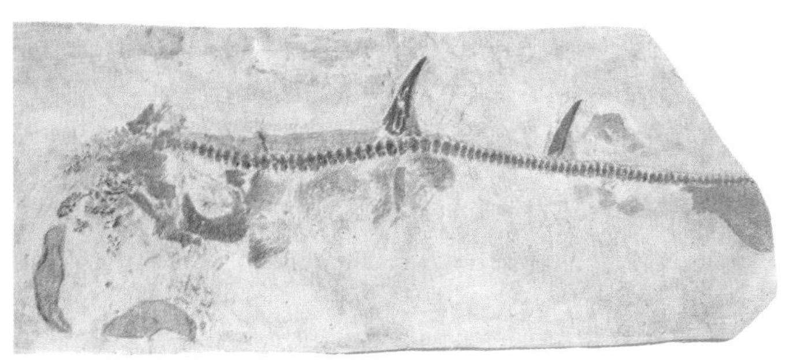

◁◁The Coelacanth fish was thought to have been extinct for 70 million years.
△ Sea Squirts are very primitive marine chordates, showing little similarity to the more advanced forms that probably gave rise to vertebrates.
◁▽ Star Sea Squirts are composite structures, being a number of animals embedded in jelly and arranged radially round one outlet.
△ The Port Jackson Shark is a survivor of forms found as fossils dating from about 200 million years ago.

By the end of the Triassic period, the modern bony fish, Teleosts had evolved from Holostean ancestors. Several different lines of evolution had begun and today approximately 20,000 species of bony fishes exist in the waters of the world. The Teleosts have adapted themselves so that a species of fish exists for every type of aquatic home.

The air-breathing fish are in the Crossopterygian division of bony fish. The surviving species are often called 'living fossils', the term covering the Lungfish and the Lobefin Coelacanth. Fossil Lobefins show that they lived in fresh water and they appear to have become extinct quite early. It was assumed that the group was completely extinct. However, the amazing discovery of a Coelacanth, caught off East London in South Africa in 1939, gave proof that these fish had been living almost unchanged for 70 million years. Many specimens have since been caught and the fish studied extensively.

Lungfishes are interesting because they possess some of the features which enabled some of their evolving ancestors to emerge from water and become land-dwelling vertebrates. As their name implies, they possess lungs and gulp air from the surface of the water. The Australian Lungfish has limb-like lobes as fins, while other lungfishes have paired fins that are reduced almost to feelers.

From outward appearances, the senses of smell, touch, sight, hearing and taste are not very apparent in fishes. Usually, however, a fish has all five senses. The sense of smell is quite acute and it is well known that Sharks and Piranha can smell blood and flesh from some distance away. Migrating Salmon find their original birth places by smell. The closely related sense of taste is well evident with taste buds being found on the tongue, the head and the body of species such as Cod and Sturgeon. Taste buds are also numerous on fishes such as Catfish and Carp which have barbels or feelers.

The eye of a fish is very like the human eye but modified for seeing under water. Most fish eyes are on the sides of the head which does not give overlapping or binocular vision. Most bony fish have colour vision while the cartilaginous fishes, Rays and Sharks, are probably colour blind.

Fishes do not possess outer ears or ear drums. The hearing organ is the inner ear which is simple in structure. The fish detects sound and gains a sense of balance through this organ.

Fishes can produce sound and communicate with other fishes. There are also pressure-sensitive nerves along the body with branches to the head region. This system allows the fish to detect movements or obstacles in the water.

Primitive Parasitic Fish

The Jawless Fish, the Lampreys and the Hagfish are parasites and scavengers. With sucker mouths surrounded by horny teeth, they rasp their food from the flesh of other fish.

Jawless Fish, the smallest class of all fish have only one remaining order, *Cyclostomata.* These are the only existing representatives of the primitive vertebrates, the *Agnatha,* fish-like animals without jaws, which fed by sucking food through their mouths. Their limited feeding mechanism did not enable the animals to develop into active predators and all but two orders are extinct. These are the Hagfishes *(Myxonoidea),* and the Lampreys *(Petromyzontia).*

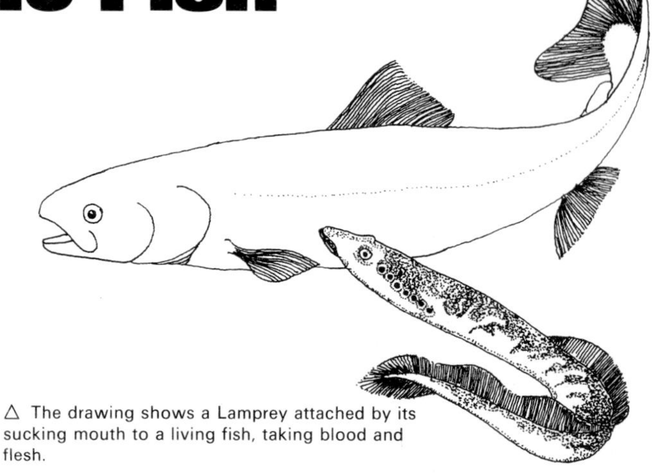

△ The drawing shows a Lamprey attached by its sucking mouth to a living fish, taking blood and flesh.

Hagfish

(Myxinoidea)

The eel-shaped **Hagfish** (not a fish but often so-called) are jawless and are scavengers and predators. They lie buried in the seabed during the day, emerging at night to feed on dead fish or waiting to burrow into the tissues of living fish. They are sightless, hunting food with the aid of sensitive tentacles on their snouts, and by smell. Horny teeth on the lips and tongue enable the Hagfish to bore into their prey, where they feed from the inside, secreting quantities of slime as they do so. Hagfish differ from Lampreys in that they have six to fourteen gill clefts on each side and only a narrow caudal fin.

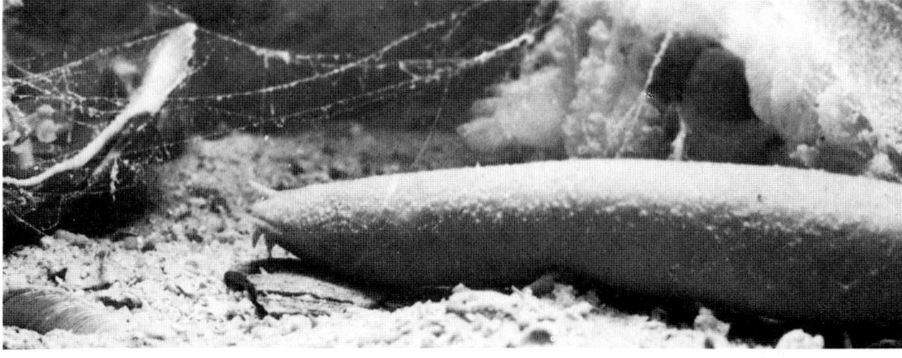

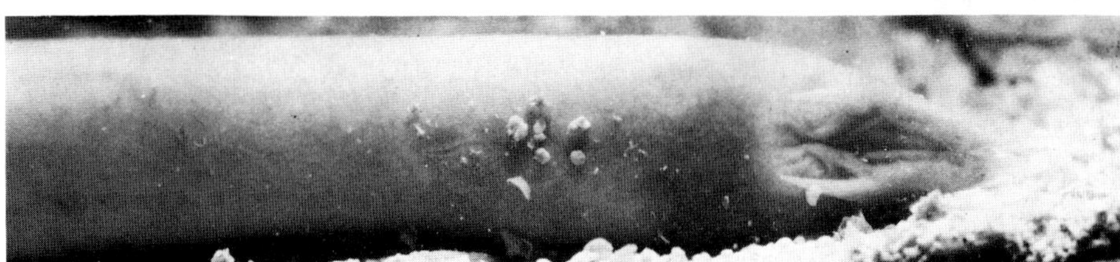

Lamprey

(Petromyzontia)

Lampreys, of which there are both freshwater and marine species, resemble the Hagfish generally but are distinguished by seven pairs of gills on each side and two dorsal fins. Though not all adult Lampreys feed after the larval stage, some attach themselves to rocks or stones, waiting for passing prey. They quickly attach their circular sucking mouths to the living fish, rasp-like tongues and ring of horny teeth tearing at the host's flesh, to suck out the blood.

Dramatic metamorphosis

During the spawning season, the Lamprey ascends the rivers to streams. Where the waters are swift-running, the female Lamprey makes furrows on a gravel bottom and deposits her eggs—about 60,000 of them. Several males may attach themselves by their mouths to the head of the female and they fertilize the eggs and bury them into the gravel. This is the end of the adult Lampreys and they die, their lives having lasted from twelve to fifteen months.

A very small proportion of the eggs hatch—about one in a hundred—and after ten days or so, the tiny larvae, having drifted from the original nest, burrow into the muddy river bottom. There they remain, feeding on organic matter for two to five years. The larva metamorphoses slowly into a six-inch long adult Lamprey, and moves back down-stream and into the sea to begin its life as a predatory marine animal.

◁ Head of a river Lamprey showing horny 'teeth'.
▷ Lampreys spawning: the female clamps herself to a stone while one or more males become attached to her head, twining around her. She releases eggs and he his sperm.
△▷ The Lampreys and the Hagfish are the only survivors of the cyclostomes or round-mouths.

Sharks, Rays and Chimaeras

Sharks and Rays are found in all the world's seas and these fishes are either scavengers or carnivores. But unlike the popular folklore, not all of the sharks are dangerous man-eaters. The biggest, the Whale Shark, is harmless to Man and feeds on tiny plankton. But others, particularly the Great White, the largest carnivorous fish in the sea, are ferocious and will attack without reason even killing on one occasion an elephant!

Existing cartilaginous fishes, class *Chondrichthyes*, may be separated into two smaller groups: *Euselachii*, the modern Sharks and Rays and *Holocephali*, the Chimaeras. The Sharks and Rays can be further separated into the orders *Pleurotremata* and *Hypotremata*. There are about five different groups of extinct relatives.

Cartilaginous fishes all have jaws and their skeletons are made of cartilage or gristle rods, rather than bone. Their bodies are covered with placoid scales—small tooth-like projections—covered with enamel. But apart from these common features, cartilaginous fishes are distinctly different from each other.

Rays and Skates, the largest order, of which there are 350 species, have flat bodies with very large pectoral fins and long thin tails. They live just above the ocean floor (with the exception of the Manta Ray which lives near the surface). The large fins, which extend almost the length of the body, are flapped to propel the fish along.

Rays feed on plankton, small fishes and molluscs. Some species can give electric shocks, but the shock is mostly used for defence or for stunning prey into immobility. Sharks, of which there are approximately 200 species, have cylindrical, streamlined bodies and although cartilaginous fishes have no swim bladder to keep them buoyant, they are a swiftly swimming fish and the necessary lift is provided by the shape of the strong tail and the pectoral fins. Sharks swim by undulating the whole of their bodies but, in spite of their streamlined looks, are not particularly agile.

Most species of Shark are carnivorous and some, the Great White Shark *(Carcharodon carcharias)*, and the Tiger Shark *(Galeocerdo cuvieri)*, in particular, are dangerous man-eaters. Sharks' jaws are strong and their rows of sharp teeth develop in succession as the front teeth wear out. The teeth vary in shape between the different species and only rarely are the teeth shaped for crushing. Generally, two main forms of teeth are found in Sharks; sharp, triangular plate-like teeth which are used for tearing and cutting, and long, pointed teeth used for spiking. Mackerel Sharks *(Lamna oxyrinchus)*, and the Blue Shark *(Carcharhinus glauca)*, have ventral crescent-shaped mouths and large, spear-shaped teeth. The Greenland Shark has oblique teeth in the lower jaw and smaller teeth in the upper jaw.

The shark's sense of smell is very highly developed and it can scent blood in the water from more than a quarter of a mile away. It can also detect changes in pressure, often the direct result of the struggles of wounded fish, and is speedily in the vicinity. The Shark does not appear to have an acute sense of hearing but can detect vibrations in the water by its lateral line system.

In reproduction, sharks fertilize internally. Some species bear their young alive while others lay eggs, ejecting them in flattened cases. These cases harden on contact with the water and are known as 'mermaid's purses'. Each of the eggs has a single embryo which—according to the particular species—takes up to 15 months to develop. When the baby shark is eventually born, it resembles its parents in miniature.

Freshwater sharks are found in two kinds of fresh water; in rivers which have direct access to the sea and, surprisingly perhaps, occasionally in inland lakes without sea outlet.

The Ganges River in India and the Zambesi each have their own named species, the Zambesi Shark being particularly ferocious. Sharks have also been noted in many of the large rivers in the world; the Fitzroy River in Australia, the Sarawak River in Borneo and the Amazon River, among them.

Chimaeras, the third and smallest group of the Chondrichthyes, are strikingly shaped and have a fixed upper jaw and three pairs of grinding teeth. These fishes have thick bodies, long tails and large pectoral fins which are used in swimming.

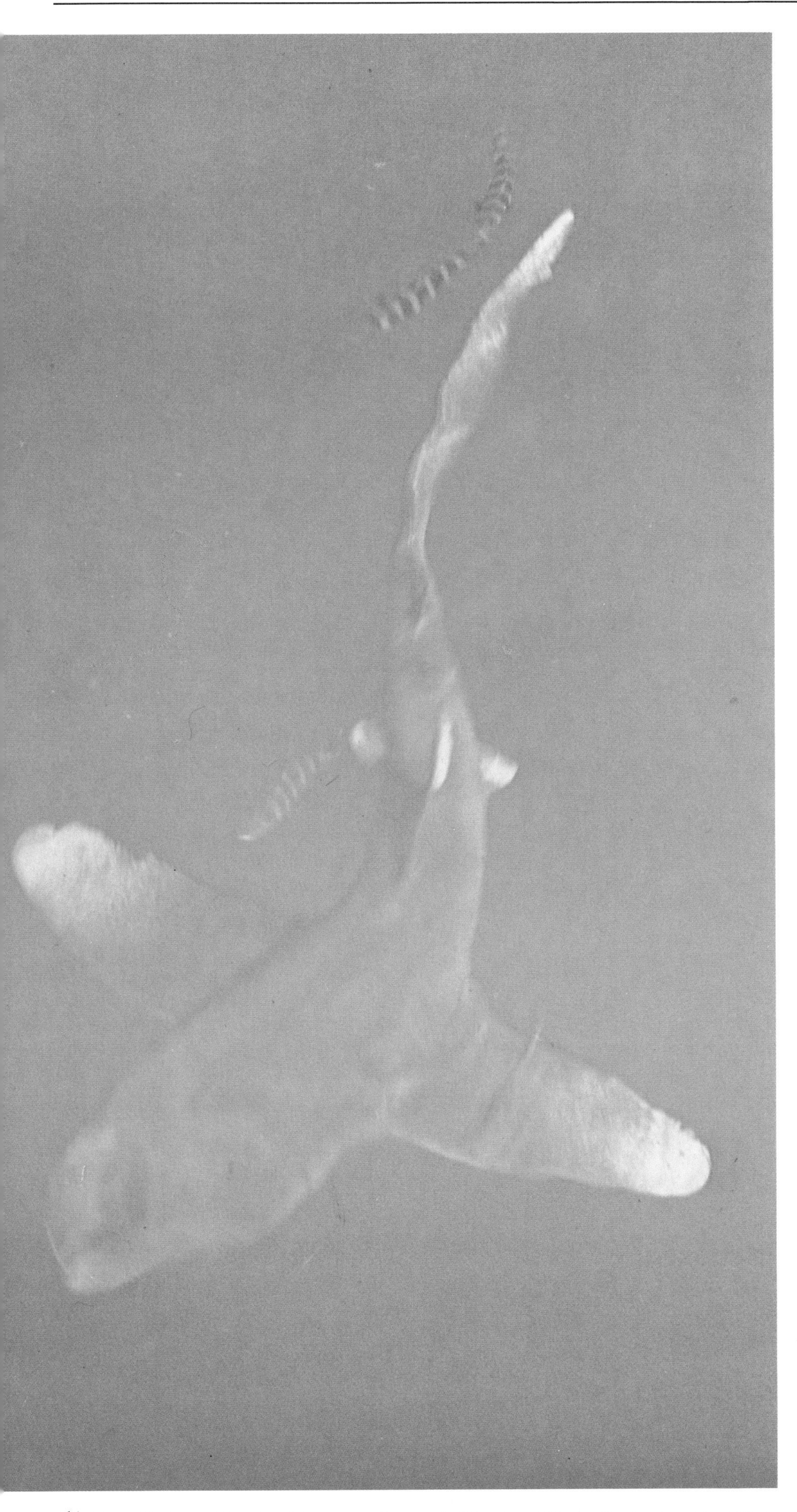

◁△ Mermaid's Purses: These Dogfish egg capsules are hollow, with slits in them to let currents of water pass through to the developing embryo inside. When fully developed, the little Dogfish escapes through a slit at one end of the case.

△ White-tipped Shark with attendant striped Pilot Fish. These fish were once believed to guide the Sharks towards suitable prey, in return receiving protection from predators. In fact, the Pilot fish probably accompany a Shark so that they can pick up the debris from its killings. It has been observed that Pilot fish are careful not to swim too close to the jaws of the Shark they are attending, which seems to disprove the myth that a protective relationship exists.

Port Jackson Shark

(Heterodontidae)

As well as being one of the strangest looking sharks, with its broad, snout-shaped head and two spine-edged dorsal fins, the **Port Jackson** *(Heterodontus)*, is also one of the oldest. *Heterodontidae,* primitive sharks, date back 180,000,000 years to the Upper Jurassic period, and the rarely seen Port Jackson could be described as a living fossil. Very few remain and those which do frequent the warm coastal waters of the Pacific and Indian Oceans. They reach lengths of 2 metres and feed mainly on molluscs. The Port Jackson is unique in that it has the equivalent of both incisor, or biting teeth and molar, or crushing teeth.

Sand Sharks

(Carchariidae)

Sand Sharks, which have brown-flecked or spotted upper surfaces, have a disconcerting habit of resting on the sea bed in shallow waters and are sometimes seen quite close to the shore. Sand sharks, even when small, can inflict severe injury to man and were once feared, in some parts of the world, as man eaters.

Nurse Sharks

(Orectolobidae)

Nurse Sharks *(Ginglymostoma cirratum)* are allegedly so-called because they retain their eggs until the moment of hatching. They are seldom a menace to man in spite of their size, which can be up to 18 ft (6 m) long. Closely related to the Sand Shark, Nurse Sharks are extremely supple and swim almost like snakes.

Wobbegongs

(Orectolobidae)

Wobbegongs, or **Carpet Sharks** *(Orectolobus)*, live much of their lives on the sea bed. The colour and texture of their skin is such that they blend inconspicuously with the sea bottom where they await their prey. They reach lengths of 15 ft (4–5 m), and are found mostly in seas off the shores of Australia and Japan. Carpet Sharks are a viviparous species, the females giving birth to live young.

Great White, Tiger, Hammerhead and Sawfish Sharks

(Lamnidae; Carcharhinidae; Pristiophoridae)

Perhaps the most ferocious of all sharks is the **Great White Shark** *(Carcharodon carcharias).* It is the largest carnivorous fish in the sea, exceeding lengths of 25 ft (9 m) and weighing more than 100 cwt (5,000 kg). The White Shark is a particularly dangerous specimen because it becomes so frenzied in a kill that it is likely to attack anything else near it.

△ Shark with Remora: Remoras have a dorsal fin which is modified as a sucker. It can disengage itself from the host fish if required. Remoras feed on debris from the Sharks own food.

The stomach of a **Tiger Shark** *(Galeocerdo cuvieri),* was once found to contain a horse's head and bits of a bicycle while a petrol can and a cow's skull were found in another. **Tiger Sharks,** like the Great White Sharks, are a ferocious species and will attack quite large fish, including other sharks. Their massive jaws are packed with razor teeth and over a ten year period a Tiger Shark will grow, use and shed 24,000 of them. Tiger Sharks are viviparous and it is reported, grow up to 25 ft (4 m) in length.

The **Common Hammerhead Shark** *(Sphyrna zygaena),* takes its name from its extraordinary appearance. The head broadens into the shape of a hammerhead with eyes and nostrils at each end. It has sometimes been suggested that this development has given the Shark better vision and directional sense of smell. Hammerheads are found all over the world. There are nine species altogether and all are considered highly dangerous. They often hunt in packs and are scavengers, following ships and waiting for waste to be thrown over-board.

The snout of the **Saw Shark** *(Pristiophorus),* is elongated and studded with teeth down each side. The fish uses this bill or 'saw' to catch its food, swimming into a school of fish and swinging it rapidly from side to side in a slashing motion. To avoid injury to the mother, the Saw shark's teeth are folded flat before birth.

Thresher Shark and Basking Shark

(Lamnidae; Cetorhinus maximus)

The tail of the **Thresher Shark** *(Alopias vulpinus),* is often as long as the rest of its body and has evolved into a most efficient hunting aid.

The Thresher uses its scythe-like tail to flail or 'thresh' the water, either herding small fish together or stunning them so it may return later and eat at leisure. It also attacks sea-birds floating on the surface.

It is otherwise known as the **Fox Shark** and grows to a length of up to 6 metres. Mainly a surface fish, the Thresher is distributed throughout the temperate and tropical waters of the world.

Basking Sharks *(Cetorhinus maximus),* have occasionally been struck by passing ships as they lie near to the surface, 'basking' in the sun.

Some authorities have concluded that the disappearance of the fish in winter is explained by the fact that plankton is scarce. The Basking Shark, it is suggested, lies on the ocean bed and 'rests'—rising to the surface again in the spring when food becomes more plentiful. They are sluggish creatures and quite harmless to man. Their diet consists mainly of plankton which is strained through filters in the same way as whales. Basking Sharks can grow to lengths of 36 ft (12 m), and weigh up to four tons (tonnes). During spring and summer, the great fish move in schools. But the rest of the year they remain alone.

The Whale Shark

(Orectolobidae)

In spite of being the largest fish known —it can grow more than 50 ft (15 m) long—the **Whale Shark** *(Rhincodon typus),* is one of the most docile. Divers have been known to have clung to its dorsal fin without the giant creature making any attempt to harm them. It should not be confused with *Carcharhinus macrurus,* a man-eating shark of Australian waters, which is sometimes called the **'Whaler Shark'.**

The Whale Shark frequents warm and deep waters and is rarely seen near the surface. However, it will sometimes seek out boats of a size similar to its own.

Whale Sharks feed on plankton and small fish, which they filter through a gill apparatus as they cruise through the water.

Dog Fish

(Scyliorhinidae)

Dogfish are all small, seldom exceeding 3 ft (90 cm) in length. They are also referred to as **Rock Salmon** and **Rock Eel.**

The **Common Spiny Dogfish** *(Squalus)* is a variety of shark.

The two other most common species are the **Greater Spotted Dogfish** *(Scyliorhinus stellaris),* and the **Lesser Spotted Dogfish** *(Scyliorhinus caniculus).* Their bodies are light brown in colour and covered in black flecks or spots.

Dogfishes are found in large numbers off the British coast as well as in the Mediterranean and eastern Atlantic Ocean. They feed mainly on molluscs, crustaceans and worms.

The Greater Spotted Dogfish is sometimes referred to as the **Bullhuss** or **Nursehound.**

None of the Dogfish family is considered dangerous to man although if not treated with a certain amount of respect, they are quite capable of ripping skin from a hand or a foot.

Electric Ray

(Torpedinidae)

The electric organs which lie on either side of the **Electric Ray's** body, between the pectoral fins and the head, enable it to produce a shock of up to 300 volts—enough to stun a large fish.

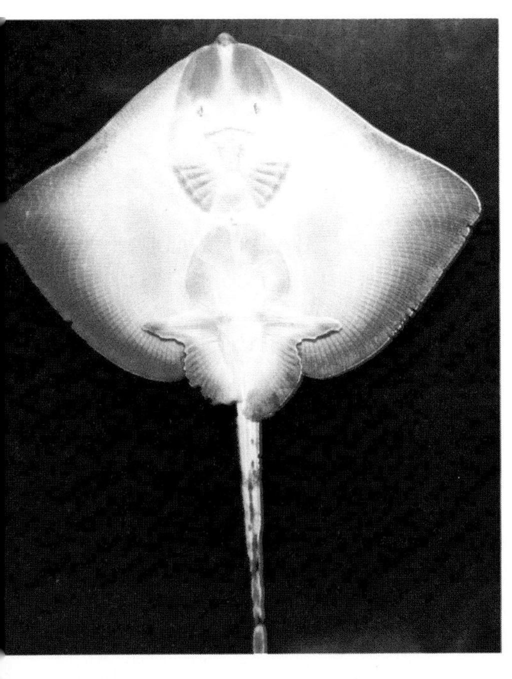

These organs are kidney shaped, and made up of muscle broken down into a honeycomb of tiny elements lying in jelly-filled tubes.

As Electric Rays spend much of their time close to the sea bottom, they draw water in through two large spiracles behind the eyes, rather than through the mouth.

There are 30 different types, most of them frequenting the warmer waters of the world although some are by no means rare off the shores of Britain. The largest is the **Atlantic Torpedo Ray** *(Torpedo nobiliana).*

A great deal of research has been carried out into Electric Ray behaviour, but it is still not yet known how the Ray first locates its prey, or at what range it can effectively stun its victim.

The **Guitar Fish** is a fish of another family, rather like a guitar in shape but with a long tail. It grows to a length of about 1½ ft (50 cm).

Sting Rays and Mantas

(Dasyatidae: Mobulidae)

The strength of the poison contained in a **Sting Ray's** tail varies from species to species, but some are certainly capable of killing a man.

The whip-like tail is usually longer than the body and is armed with one or more mobile spines. The venom is secreted by glandular organs in the skin and flows down grooves to the spines.

There are approximately 90 different species of Sting Rays and they are in all the world's seas.

The biggest of the Ray family is the **Manta Ray** *(Manta birostris)* or **Devil-fish.** This formidable creature has a breadth of more than 20 ft (6 m), and can weigh as much as one and a half tons.

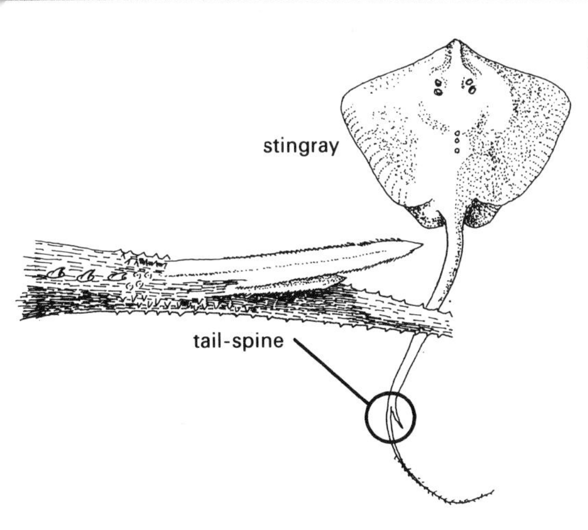

△△△ Underside view of a Ray: The claspers, associated with breeding can be clearly seen. The large fins, fully extended, fuse with the snout of the fish and to the tail. The tail is a whiplike structure used for steering only. The propelling force is provided by the large fins.

△△ Skate: The camouflage colouration enables the fish to be concealed when lying on the seabed. The rippling of the pectoral fins provides the propulsion of the fish.

△ Drawing shows a detail of the tail spine of a Sting Ray. The sting takes the place of a dorsal fin and sometimes grows up to 18 inches long. When the tail is lashed from side to side or curled around prey, ugly jagged wounds can be inflicted. Sometimes, two or three spines can be present on the tail of a Sting Ray at the same time.

Skates

(Rajidae)

Skates, in common with other flat fish, are not born flat. They hatch in a rounded shape which changes as the fish assumes its adult form.

Skates and **Rays** belong to the sub-order *Batoidea* and use their large pectoral fins to swim close to the ocean bed. These are flapped slowly, creating a wave which sweeps back, thus providing the Skate with a surface to push against.

Some members of the Skate family possess small electric organs with which they can inflict slight but perceptible shocks. These organs lie along the sides of the tail and are connected with nerves from the spinal cord.

The electric power produced by the Skate is, however, very low. The **Thornback Ray** *(Raia clovata),* for instance, which is found off the coasts of Britain, produce only four volts. This is used as a means of defence, not as a way of attacking other fish.

Skates vary greatly in size. One of the largest is appropriately named the **Barn-door Skate,** an Atlantic species.

Skates feed on molluscs, crustaceans and small fishes.

However, in spite of its name and size, the Manta Ray is a completely harmless fish and feeds on plankton and small molluscs. Mantas do sometimes leap out of the water, but usually only in play and, unlike other Rays, swim near the surface.

Holocephali

(Chimaeridae)

The third sub-class of the cartilaginous fishes are the primitive *Holocephali,* including the family *Chimaeridae,* of which only 25 species remain. One genus, the **King of the Herrings** *(Chimaera),* is found in many parts of the world, in deep off-shore waters. Chimaeras look rather like sharks but with blunted heads and small mouths. Their pectoral fins are large and in the **Ratfish** species, *(Chimaera monstrosa),* the tail is long. Chimaeras feed on crustaceans and molluscs.

All the Chimaeras have an unusual appearance but the most startling is the **Elephant Chimaera** *(Callorhynchus).* This possesses a long snout which grows downwards and then curves back up again towards the mouth.

Bony Fish

Fossils of bony fish appeared suddenly in the Devonian period, in fresh waters, and all the major forms were present. The previous stages in their development must have occurred earlier, perhaps in the upper reaches of rivers from which no deposits have been preserved. These bony fish flourished until by the end of the Paleozoic period they were the major occupants of lakes and rivers and had also entered the seas. Unlike many other groups that were prominent in geologic periods and subsequently became reduced or extinct, the bony fish have increased their importance both in species and biomass.

There are countless millions of bony fish divisible into about 25,000 species forming over half of the total 42,345 species of vertebrates. Today nearly all of the freshwater fish as well as the majority of the marine ones belong to the class Osteichthyes, the bony fish. They have penetrated a wide range of habitats from shallow coastal waters to abyssal depths of the sea. Besides rivers and lakes bony fish are found in hot springs, artesian wells, caves, torrential rivers, almost freezing water and even the gill chambers of other fish.

Most of the bony fish have streamlined bodies, with skeletal elements of bone, and are active swimmers. They have paired pectoral and pelvic fins, median dorsal fins and a vertical tail fin. The fins have a thin membrane between the hard or soft supporting rays. Swimming is achieved by lateral undulations travelling along the body, the tail fin providing the final thrust. Most obtain oxygen through paired gills with a single opening on either side covered by an operculum, a bony flap. A few freshwater fish have sacs or primitive lung-like structures and they come to the surface to obtain air. The majority of the bony fish have bilateral symmetry, at least during their early life; however flatfish, such as the Plaice, come to lie on one side after passing through metamorphosis, a change in their external form. Fertilisation is external, the eggs and spermatozoa being shed into the water at the same time. The eggs of most of the bony fish are about 1 mm in diameter. The heaviest of the bony fish is the Sunfish which can weigh over 5,000 lbs (2,240 kg), the longest is the Sturgeon, 283 inches (720 cm), and the smallest the Marshall Island Goby, 12 to 16 mm.

△ Batfish *(Platax pinnatus)*. Batfish are tropical sea fish, generally found near rocks, reefs and sand banks. They have flattened, deep bodies, small bony mouths and long pelvic fins. They are medium sized and grow to between 8 and 12 inches in length.

A very wide diversity of shape and form is seen amongst this group of fish, which include long tubular-like Eels, prickly Puffer Fish, strangely shaped Sea-horses and the more usual form of the Salmon, Cod, Mackerel or Herring.

Many-finned Bichirs

These are primitive and belong to the sub-class *Actinopterygii*, the ray-finned fishes and the infra-class *Chondrostei*, today represented by two groups of descendants: The **Bichir,** *(Polypterus),* meaning many fins, and the **Reedfish,** *(Calamoichthys).* Both live in the freshwaters of tropical Africa. The body is covered with thick shiny scales and the dorsal fin is formed by a series of small finlets or 'sails' which may be erected if the fish is alarmed or excited. It will often support itself on its fan-like pectoral fins which have a well-developed fleshy lobe. The remarkable feature of this fish is the

retention of a pair of ventral lungs in the form of a simple bilobed sac opening out of the bottom of the throat. This is of great importance in allowing the fish to survive during periods of seasonal drought. The larva has leaf-like external gills such as are found in amphibian larvae but are unknown in other bony fish.

The Sturgeons

(Chondrostei)

This is the second group of degenerate survivors belonging, like the Bichir, to the infra-class *Chondrostei.* Their ancestors had a dense bony skeleton but this has been replaced by cartilage, the scales have been lost and instead there is a series of fairly large, conical, bony plates along the body.

Some 25 species of Sturgeon are found in the cold temperate waters of the northern hemisphere. Some spend much of their life in the sea and migrate into rivers to spawn, but others live entirely in freshwater. They are slow-moving and browse along the bottom; the mouth, on the underside of the head, is protusile and sucks in food material. Fleshy barbels around the mouth aid in the detection of prey which includes molluscs, worms and other small invertebrates.

The Sturgeons include the longest bony fish, the **Beluga,** which may reach a length of 283 ins (730 cm), and a weight of some 2,600 lbs (1,200 kg). Sturgeons are valued for their eggs, the gourmet's caviar, as well as the swimbladder from which isinglass is obtained, while the skin is tanned.

The Bowfins and Gars

(Holostei)

The **Bowfin** is a cylindrical, solid-looking fish with a long dorsal fin, and armour-like scales. It usually swims slowly, movement being by means of a series of waves passing along the dorsal fin; it can also swim faster by undulations of the body. During the breeding season the male makes a rough nest on the bottom, and, after mating with several females, will guard the eggs. After hatching the male continues to care for the young until they can swim well.

The **Gars,** or **Garpikes,** have a long, thin body covered with thick, shiny, diamond-shaped scales and a dorsal fin close to the tail. They live in weedy waters and spend much time moving slowly through it but accelerate rapidly when prey is sighted.

Herring and Relatives

(Teleostei)

These, and all the fish in the following sections belong to the great infra-class **Teleostei**, the dominant group of fish since Cretaceous times. The **Herring** belongs to a primitive but extremely successful order in modern times—the Clupeiformes. Besides the Herring it includes the **Anchovies, Sprats, Pilchards** and also the lesser known **Shads** which enter freshwaters during their spawning period. All are silvery fish with scales that are easily lost.

The Herring has been fished off East Anglia since the 5th century and today it is still caught in large numbers. A female lays 5 to 200,000 sticky eggs on the sea bottom where they adhere to stones. At 10 to 20 days after hatching, the young fish resembles the adult and they form shoals that move towards the coast. When about 6 inches (15 cm), long they move out into deeper waters where as adults, of up to 10½ inches (27 cm), they form great shoals. Some shoals have been estimated to contain some 150 million Herrings.

Sprats, Anchovies and Pilchards, smaller fish than the Herring, are also generally found in shoals. In contrast, the Shads are solitary fish and are larger, reaching a length of about 19½ inches (50 cm).

▽ The female Salmon lays her eggs in river gravel and covers them up. When the larvae are hatched, they carry yolk in a sac on the belly and feed on this. When the yolk has been consumed, they begin to feed by mouth and are called alevins.

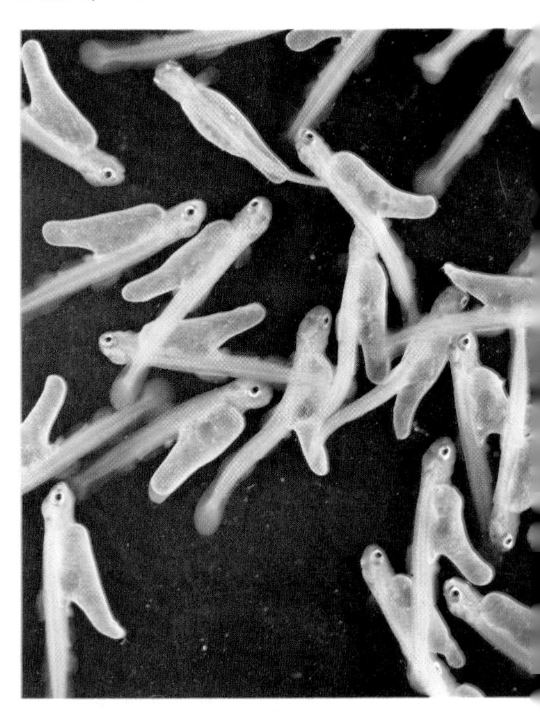

Trout and Salmon

(Salmoniformes)

Slender predatory fish of fresh and salt waters, the characteristic feature of fish of this order is the presence of two dorsal fins, the posterior one being adipose, or fatty.

The **Salmon,** *(Salmo salar),* and the two forms of the Trout *(Salmo trutta),* the **Brown Trout** of rivers and lakes and the **Sea Trout,** all have migratory habits. Their eggs are always laid in fast-running fresh water, the young of the Brown Trout then migrate downstream or into lakes, or in the case of the Sea Trout and the Salmon, to the sea, where they feed well and grow to adulthood before returning to their natal river to reproduce. When Salmon enter rivers they are beautiful, sleek, silver creatures that delight the angler.

Grayling and Smelt

(Salmoniformes)

Both these fish have the characteristic adipose dorsal fin. The **Grayling** belongs to the family Thymallidae, characterised by the long-rayed dorsal fin—the last of which is as long as those in the middle of the fin. It is a freshwater fish found in fast-running rivers of the Midlands and south-east of England. At spawning the eggs are buried in the gravel. The fish lives for about 6 years.

The **Smelt,** or **Sparling,** lives in inshore waters off the east coasts of Scotland, England and Ireland and congregate in the river estuaries in spring. They move up river to spawn in their second year. The eggs adhere to stones and hatch in about 20 days. Usually a length of 7 to 8 inches (20 cm) is reached, and a weight of about 8 oz (225 gm), but one of 12½ inches (32 cm) has been recorded. They feed on a wide range of fish including Herrings, Sprats, Whiting and Gobies.

Deep-sea Fishes

Bristle Mouths are small fish, about 3 inches (7.5 cm) long, with thin fragile bodies; they are also called **Lightfishes** since they have luminous organs. They belong to the family Gonostomatidae and the order Salmoniformes. They are abundant in the depths of the sea but are usually only seen by people on research ships. One genus, *Vinciguerra,* produce eggs which float but it is difficult to distinguish their larvae from those of Sardines.

Hatchet fishes of the family Stomiatidae, and order Salmoniformes, are found at depths of 750 to 1,500 feet (250 to 500 metres). They are deep-bodied, compressed and when seen from in front, little, but the large gaping mouth and bulbous, tubular eyes can be seen. The lower part of the body has many light organs.

Viperfish are active swimmers found at between 5,000 and 6,000 feet (1,800 metres), rising at night to the surface. They have large heads and the elongated body tapers towards the tail. The mouth can be very widely opened and, with its long, dagger-like teeth is capable of taking and holding large prey. The transparent larvae of these fish have eyes at the end of long stalks, but as they grow so the eyes form an integral part of the head.

Other Stomiatoids have chin filaments or barbels, used as feelers. One fish, *Ultimostomias,* has a chin filament ten times the length of its body.

Arapaima Giant

The **Bony Tongues,** of the order Osteoglossiformes, are primitive fish whose members are found in the freshwaters of Africa, Australasia and South America. They are characterised by their use of the teeth borne on the tongue and roof of the mouth rather than those of the upper and lower jaws.

The **Arapaima** of South America is a large freshwater fish reaching a length of about 10 feet (3 metres). It lives in murky waters and feeds on fish, even taking the armoured Catfishes. The swim bladder has been adapted for breathing air and an adult will surface once every 12 minutes or so in order to breathe. Its jaws are upturned so that the fish can breathe while virtually submerged. The eggs are laid in holes dug into the river bed and guarded by the male who continues to care for the young fish after hatching. The **Arawana,** also of South America, swims at the surface of the water. It has two leaf-like processes on the end of its lower jaw, which is uptilted. These are held stretched out in front of the fish and probably indicate the presence of insects, its food.

Pike and Relatives

(Salmoniformes)

These fish live in freshwaters of North America and Eurasia. The **Pike** is found throughout the British Isles and Europe except Portugal, and in parts of Russia. **Mud Minnows** are found in North America and in western Australia and Hungary. The **Alaska Blackfish** lives throughout northern Canada as well as in Alaska and also in Siberia.

The Pike is a solitary predator, stalking its prey and darting rapidly towards it when close enough. It can take fish almost as large as itself, indeed one of 9 lbs (4 kg) has been recorded as taking a Salmon of almost the same weight, but 3 days passed before the entire body of the Salmon disappeared from the mouth of the Pike. Most Pike begin to breed at two years and the eggs adhere to plants. After hatching the young remain attached to the plants by a fastening organ and they soon begin to feed on small crustacea. Large females reach some 37 inches (95 cm) although the males are smaller. Most survive about 12 years although ones of 17 years are known.

The Mud Minnows are small fish as is the Alaska Blackfish which reaches about 8 inches (20 cm). The Blackfish can withstand being frozen up in the icy rivers or in the partly dried bogs of northern Canada for the entire winter and still survive.

◁ Northern Pike *(Esox lucius)*. Pike are voracious feeders and can take fish almost as large as themselves. They have binocular vision, important for snatching prey and back-slanting upper teeth which traps the prey within its mouth.
▽ Drawings show the position of electric organs on the Electric Eel, the Catfish and the Ray.

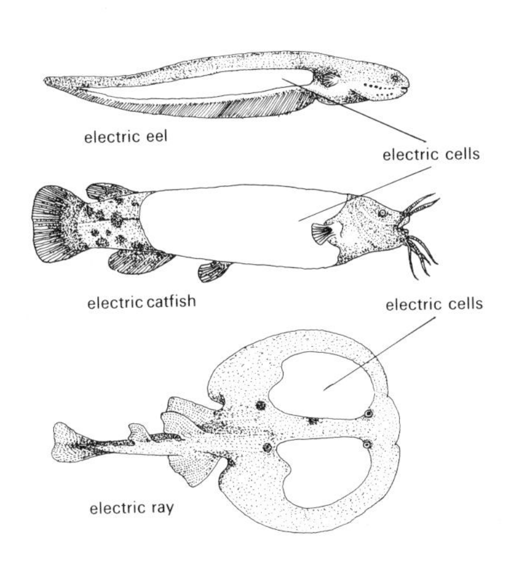

▷ The deadly Piranha fish swim together in schools, acting together on their prey. When blood is released into the water, the whole school becomes activated and excited. *Serrasalmus piraya*, can grow up to 15 in in length and a school of these large Piranhas can consume the flesh of quite a large animal or fish within a very short time.

Lantern Fish and Gulper Eels

Lantern Fish form a large family of small fish. Most are soft-bodied and silvery black in colour. They have many small luminous organs, the arrangement of which varies greatly between different species. In some, they are in rows running along the length of the body while in others they form complicated patterns. Most are deep-sea fish that come to the surface at night, and they have large eyes. One genus, *Diaphus*, has large luminous organs at the end of the snout. All have large mouths with arrow-shaped teeth.

The **Gulper-Eels** are deep-sea fish with thin elongated bodies up to 67 inches (170 cm) in length. They live at depths of 750 to 21,000 feet (250 to 7,000 metres). The most notable feature is the large mouth provided with very sharp teeth. One genus, *Saccopharynx*, has a sac-like mouth and a light organ on the tip of the tail. One of the most common of these Eels is *Eurypharynx peleconoides* which has the largest mouth of all the Gulper Eels. Its teeth are fragile and are perhaps used for straining small organisms.

Characins

Small, fairly primitive freshwater fish of the family Characidae which contains about 1,500 species, **Characins** are found most commonly in southern and central America with some genera in Africa. They have a wide variety of forms including small shoaling ones; pike-like forms, and deep bodied bream-like ones, as well as highly coloured forms and others which are almost transparent. The family shows adaptations to many different modes of life.

The well-known **Piranhas** are small, the largest reaching some 15 inches (38 cm), but extremely ferocious fish whose main diet is other fish or mammals. Indeed travellers say even Man has been attacked by Piranhas. The **Tigerfish** of Africa are equally voracious but grow to a larger size.

The genera *Hemigrammus* and *Hyphessobrycon* contain most of the small, attractive Characins well known in aquariums and referred to as **Tetras.** They are lively but peaceful shoaling fishes with a small, erect dorsal fin. The **Glowlight Tetra** has a glowing ruby-red stripe along its grey-green body; others include the **Frame Tetra;** perhaps the best known is the **Neon Tetra** with an iridescent stripe along its body.

Electric Eel

Although eel-like in appearance, this is not a true Eel but a member of the family Gymnotidae. It can grow to about 6 feet (180 cm) in length and seven-eighths of the total length is tail which consists largely of electric organs. The Electric Eel lives in the Amazon basin in South America and has very powerful electric organs. These are divided into three parts on each side of the body and are capable of producing a discharge of 550v. This is quite sufficient to stun prey and even to give Man a severe shock. The electric organs have another function when discharges are produced from the electric organ at the posterior part of the tail. These discharges are probably used in orientation, for keeping the body axis along a straight course when it swims—an important function since the animal lives in muddy water. It could also be used for the detection of prey in the vicinity of the fish.

Minnows and Loaches

These fish belong to the super-order Ostariophysi, the second largest group of primitive bony fish, living mostly in freshwater. The **Minnows** and **Loaches,** of the order Cypriniformes, are found in northern temperate regions. All live in slow-moving waters of lakes, rivers and occasionally ponds.

The Minnow is found in shoals of about 100 fish swimming near the surface between April and December. They breed from May to July and the eggs are attached to stones. They grow slowly and live for about 3 years.

The Loaches are characterised by a worm-like shape and a small mouth, without teeth, on the underside of the head, surrounded by barbels. Much time

is spent dormant, moving only at night in search of bottom-living invertebrates, such as insect larvae, worms and shrimps, upon which it feeds.

The **Carp** is Asiatic in origin; it lives in small lakes or ornamental ponds where the bottom is muddy. Food is obtained by sucking up mud, digesting the contained detritus and small animals, before finally ejecting the inorganic material. They generally live for 12 to 15 years.

Catfish

There are over 2,000 species of **Catfishes** known, adapted to many habitats ranging from cave-dwellers to parasites. They have barbels around the mouth and the body of some is without scales while others have thick, armour-like scales. With the exception of two families that live in the sea, all the other members of the remaining 29 families live in freshwater. They are found in all parts of the world except the colder regions of the northern hemisphere.

Two Catfish are found in Europe, where they arrived from Asia after the last Ice Age. The largest of these, the **Wels,** was introduced into Britain by the Duke of Bedford. The Wels spends the day at the bottom of deep, slow-running waters and emerges at dusk and the early morning in search of prey. The other European species is *Parasilurus aristotelis,* also from Asia, found in a few rivers in Greece. It was named after Aristotle who gave an accurate account of its biology 2,000 years ago.

The **Glass Catfish,** or **Ghostfish,** from Burma, has a transparent body and is well known to aquarists. One of the African species is the **Upside-down Catfish,** named from its habit of swimming on its back. This reversal has resulted in its belly being darker in colour than its back.

Eels

(Anguilliformes)

These fish are distinguished by their serpent-like form, confluent dorsal and anal fins, small gill-openings, and a slimy skin either naked or with minute scales. The majority of Eels are marine, some living at great depths, and without exception all spawn in the sea. However, the Eel, *(Anguilla anguilla),* spends most of its life in freshwater travelling some 3,500 miles (5,600 km) to its spawning grounds in the Sargasso Sea. The eggs hatch into a leaf-like larva and as they travel towards their adult habitat they grow. Metamorphosis into the adult form occurs near the coasts.

The **Conger Eel** is a large, thick fish with a somewhat flattened head and is often seen lurking amongst rocks or in shipwrecks by divers. It hunts for food at night, taking chiefly fish and cephalopods. The Conger has a wide distribution in the Atlantic, Mediterranean and Indo-Pacific regions but all go into deep waters to breed. The **Moray Eels** are much more colourful than most of their relatives and are very voracious feeders. They are found in the Mediterranean and tropical seas. Other members of the order include the deep-sea **Snipe Eels,** the brightly coloured **Snake Eels** and the burrowing **Worm Eels.**

Needlefishes and Flying Fishes

Most of the **Needlefishes,** or **Garfishes,** are marine but there are two freshwater species in Asia. These are elongated fish with very long, slender jaws bearing needle-like teeth. The **European Garfish,** *(Belone belone),* has an electric-blue back, silvery belly and flanks, while the pectoral fins are pinkish. It is a pelagic fish and arrives off the coast of the British Isles in April or May, some weeks before the **Mackerel** and is thus often called the 'Mackerel-guide', The Garfish is very good to eat but is notable for its translucent, green-coloured bones.

The **Flying Fishes** have streamlined bodies and very large pectoral fins which can be spread like wings. The tail fin has an enlarged lower lobe to provide more thrust as the fish leaves the water. The speed reached before the fish takes to the air has been known to reach 35 miles per hour (56 kph). Flight can last up to half a minute and the fish may cover a distance of as much as 1,000 feet.

The Needle fishes and the Flying fishes both belong to the order Atheriniformes.

Sea Horses and Relatives

(Syngnathidae)

These fish have a small mouth at the end of a tube-like snout. The body is encircled by bony rings and there is a reduction in the number of fins. They live in the sea and are usually well con-

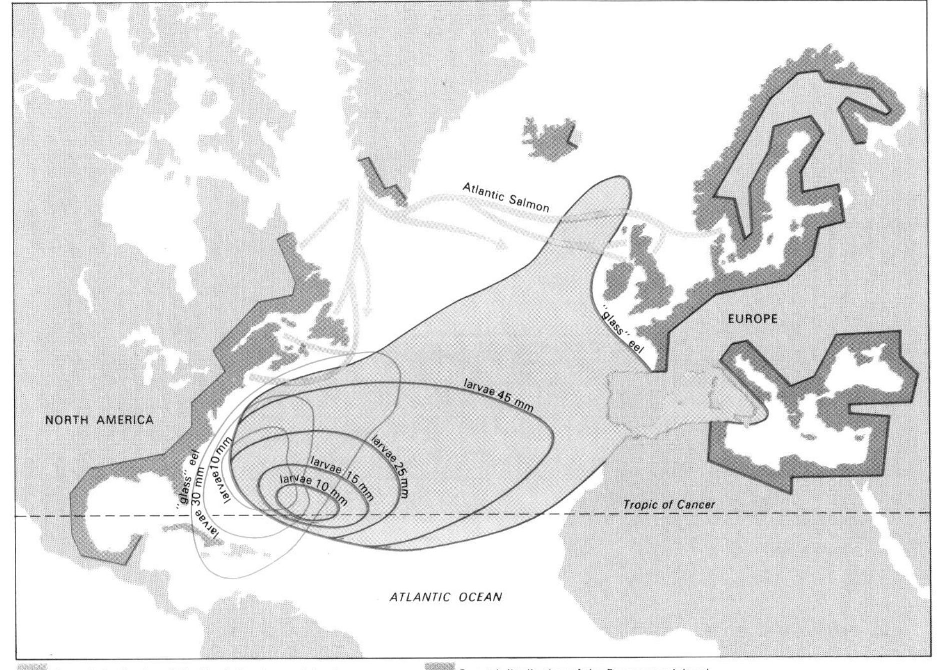

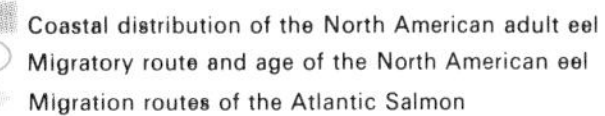

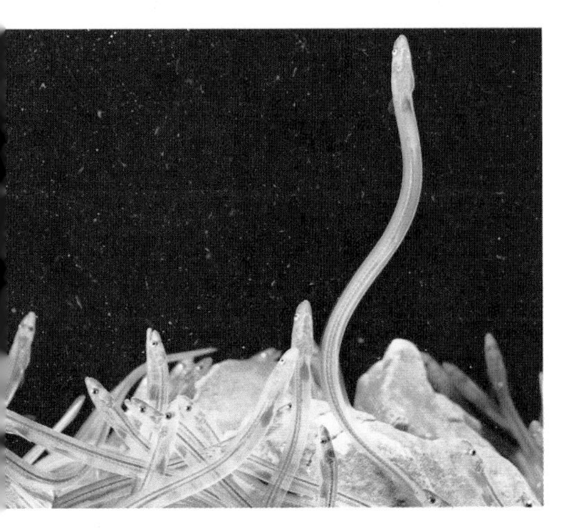

◁ ◁ Catfish (Ictalurus nebulosus), has been introduced into Europe from eastern USA. It is a popular aquarium fish. The barbels on the Catfish head help it to locate food in muddy river water.

▽ Moray Eels grow to lengths of over 6 ft and are a ferocious fish. They wait in rock crevices, darting out to sieze their prey in powerful jaws. The teeth of the Moray Eel are large and pointed either backwards or forwards.

◁ Elvers of the Common Eel, (Anguilla anguilla). At this stage they are known as 'glass' eels and have drifted to European shores as larvae from the spawning grounds in the Atlantic Ocean, near the Sargasso Sea. The following spring, the elver ascends to freshwater rivers and colour begins to enter the skin.

△ Map shows the migratory routes of the North American Eel *(Anguilla rostata)*, the European Eel *(Anguilla anguilla)*, and the Atlantic Salmon.

▷ The Seahorse can hold onto seaweed with its prehensile tail, the only fish which can do this.

cealed. The **Pipefish** adopts a vertical posture closely resembling the eel grass amongst which it lives. The **Sea Horse** has a broken outline and holds on to sea weed with its prehensile tail, indeed the **Leafy Sea Horse** has fleshy appendages resembling the leaves of weeds amongst which it lives. The **Shrimp Fishes,** or **Razorfishes,** look much like the blade of a knife and spend most of their time head downwards, again making them difficult to find amongst weeds. The males have a fold of skin or pouch on their belly in which the eggs are placed and they take 4 to 5 weeks to develop. When they hatch the male expels the young, looking like miniature adults, by flexion of his body.

The Sea Horse swims by oscillations of the dorsal fin and it can manoeuvre slowly but precisely to come close to its prey. The tiny crustacea are then sucked into the small mouth and swallowed.

Squirrelfishes and John Dories

Squirrelfishes, of the order Beryciformes and the family Holocentridae, have a world-wide distribution in tropical waters. Their head profile somewhat resembles that of a squirrel and they have large eyes. Inhabitants of shallow water, particularly around coral reefs, they hide during the day coming out at night to feed. Usually bright red in colour, with silvery spots or bands on the flanks, they have large scales which have serrations or sharp points making the fish rough to handle. When courting, most of the Squirrel fishes make noises and under some conditions can even be heard out of water.

The **John Dories** belong to the order Zeiformes and are found down to depths of 600 feet (200 metres) in temperate oceans. The John Dory is also known as **St. Peter's Fish,** the dark spot on the flank being alleged to represent his thumbprint. It swims slowly in midwater or near the bottom, undulating the dorsal and anal fins. A predator, it takes small fish including Pilchards, Herrings and Sand Eels. Prey are taken after the very slender Dory has approached closely and then with great rapidity it extends its jaws forward to take the fish. Usually 10 to 18 inches (25 to 45 cm) long, the John Dory may reach 22½ inches (57 cm). Other species live in deeper water and are infrequently caught.

Perch and Relatives

All these fish belong to the order Perciformes, a group which showed rapid radiation in the early Tertiary Period. Today the great majority of spiny finned fish are currently regarded as belonging to this order, and they occupy an enormous range of habitats.

The **Perch** *(Perca fluviatilis)* is widespread throughout Europe including England and Ireland although it is rare in Scotland and Norway. A deep-bodied fish with two barely separate dorsal fins, the first one being spiny, it is a fish of slow and sluggish waters which feeds on small fish and invertebrates. It is much loved by anglers, and easily taken.

Dolphinfish, found in tropical and sub-tropical oceans, have a large head with a body tapering to the strongly forked tail. The long dorsal fin and back are green while the flanks and tail have an orange band. A length of 59 inches (150 cm) may be reached. With a swimming speed of up to 37 miles per hour (59 kph) it can catch easily the Flying Fish on which it likes to feed.

The **Grouper** is a fish of temperate and tropical seas. A large and bulky fish up to 11½ ft (350 cm) in length, it has an enormous mouth. It is capable of varying its colour with some rapidity although they are rather drab. It is a valuable food fish.

Sunfishes are fairly common freshwater fishes found in North America and includes **Crappies, Bluegills,** and **Black Basses.** All are nest builders, the male making a hollow for the eggs and which he subsequently guards.

The **Bass** is a shoaling species of warm waters, being found usually along rocky coasts. They are long-lived and a female 30 inches (76 cm) long, 11½ lbs (5.3 kg) was found to be 21 years of age.

Archerfish

(Toxotidae)

These fish are found in the freshwaters of south-east Asia. They are small fish, up to 8 inches (18 cm) with deep compressed bodies and dorsal and anal fins far back on the body near the tail; generally silvery in colour, it has 3 to 5 dark bars along the flanks. It is the means by which the Archerfish obtains its insect prey that is so fascinating to watch and is possibly unique in the animal world. They swim near the surface searching for insects on overhanging foliage. When an insect is seen, the Archerfish orientates itself in order to eject a jet of water at it. On being hit the insect drops into the water to be eaten by the fish. Adult Archerfish can eject water over a distance of 35½ inches (90 cm) with a remarkable degree of accuracy. The water is taken into the mouth and squirted from between the roof of the mouth and the tongue which has a groove, thus forcing it along a narrow channel to emerge as quite a powerful jet.

Butterfly Fishes or Angel Fishes

The common name of these fish is **Angel fishes** in America but they are also known as **Butterfly** fishes. They are small, often brilliantly coloured marine fish found mainly on coral reefs. Deep-bodied and compressed they have long anterior dorsal and anal fins. They are agile fish and give the appearance of fluttering around the coral reefs into which they disappear at the first indication of danger. Many have dark vertical bars across the eyes while there is a large eye-spot near the tail. The mouth is small and the jaws have sharp teeth for taking small worms and other invertebrates from the cracks in the coral. In some species the snout is elongated with a small mouth at its extremity, allowing the fish to take animals from greater depths in the crevices. They are territorial in habit and defend it vigorously. In some regions these fish appear to work in shifts, some species occupying areas during the day while others move in at night.

Cichlids and Wrasse

Both **Cichlids** and **Wrasse** belong to the order Perciformes. There are some 600 species of Cichlids, found chiefly in lakes or sluggish waters of Africa and South America, with one genus in India and Ceylon. Some show interesting breeding habits. Many exhibit territorial behaviour while guarding the nest, usually by the male, and its eggs. Other species are the so-called mouth-brooders, in these the eggs are sucked into the mouth, usually that of the female, where they remain and even after hatching the young stay until they are able to look after themselves. Many Cichlids are brightly coloured and this, together with their breeding habits, makes them popular with aquarists.

Wrasses have thick lips and strong conical teeth while hard scales cover the operculum and cheeks; they inhabit shallow water near the rocks. Some species are brightly coloured as some of their names indicate, for example the **Rainbow Wrasse.** They swim by synchronous movements of the pectoral fins, the tail fin being used as a stabiliser and rudder. The males of some species build nests and guard the adhesive eggs while other species produce eggs which float.

Moorish Idol and Surgeon Fishes

Both the **Moorish Idol** and **Surgeonfishes** belong to the order Perciformes; they are found on coral reefs. The Moorish Idol, of the family Zanclidae, is a highly coloured fish of the Indo-Pacific region. It has a very distinctive shape; the depth of the body being about the same as its length, and the snout is elongated. The first spiny ray of the dorsal fin is of enormous length, the remaining rays tapering from the second ray, which is a little less than half the length of the first ray. In large adults a horn-like protuberance develops on the head just above the eyes.

The **Surgeonfishes,** of the family Acanthuridae, derive their name from little keels on either side of the base of the tail. This keel is very sharp and blade-like making the fish difficult to handle. The mouth is small and has a single row of teeth used for scraping algae and other encrusting organisms from the coral reef. The five-banded Surgeonfish reaches a length of about 10 inches (25 cm) and is also known as the **Convict Fish.** Larval Surgeonfishes have vertical ridges on the body and bear little resemblance to the adult.

◁ Butterfly fish. These are small fish with deep, flattened bodies and are characterised by the small brush-like teeth on the jaws.

▷ Mudskippers are adapted for life both in water and on land, absorbing air through modified gill chambers. Pectoral fins help them to climb out of the water onto rocks, sand or muddy shores, but they must return to the water to feed and breed.

Mackerel, Tuna and Relatives

(Scombroidei)

Mackerel are shoaling fish and during spawning numbers may be seen at the surface of the waves some distance from the shore. An average size female may produce up to 450,000 eggs but sheds only about 40,000 at a time, as they ripen. Mackerel grow rapidly and may be 8 inches (20 cm) long at 1 year with a maximum length of some 23½ inches (60 cm). They feed on Shrimps and other crustaceans, polychaete worms and small fish.

Tuna or **Tunny** are larger and some, 9 feet (274 cm) in length, have been caught in British waters. An oceanic fish, it migrates over considerable distances, as the home of the Tuna is the Mediterranean, and this is where they collect in shoals prior to mating and spawning.

The **Sailfish** derives its name from the very long and large sail-like dorsal fin. It is a common fish in tropical Atlantic waters, and in the Gulf Stream off the American coast. It is popular with fishermen because of its spectacular fight once hooked.

Marlins, or **Spearfishes,** are allied to the Tuna but have an elongated upper jaw. Distributed in the tropical seas of the world, they are large fish and, like the Sailfish, leap and fight once hooked.

Gobies and Blennies

(Perciformes)

The Marshall Island **Goby** is the smallest of the bony fish and indeed most of the Gobies are small. Some inhabit the sea-bed in offshore waters, except the transparent Gobies which are pelagic, while others live in coastal waters. The distinctive feature of these fish is the fusion of the pelvic fins by a membrane extending across them anteriorly—this forms a mechanism of attachment to the bottom. Gobies lay rather large eggs which stick to the underside of stones. In some species the male guards the eggs until the young hatch.

Blennies are also small and inhabit shallow, rocky inshore waters. They have an elongated body with a somewhat

slimy skin, lacking scales. The mouth is usually large with sharp, close-set teeth. Many of the Blennies exhibit territorial behaviour during the breeding season. The male usually occupies a hole or empty mollusc shell and after the eggs are laid aerates them for several weeks until they hatch. Blennies feed on crustaceans, worms, molluscs and barnacles.

Fighting Fishes and Kissing Fishes

(Perciformes)

All of these fish live in freshwaters. Of the fighting fish the best known is probably the **Siamese Fighting Fish,** found chiefly in Thailand and the Malaysian Peninsula. These fish have been 'domesticated' in Thailand and used for sport because of their territorial and fighting habits. During a fight between two males the fins are spread out and the mouth and gill covers opened wide. In wild populations the fins are short and the colour of the body is rather dull. Because of domestication special varieties with vivid colours and long fins have been bred. These fishes are notable for the production, in many cases, of froth nests made by the male of mucus-covered bubbles. In many species the eggs and young are protected by the male.

Gouramis are found from India to Malaya, the best known being the **Kissing Gourami** from its habit of applying the broad lips to another fish. They include the **Dwarf, Pearl** or **Mosaic,** and the **Croaking Gouramis.** Males of the Croaking Gourami surface at night for air and in using their accessory breathing organ make croaking noises.

The **Climbing Perch** can survive out of water for some time as it has an accessory breathing organ, a series of plates in the upper part of the gill chamber. These fish will move from one pool to another travelling overground with the use of their spiny gill cover as 'legs'. These are spread out to anchor the fish while the pectoral and caudal fins push it forwards.

Barracudas and Mullets

(Perciformes)

The **Barracudas** are known from their fierce predatory habits. Their relatively enormous mouths and sharp dagger-like teeth make them formidable enemies. They feed on fish and will herd a shoal of fish until ready to attack. The smallest of the Barracudas reaches about 18 inches (46 cm) while the **Great Barracuda** reaches 95 inches (240 cm). Inhabitants of warm waters, they are found in the tropical Atlantic and Pacific as well as the Mediterranean.

The **Grey Mullets** have almost world-wide distribution. They have long, somewhat cylindrical bodies and mouths adapted for bottom-feeding. They suck food and mud into the mouth and, after sieving, the mud is subsequently expelled. They also scrape algae and diatoms from rocks. Shoaling fish of coastal waters, they will also enter estuaries. A length of some 24 inches (60 cm) can be reached.

Red Mullets, or **Surmullets,** are inshore fish of the Mediterranean and east Atlantic. They have two barbels below the chin and the body is of shades of red. They feed on shrimps, molluscs and worms and may reach a length of 18 inches (45 cm).

Scorpion Fishes

(Perciformes)

These marine fish have characteristic bony plates and spines and a thin tapering body covered with scales, while the head is without scales. The fins have robust spiny rays, and 2 to 3 free rays of the pectoral fins are used as feelers to explore the bottom for prey and also to 'walk' along the bottom. Many of the spines of the fins are pungent or poisonous. The most poisonous of the Scorpionfishes, and indeed perhaps of all the venomous fish, is the **Stonefish.** As its name suggests it closely resembles the stones and rocks of its habitat and on which it lies motionless for long periods. A swimmer stepping onto a Stonefish, if poisoned, may shortly afterwards die. The breeding habits of the Scorpionfishes are varied. Some produce floating eggs, others eggs embedded in a gelatinous matrix, while others retain the eggs either until just prior to hatching or until they are liberated as young larvae.

Sticklebacks

Common fishes of fresh and salt waters of the northern hemisphere, **Sticklebacks** have isolated dorsal spines that can be erected and have a locking mechanism. This provides them with effective protection against would-be predators including the Pike. Some of the Sticklebacks live entirely in freshwater, such as the **Ten Spined** which is widespread across Europe. The **Three Spined Stickleback** can live in freshwater, brackish or seawater and is found in England, Europe, Asia, Japan and North America. The **Fifteen Spined Stickleback** is entirely marine, living in coastal waters. All show migratory and territorial habits prior to spawning and it is the male who builds the nest and guards the eggs until hatching. Just prior to the spawning season the male Three Spined Stickleback becomes bright red below the chin and during courtship he swims in a series of zig-zags when a silvery female, swollen with eggs, enters his territory. Once she has laid eggs the male chases her from the nest and then continues to attract other females to lay eggs. As many as five females may deposit eggs in one nest. The freshwater Sticklebacks stalk their prey taking small crustaceans, aquatic insects, larvae and worms all of which are swallowed whole.

Flatfishes

The bottom-living bony fish all begin life as planktonic, bilaterally symmetrical larvae with eyes on each side of the head. During metamorphosis one eye gradually migrates over the head to lie close to the other, the head bones become twisted and asymmetric. As this takes place so the fish comes to lie on one side while the swimming posture changes from vertical to horizontal. The **Plaice, Halibut, Flounder, Sole** and **Dab** all rest on their left side while the **Turbot, Brill, Megrim** and **Topknots** all lie on their right side. The blind or lower side is usually white or pale whereas the upper side can change its colour or pattern to come to resemble the background on which the fish is lying. This provides excellent camouflage together with their habit of covering much of the body with sand or gravel from the bottom. The adult flatfish feed on bottom-living invertebrates such as bivalves, crustaceans and Polychaete Worms, as well as small fish such as Sand Eels. The Halibut is one of the largest of the flatfish and can reach a length of 270 cm. The Flounder and a few other species can live in brackish waters. Many of the flatfish are of great commercial value. The Turbot, Halibut, Plaice, etc., belong to the suborder *Pleuronectoidea* and are found in the Pacific and Atlantic Oceans.

Triggers and Remoras

Triggers, of the order Tetraodontiformes, are coastal fish usually inhabiting rocky areas. The Trigger fish is so-called because of the erectile mechanism of the three dorsal spines which can be raised if an enemy is sighted. If captured with the spines erect they can be released by pressure on the base of the small third spine. The young drift with the currents and are more pelagic than the adults which are rather poor swimmers.

The **Remoras,** of the order Perciformes, are oceanic and have world-wide distribution. The distinctive feature of these fishes is the modification of the dorsal fin into a flat laminated sucking disc on top of the head. It is used for attachment to other animals, usually Blue Sharks, but also Turtles. This allows the Remora to travel long distances to new feeding grounds without any expenditure of energy. On arrival it detaches itself and swims actively in pursuit of prey, later again finding a suitable anchorage.

◁ Scorpion fish, also known as Lion fish, Dragon fish and Turkey fish, are found on coral reefs in Indo-Pacific waters. The Scorpion fish shows extreme modification of fins. The dorsal fins have associated poison glands.

▽ Like all bottom-living fish, Plaice larvae are bilaterally symmetrical with an eye on each side of the head. As the larva develops, one eye gradually moves over to the top of the head, the head bones becoming twisted and asymmetrical.

▷ Triggerfish have a trigger-like spine in front of the dorsal fin. One species preys on Sea Urchins, pulling the spines off to get at the animal. Other Triggerfish feed on coral and invertebrates.

▷▷ The pectoral fins on the Lungfish are reduced to threadlike feelers and have no locomotory function. Lungfish have adopted air breathing abilities.

△▷▷ Lungfish which were once widely distributed are now found only in tropical Africa, South America and Australia.

Puffers

Fish belonging to the order Tetraodontiformes have derived this name from what appear to be four teeth in each jaw, as the teeth have fused leaving a gap in the centre. The **Puffers** are carnivorous and depend upon their camouflage while approaching prey since they are clumsy swimmers. They use their dorsal and pectoral fins as main propulsive forces while the small anal fin and tail act as rudders. The common name of these fish indicates their habit of inflating their bodies with air or water, thus providing them with a means of defence as the Puffer becomes too large to be swallowed by its predator. When not inflated they look rather clumsy, often with scales modified into spurs. The head and chest are large but the body narrows towards the tail. Some of the Puffers are brightly coloured and it is usually these that are also poisonous. The Puffers are a major source of fish poisoning in Japan in spite of elaborate care when preparing a dish of fugu.

Anglerfishes

A group of highly specialised fish with a world-wide distribution, they are found at all depths of the sea. The feature which gives these fish their name is the abnormal development of the first ray of the dorsal fin into a long lure, with which the fish 'angles' for its prey. There is a small flap of tissue at the tip and as the prey investigates this so the lure is brought in towards the huge mouth which engulfs it. The jaws are lined with sharp, needle-like teeth and food taken can be stored in the mouth before being swallowed. Aristotle recorded the feeding habits of this fish over 2,000 years ago. **Anglers** seem to feed on anything that is available since not only have fish been found in their stomachs but it is not uncommon to find seagulls too. In the deep-sea Anglers the male is small and becomes attached to the very much larger female. The Anglers lay long ribbons of spawn which are often seen at the surface of the sea. The young spend their larval life in surface waters and much resemble other larval bony fish, only developing their somewhat grotesque forms at metamorphosis.

Lungfishes

These are primitive fish that during the Devonian and Triassic periods had world-wide distribution. Some species have survived until today but are now restricted to Africa, Australia and South America. Their characteristic feature is the presence of either single or paired lungs which are used for air breathing. The **Australian Lungfish,** *Neoceratodus,* is the rarest of the Lungfishes and is confined to a few rivers in

Queensland and shows the least external changes from its fossil ancestors. It has large scales, paired fins with fleshy bases and the dorsal and anal fins are continuous around the hind end of the body. It has a single lung and in well aerated waters uses its gills and does not tolerate foul conditions as well as the other Lungfishes. The **South American** and **African Lungfishes** are more elongated and the paired pectoral fins have developed into fleshy 'feelers'. They have paired lungs and air-breathing is important as they often live in foul or drought conditions. Both species aestivate, that is they are dormant during the dry season; the South American Lungfish makes a mud tunnel and the African ones make a hard cocoon in which they remain during the drought. The African Lungfishes make nests and care for up to 5,000 eggs; after hatching the young remain in the nest for as long as 8 weeks. The young Lungfish have external gills which are gradually lost.

Adult amphibians, as they exist today, have smooth skins and are capable of living on land or in the water, after passing through an aquatic larval stage during which they cannot survive on land. The vast majority of adults have four legs, although the small order of about sixty species of Caecilians are legless. There are about 300 tailed amphibians (Newts and Salamanders), and roughly 2,000 tail-less ones (Frogs and Toads), known today.

By the end of the Palaeozoic era, 280 million years ago, five orders of amphibians had evolved, all of which were extinct by the end of the Triassic (190 million years ago). Two of these are members of the primitive forms known as labyrinthodonts (from the tortuous surface of their teeth), the earliest known being the large *Ichthyostega* up to 3 ft (1 m) long, which still bore a fin on the upper aspect of its tail. The first were the Temnospondyls, many bearing gills as a mark of their aquatic habits. Some, such as the Trematosaurs, occurred in salt water.

The largest known is the 14 ft (4.5 m) long amphibian called *Cyclotosaurus*. This was one of the Stereospondyls, a late group of the aquatic Temnospondyls occurring during the Triassic. The other group formed before the Temnospondyls became extinct was the Plagiosaurs, squat creatures up to almost 3 ft (1 m) long, which were probably wholly aquatic.

The other order of labyrinthodonts are known as the Anthracosaurs. One group was the Embolomeres, aquatic fish-eating forms several feet long, and the other the Seymouriamorphs, many of which were terrestrial, and under 12 in long. Although the land vertebrates probably evolved from the early Seymouriamorph line, no present-day amphibians are descended from the labyrinthodonts.

The other main group of early amphibians, beside the labyrinthodonts, was the Lepospondyls. These were scaly, many were aquatic or semi-aquatic and they ranged in size from a few centimetres to just under 3 ft long. It is likely that one group, the Microsaurs, developed and diverged to give rise to the three orders of present-day amphibians. Thus, although the Lepospondyls are now all extinct, the present-day tailed and tail-less amphibians and Caecilians have evolved from them.

Many amphibians are attractive food-items for a wide range of predatory species, ranging through fish, snakes, birds and mammals, including man. Cannibalism is a commonplace in some species. The fact that amphibians still survive in numbers means that they have successful defences. For some it lies in toxic skin secretions, as in some newts and toads, for others in evasive action, in camouflage or in the sheer numbers produced. When a clump of frogspawn, containing around 2,000 eggs is laid, this means that some 1998 tadpoles and froglets are expendable.

Some amphibians which have toxic skin secretions also are brightly coloured—this is known as warning coloration. Predators soon learn to associate these colours with distasteful prey, and soon, less common but harmless species, mimic those with warning colours (or sometimes warning movements), and so avoid predation. This is referred to as Batesian mimicry. Sometimes a group of species which are all distasteful resemble one another (Mullerian mimicry).

The distribution of modern amphibians must reflect their time of origin from the ancestral stem and the wanderings of earlier generations as they survived all the threats of weather, predators and other ecological factors. The tail-less anurans are the most widely spread, through all the continents except Antarctica, extending North as far as the Arctic Circle. Tailed amphibians are only found in the Northern Hemisphere, apart from a few which have managed to cross into South America. The Caecilians occur mainly in the tropics, but have spread a little outside in South America and Asia.

The long hind legs of the Treefrog are specialised for jumping and they can launch themselves accurately at a sitting insect, so that the insect is engulfed as the frog lands. Their digits have spatulate ends which have discs on them which helps the frog to cling to a variety of different surfaces. Male Tree Frogs have a vocal pouch which the females lack.

Amphibians: Survivors of an Ancient World

The Evolution of Amphibians

The origins of the amphibians are not clear because the fossil record is incomplete. There is no doubt that their origin lies among the Rhipidistian fishes, a now extinct family, and perhaps from two distinct groups, the Osteolepiformes and the Porolepiformes. The tail-less amphibians come from the former and the Newts and Salamanders from the latter.

The typical amphibian lives two different types of life, a larval or tadpole stage in water, which is followed by adult life on land. But many of the modern amphibians are not typical, because they either live entirely on dry land or completely in the water. In the case of the water-livers, it is not that they have never left their aquatic environment; after millenia on land, their anatomical structure shows that these amphibians have returned to an aquatic existence. Species now extinct, such as the Trematosaurs of the Triassic period, over 225 million years ago, had even adapted to life in salt water.

Modern amphibians comprise three orders. The earliest to appear were the tail-less amphibians or anurans, which can be traced back over 135 million years, to the Jurassic period. The tailed Newts and Salamanders appeared very much later, during the Triassic period, roughly 180 million years ago; while the legless Caecilians are only known from the past few million years of the Pleistocene age.

△ The Clawed toad *(Xenopus laevis)* spends all its life in the water, and only comes on land during wet nights. The eyes look upwards instead of forwards.
▽◁ This 'cleared' specimen of a salamander shows the outline of the body as well as the skeleton. In contrast with anurans, the body is slender and can readily be thrown into curves. There are numerous tail vertebrae.
▷▽ The 'cleared' specimen of a frog shows the squat, broader body, with only a few vertebrae in front of the large pelvis. There are no tail vetebrae, but the hind limbs are long, being used in jumping.
▽ The Caecilians are worm-like amphibians; some live in the eath whilst others spend their lives in water.

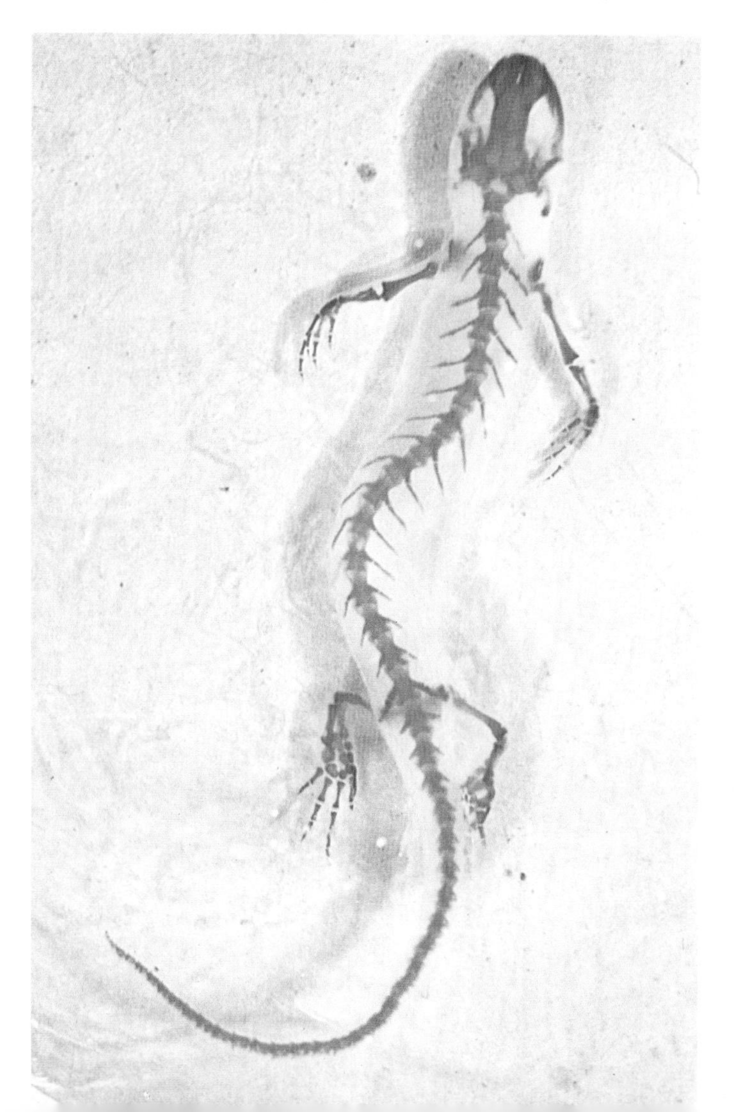

Several older orders vanished during the long ages from the Devonian period. These had their origin in the Carboniferous period and had gone by the end of the Jurassic. The main line of descent of the modern orders comes via the small, scaly Lepospondyls, through the little Microsaurs, which lived salamander-like in muddy pools, to the tailed amphibians and Caecilians of the present day. The origin of the anurans is less certain. Some of the early amphibians which lived 300 million years ago, were ancestral to the first reptiles, from which in turn came the mammals and birds, as well as the modern reptiles.

Problem of survival

The major problem which the earliest amphibians had to overcome to survive was their emergence on to the land. Doubtless many were stranded and died in times of drought, until forms evolved which could survive on land. To do this, they had to develop a breathing mechanism independent of gills, and a circulatory system to supply these new lungs. Some of the necessary changes can be seen as tadpoles develop into froglets.

The ancestral fishes had lobe-fins which became the limbs to support their weight on land. Mobile fins had to develop into pillars which could also be moved, with fin-rays changing to a series of digits. Enabling the head to be held in the air, away from the supporting water, the skull bones became lighter. While these changes were gradually occurring through natural selection, the middle and inner ear were developing, while the eyes were becoming better positioned for life on land rather than in the water. But standing still until the rains came and the water level rose was not enough. The successful amphibians were those which could move ashore at will and go about on land, feeding on the abundant invertebrates.

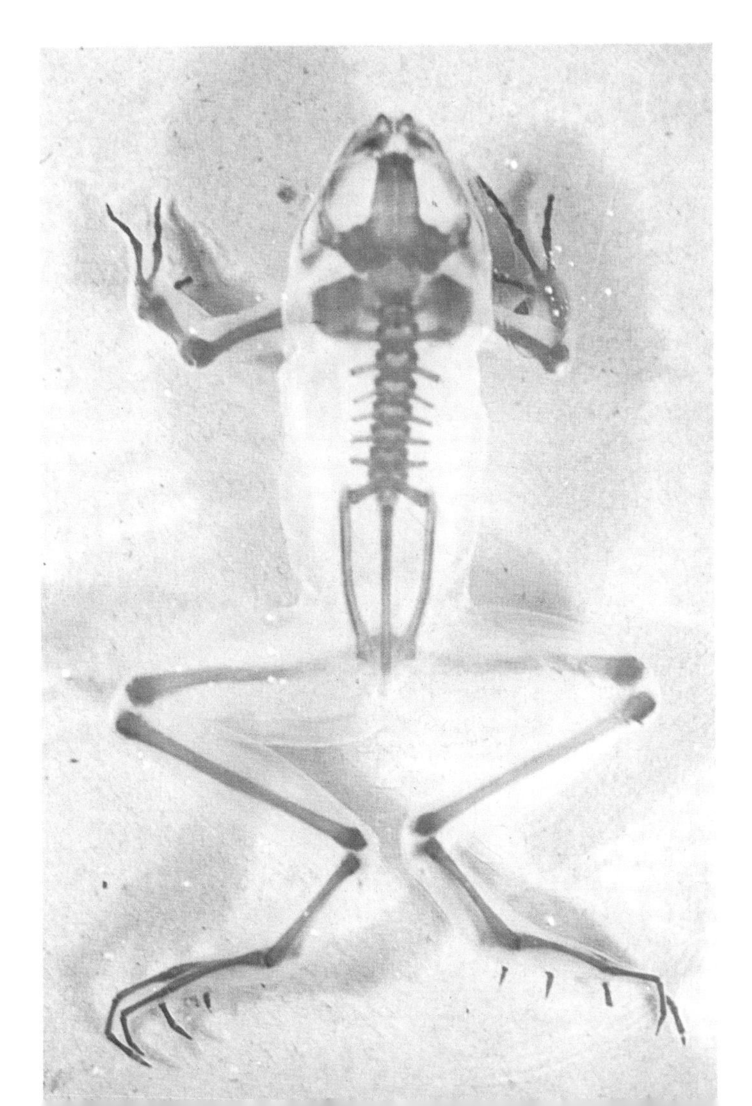

The life of the amphibians lies between the land and the fresh water. Just as the earliest forms learnt to come ashore, so they had to come back to the water to breed. With the transition to land, they eventually lost both the fins and the scales of the fishes, though the tail-fin was still present in *Ichthyostega*, the earliest amphibians known. As the climate changed, so the amphibians had to combat the danger of drying up, either by hiding in damp places or returning to the water. Their eggs and young, smaller and therefore more at risk of drying up, remained in the water and did not undertake life on land. To compete successfully in the water, the young retained many fishlike characteristics, and this meant that at the time when they left the water for the land, they had to undergo a change of form, to take on the adult characteristics. This is known as metamorphosis (from the Greek, meaning change of form).

The adult modern amphibian has developed to the stage where it has a moist cool skin, plentifully besprinkled with glands, some of which produce secretions that are toxic to predators. The skin also contains pigments which are responsible for the colour and markings. The larvae retain the lateral line organ found in the fish, which gives warning of pressure changes in the water and so of the approach of predator or prey, but this is lost at metamorphosis, though kept by some species which are wholly aquatic.

Three respiration methods

The skin of the modern amphibians, unlike that of mammals, is permeable to water in both directions, and also to gases. This means that where fine blood vessels known as capillaries are near the surface, gases can pass between these and the outside surface. Amphibians thus have three possible pathways for respiration—lungs, gills and through their skins.

The larva still bears the gills of its primitive fish ancestor. These comprise a series of gill arches just behind the head, in which run fine blood-capillaries, where gases are exchanged with the outside water, oxygen being taken in and waste gases excreted. At metamorphosis, some of the gill-arches degenerate, while others remain as an integral part of the new system of blood vessels, used to enable blood to be pumped to the lungs instead, where gas-exchange with the outside now takes place. The lungs are the ultimate development from the former air-sacs of the primitive fish ancestor.

Caecilians, Newts and Salamanders

▽ This Salamander is eating a nestling mouse, showing that they are carnivorous. Salamanders live in wet places and eat slugs, worms and small insects—anything in fact which moves slowly.

Among the Newts and Salamanders are found fascinating mating habits which are different in some respects from those of other amphibians. Newts for instance are voiceless and have no song but the male assumes wonderful breeding colours to attract the attention of the female. In one species of Salamander, both the male and the female care for the egg sacs, the female laying them in the water and the male subsequently spending the winter in the water guarding them—while the Lamper Eel male builds a nest for his mate so that she can lay her eggs in it.

Those amphibians commonly called Newts and Salamanders are sometimes known as the *Caudata*. They generally have four legs and a tail, although in some species the legs may be reduced to only a single pair, or even be completely absent. *Caudata* vary in length from 4 to 8 ins (10 to 20 cm), with some extending up to 3 ft (1 m) long. Some of the species, where the adults retain the larval form, have external gills, as part of their re-adaptation to a permanent life in water.

The *Caudata* probably originated in the northern hemisphere, judging by the present-day distribution of the 300 species which have survived. Newts and Salamanders are found through much of this part of the world; in the north-west of South America, in North Africa and on a corresponding level in South China. While the animals can apparently survive easily in cooler climates, including mountain regions, tropical areas seem to be a barrier to their spread southwards.

Many species of Newts spend the greater part of their adult life on land, returning to the water to breed, sometimes staying there for months at a time. Salamanders are frequently more terrestrial, and may even spend their larval stage on land. In order to survive periods of drought on land, both Newts and Salamanders spend much of their time underground or beneath surface debris, while some use rock crevices or even caves as a regular habitat. One or two species are found in holes in trees or in the axils of leaves of certain plants.

However, agile as they are in water, many *Caudata* are rather lethargic on land, yet others are remarkably lively, even jumping when approached. Many have skin secretions which are toxic, or at least, unpleasant to predators, sometimes associated with warning coloration. Their food is mainly small animals, frequently insects or other invertebrates. When they are in water, tadpoles or fishes may be eaten and hole-nesting amphibians may even eat some of their own eggs.

Caudata have a remarkable facility for returning accurately to their breeding site. Marked individuals of one species of American Newt travelled thousands of yards to their home range from a release site although this involved crossing hilly terrain.

The Fire Salamander

The brilliant black-and-gold **Common Salamander** is widely distributed throughout Europe. It is sometimes called the **Fire Salamander** from the ancient belief that it could live and thrive in fire. During the day, the animal stays under stones, coming out to feed at night. Clumsy in water and liable to drown, the Fire Salamander spawns in shallow water, the female being only partially immersed. The closely allied **Black Salamander** *(Salamander atra)* of the Alps, produces two young which are developed in the mother's oviduct.

The warning coloration of the Fire Salamander is associated with a toxic skin secretion. The amount of yellow coloration is variable, and selective breeding in captivity has produced almost pure yellow or black forms. In parts of Spain, the species have red spots on the head.

Among the other genera, the American **Plethodontid Salamanders** are noteworthy, being lungless. Respiration is wholly through the skin.

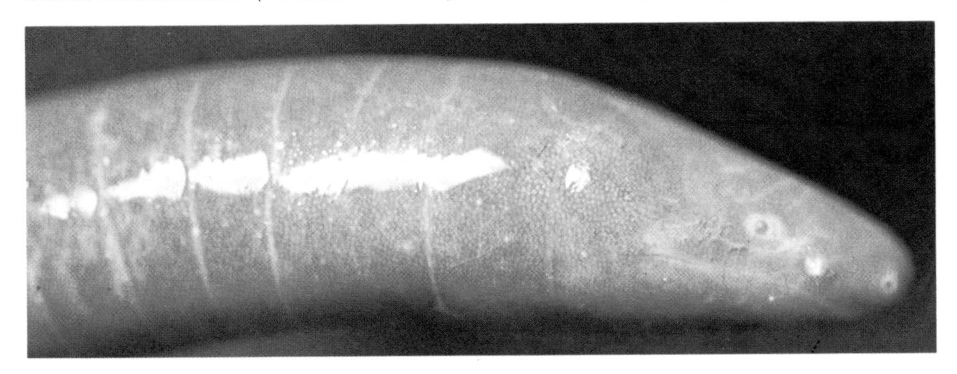

△ Caecilians are burrowing animals and are rarely seen above ground. The retractible tentacle which lies between eye and nostril and helps the animal to move underground, can be seen.

▽ A pair of Great Crested Newts during courtship. The splendid crest on the male disappears when he leaves the water in the autumn for hibernation.

Caecilians

These comprise a little-known order with about sixty tropical species ranging through South and Central America, Central Africa, the Seychelles, and South and South-east Asia, occurring some way above sea-level. They are sometimes known as *Apoda* (footless), since they have no legs, but a wormlike shape which enables easy passage through moist earth or in water. Many live in the earth, though some are aquatic. Some **Caecilians** have tails, while other species are tail-less, but all can be identified by a small retractible tentacle found between eye and nostril, which is an organ of sensation. The left lung is normally degenerate. Some Caecilians, which live wholly in water, have fish-like dorsal and ventral fins, and some other aquatic species still have scales on their body surface. At least one uses its anal disc as a sucker to anchor itself in areas of turbulent water.

Some Caecilians lay small batches of eggs, while others give birth to one or more young which have developed inside the mother. Depending on their habitat, aquatic larvae may have gills, which are lost at adulthood.

Newts

These vary from the primitive forms found in cold waters of northern Siberia, such as *Ranodon* (whose larvae take two or more years before metamorphosis, and whose males fight over the egg sacs rather than for breeding females), to familiar species such as the three British ones—the **Common, Palmate** and **Great Crested Newts.** Their requirements in breeding pools are obviously different since a quick survey of a dozen ponds or so will show the Palmate in most of them, the Common Newt in about half, and the Great Crested in only about two or three. In the ponds where the Great Crested Newt is found, one or two of the other species will invariably be discovered. Like other amphibians, Newts can take in oxygen through their skins.

The distribution of species, subspecies and races of Newts in Europe has obviously followed the movements of the great ice sheets during the Ice Ages. Communities of Newts developed in isolation in sheltered pockets, while those in intervening areas died out. An even more startling example of such environmental effects is shown by the present-day distribution of the **Salamander** genus *Hydromantes*, now found only in southern Europe and in south east United States of America.

British Newts

The three British species belong to the genus *Triturus*, which is confined to Europe. These are the **Great Crested,** *(Triturus cristatus)*; **Common Newt** *(Triturus vulgaris)* and **Palmate** *(Triturus helveticus)*. Unlike some of the American species, these show no major differences in appearance between land and water phases, although when they are found under stones on land, they are often rather dried up and wrinkled. There is a marked difference between the two sexes. The female Common Newt, for instance, which is about 3 ins 8 cm) long, is a drab brownish colour, but the smaller male has large black spots on a paler ground and with a brilliant, orange belly. During the breeding season, the male develops a crest along his back and the whole length of the tail, on both upper and lower surfaces.

Newts return to the water to breed during the spring in Britain, and some stay there until the autumn. Hybrids between the British species are known and, rarely, Newt hybrids have been recorded in the wild.

▷ The external gills which the Axolotl retains into adult life show clearly here. The Salamander is sexually mature and ready to breed even though it still seems juvenile in appearance.

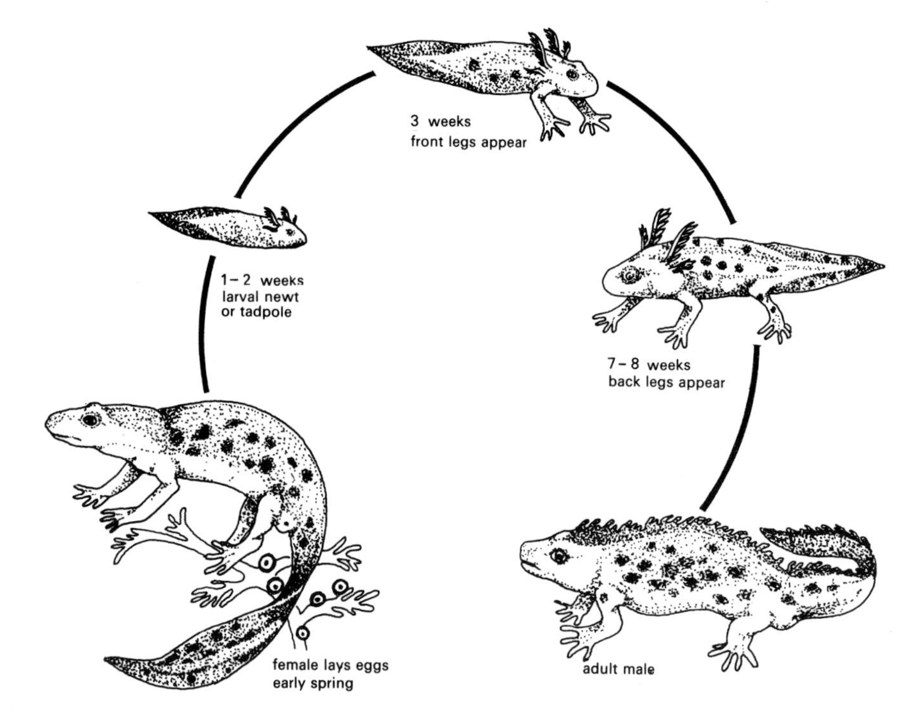

▽ Life cycle of a Newt: from the laying of the eggs to the appearance of all four legs when the Newt is an adult takes 10 weeks.

Newt breeding cycles

Newts come out of hibernation as the weather becomes milder in spring, and by night find their way to the breeding ponds. The males display in the water before the females. Some of the American species perform a nuptial walk on land with the female's nose pressed at the base of the male's tail. The male then deposits a packet of sperms in jelly from glands in his pelvic cavity. This is taken up by the female and the sperms stored and later released to fertilize the eggs when laying.

The eggs are laid singly on the leaves of underwater plants and soon hatch into tadpoles with external gills on a series of gill-arches. As the tadpole feeds and develops, so fore and hind legs grow. The changes of metamorphosis take place, the gills disappear and the lungs develop, with corresponding changes in the circulation. The young Newts develop into miniatures of the adults with ovaries and testes starting to grow, preparatory for the breeding season in the spring. The young Newts come ashore ready for hibernation in the autumn.

The Axolotl

The Mexican name **Axolotl** is given to one or more species of Salamanders of the genus *Ambystoma* which lives in cold, upland lakes. The true Axolotl is *Ambystoma mexicanum*. The adult of this species normally retains the external gills and flattened shape of the larva, though it breeds normally in this form. This occurrence of breeding in the larval form is known as neotony and is reported from time to time even in British Newts. Some species, such as the little-known **Typhlomolge,** from deep artesian wells in Texas, are completely neotenous, but on rare occasions the Axolotl can develop into a land-living adult Salamander, having lost its gills.

Axolotls, like many larval amphibians, readily regenerate limbs which have been lost as a result of fighting or of some accident. In captivity they have a reputation for being difficult to breed, but often, placing a pair together in fresh tapwater will result in production of eggs overnight.

▽ Marbled Newt swimming underwater: both the Newts and the Salamanders have kept the elongated body and tail of their ancestors and these are highly effective when the creature is in water, providing it with considerable agility. Some species are even adapted for different kinds of water, such as mountain streams and deep, still pools.

Congo Eels

(Amphiuma)

The **Congo Eels** are 20 to 30 ins (60 to 90 cm) long, with only rudimentary limbs. The gills have degenerated and they have fully functional lungs. Congo Eels live in burrows under water and the female stays with her eggs.

The Giant Salamander

(Megalobatrachus)

The **Giant Salamanders** of China and Japan are up to 3 ft (1 m) in length and the **Hellbender** *(Cryptobranchus)*, of the United States of America is allied to these. Adults of both genera have corrugated skins, well supplied with capillaries, believed to be important in the animal's respiration.

Mudpuppies and the Olm

(Proteidae)

This covers two distinct genera; **Mudpuppies** *(Necturus)*, found in parts of the United States of America, and the **Olm** *(Proteus)*, found only in deep underground cave waters in Jugoslavia. Both genera comprise large Salamanders measuring up to 12 ins (30 cm) long or more, with plumose external gills. Of the half dozen species of Mudpuppy, the common *Necturus maculatus* is widely distributed in North America, where it is found in both clear and muddy water. The species has both gills and lungs, as well as taking up oxygen through its skin. The lungs are inadequate for the passage of gases, having no alveoli, so that the animal cannot survive in air. They seem instead to be used as buoyancy tanks to lower the specific gravity of the whole animal. Mudpuppies can get enough air for their needs through the skin alone, if the water is well aerated. In these circumstances the animal stays on the bottom, but in poorly aerated water, it surfaces. The whole body of the Mudpuppy is light-sensitive and it bears lateral line organs which seem to be stimulated by sound. Mudpuppies spawn in the spring or summer, laying their eggs in small batches in hollows under stones or debris on the bottom. The female of the common species is said to remain with her eggs.

The Olm *(Proteus anguineus)* is normally only found in deep underground waters, though it may be swept outside when the river is in spate. The adult is a pale, white colour, but if reared in light will grow up wholly black. Light is lethal to the eggs, which are laid on the undersurface of stones, where the male uses his tail to direct a current of water around them. Without this, the eggs fail to hatch.

Sirens

The family *Sirenidae* comprises two genera; those of the **Sirens** proper and the **Mud-Sirens** *(Pseudobranchus)*. The **Great** and **Dwarf Sirens** are completely aquatic, living in muddy ditches and ponds, where they hide by day in thick waterweed or under debris on the bottom. In the heat of summer, they may aestivate in burrows. The hind limbs have been lost, though the front pair remains. The animal is very eel-like, though with external gills, even although the main effective respiratory surfaces are in the lungs. Batches of several hundred eggs are laid in hollows in the mud. The Mud-Sirens are rather similar animals, but striped, and live particularly in thick beds of water-hyacinth, laying the eggs singly or in pairs on the plants.

Frogs and Toads

Toads that loop-the-loop while mating, Frogs that 'fly', species that provide poison for hunting arrows and water in the desert are just some of the more exotic species of Frogs and Toads.

The Anurans, Frogs and Toads, are relatively stouter than the Newts and Salamanders. Many can hop and others run swiftly. Some have moist smooth skins, while others have dry and warty ones. The colour of the skin is produced from chromatophore cells in the deeper layers. The dark melanophores of some species can vary their degree of contraction according to the colour of the animal's background or under stress, so that a bright green Tree-frog can turn to dark brown in a few minutes. The size of an adult can range from several small species of about 1 cm long to over 12 ins (30 cm) for the Giant Frog *(Gigantorana goliath)*. Giant Toads can easily eat adult mice, while other Anurans eat invertebrates as small-sized as ants. Normally, anything living which can be engulfed and swallowed can be regarded as an Anuran's prey. Food is taken by a sticky tongue being projected at it.

Frogs and Toads themselves form food for a very large number of creatures—snakes, birds and mammals, even man. Some species are protected by their skin secretions and these often have warning coloration. Others may be well camouflaged against predation, or well hidden in holes in the ground. Many are nocturnal.

Anurans return normally to the water to spawn and sometimes more than once annually. On these occasions, the song of the males both repels other males and attracts females of the same species. The Frogs pair, the male clasping the female either under the armpits or round the hips, and they remain together until the female produces her eggs. At this moment the male fertilizes them before releasing the female. Some species lay the eggs in a clump of jelly, others in strings of jelly which are wound round vegetation.

The eggs soon hatch into tadpoles, which have a rounded head and body, and a mobile tail used in swimming. The gills of the tadpoles are on the inside of the gill-arches, so that in order to get air the tadpoles have to gargle water round the upper throat area, though some air also diffuses in through the tail membranes. The tadpoles are vegetarians until they have reached the time of metamorphosis, with the growth of the limbs and resorption of the tail, when they take to protein food. The mouth enlarges, the gills disappear and lungs develop for air-breathing. At this stage, they leave the water for the land.

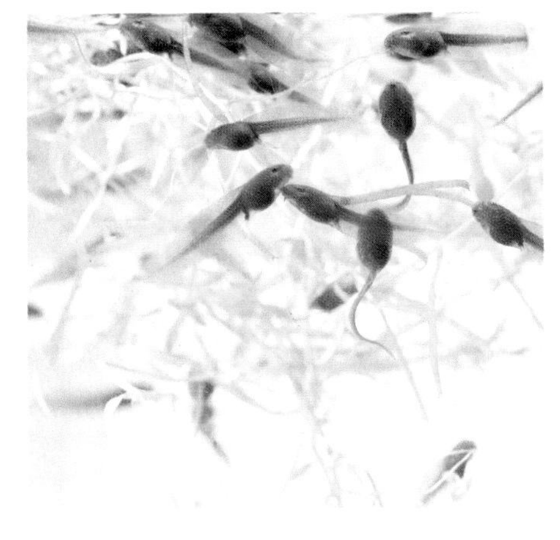

Frog or Toad?

In Britain it is relatively easy to distinguish between Frogs and Toads, since there are only three species of Frogs (two of them introduced by man), and two Toads. Generally speaking, Frogs are typical members of the family Ranidae *(Rana)*, and Toads of the family Bufonidae. The Ranidae are rather sharp-nosed Anurans with smooth, moist skins, and are good jumpers. The Toads are more slow moving, walking or running rather than jumping. Their heads are more rounded, their skins are warty, and behind the eardrums, they have a pair of large parotoid glands.

With these differences in appearance, Frogs and Toads can be distinguished fairly easily. But among other Anuran groups there is a vast diversity of colour, form, size and behaviour.

△ Top left to right: European common Frogs spawning; each year, after hibernation, frogs gather at the same ponds as they did in previous years.
2. Frog spawn, the embryos protected by jelly.
3. Gilled tadpoles eat first the jelly and then feed off the pondweed. 4. After the external gills have disappeared, the tadpole breathes on internal gills.
◁ European Green Toad.

Toads

The Toads are disposed through a number of genera, though far the largest of these is *Bufo*. The majority of the species are dark in colour, frequently brown or black, though sometimes mottled with green and in some cases carrying a dorsal yellow stripe. When disturbed, toads exude a milky skin secretion which is especially produced by the parotoid glands. This is particularly irritant, and in fact several drugs affecting the heart have been isolated from the secretion. Not only does it make the toad distasteful to many predators, but even its tadpoles are rejected. Some toads are killed at their breeding ponds, but are only eaten by predators after being skinned. Even then, the oviducts and jelly are usually left.

Of the two British Toads, the **Natterjack** *(Bufo calamita)* is a running toad which lives in sandy areas where it can burrow easily. In contrast, the **Common Toad** *(Bufo bufo)* is universally distributed over the countryside. On a world basis, toads have spread to occupy a wide variety of habitats.

Frogs

This comprises a number of genera, among which *Rana* contains the greatest number of species. These true Frogs vary externally as well as in behaviour. Some, like the British **Common Frog** *(Rana temporaria)*, have no external vocal sacs, while others, like the two introduced species, *(Rana ridibunda* and *Rana esculenta)*, have large external sacs which become inflated when the animal croaks. The Common Frog is what in Germany is called a **Grassfrog.** This species hides in moist, low vegetation. In contrast, the two introduced species, *(Rana ridibunda* and *Rana esculenta)*, are water frogs, often basking in the sun on the banks of a pond or dyke, seldom more than a frog's leap from the sheltering water. Although not confined to the water like the **Platannas,** they spend much of their time in it and will feed underwater, even eating their own tadpoles. In contrast, the American **Bullfrog** *(Rana catesbeiana)*, often sits surrounded by a shoal of tadpoles and is said to protect them against predators by its aggressiveness.

◁ Australian Bullfrogs.
▽ Paradoxical giant; the tadpole of the South American Paradoxical Frog is three times the size of the adult frog. When the tadpole decreases in size, the internal organs also become smaller.

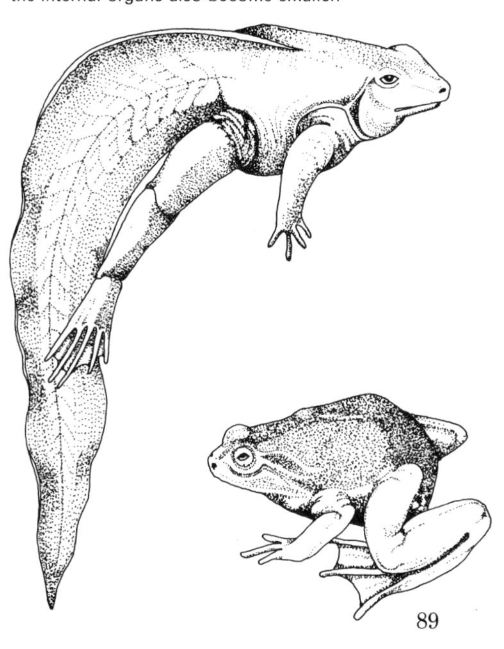

The Clawed and Surinam Toads

Some Anurans have returned permanently to the water and are seldom seen out of it unless migrating overland during a thunderstorm. Their flattened bodies are particularly adapted for swimming and their eyes remain looking upwards rather than forwards. Their hands have fingers which are particularly sensitive and are used for cramming food into their mouths.

The best known of these are the Platannas or **Clawed Toads** of the genus *Xenopus*, from Africa. Their Tadpoles are fishlike and hover in midwater, inclined to about 30° from the horizon, head down. They retain the lateral line organs through life. Less well known are the **Surinam Toads** *(Pipa)*, from South America, with fingers forking at the ends into fingerlets. The parents 'loop-the-loop' while mating and as the eggs are laid, they drop onto the corrugated back of the mother toad and sink into little pits in her skin. The tadpoles develop on the parent's skin, emerging only when they have become miniature froglets.

Arrowpoison Frogs

The South and Central American genus *Dendrobates* contains a number of colourful Treefrogs with toxic skin secretions, which are active by day. These are known as **Arrowpoison Frogs** because the local Indians have regularly used their toxins. The frogs are placed by a hot fire until the toxins are exuded: the poison is smeared on the points of arrows, causing paralysis in the bodies of shot creatures.

Treefrogs of this genus even breed up in the trees, in small pools of water in the hollows of branches. If both food and water are in short supply in one such pool, the male Treefrogs carry the tadpoles on their backs from one tree pool to another. Other species of Treefrogs use the water in the leaf axils of bromeliad plants for their eggs, but are not known to carry the tadpoles about. At least one species, where the adult lives in the water in bromeliads, has a cupped area on its back where the tadpoles remain.

Spadefoots

Spadefoots are small-sized Anurans which have developed a digging tubercle on the undersurface of the hind foot, which is used to excavate a burrow in soft earth. The **European Spadefoots** are placed in the genus *Pelobates* and the North American species in *Scaphiopus*. Spadefoots are adapted to life in dry habitats, since in their burrows they escape dehydration. Other species of Anuran in a whole series of genera through the world also burrow, but they have no special digging tubercle. Some use all four feet, others even their noses.

Spadefoots spend most of their lives in their burrows, only emerging at night when it is moist enough. Their home range is only about forty yards radius about their burrow. After torrential rain they emerge and speedily breed in vast gatherings in temporary pools. These pools are liable to dry up again in only a few weeks and the tadpoles very quickly reach metamorphosis before this happens.

Treefrogs

Treefrogs embrace a whole host of different Anurans of a wide selection of families. Their common point is that with the aid of adhesive discs at the end of their digits, they can live on vegetation, whether trees, shrubs or lower plants. In the European species, even the newly metamorphosed young cling vertically to grass blades rather than hide on the ground among the tussocks.

Prey is captured by leaping and engulfing it. The adhesive discs on the Treefrog's digits take care of landing safely on the vegetation beyond. In one or two species, the feet are enlarged so that with the body, they can form a parachute to enable the 'flying' frog to come to earth when it jumps, to escape pursuit in the treetops.

Among treefrogs of the genus *Gastrotheca*, from South America, the eggs and developing young are carried in a pouch on the mother's back. In some species they are liberated as tadpoles into a pool of water, but in at least one they develop to froglets within the pouch, escaping when the pouch splits along the midline.

△ Tree Frogs can fasten themselves to smooth surfaces by means of suction pads on their finger and toe tips. The suction pads can be seen clearly here on a frog which has been photographed through a piece of glass.

△ Spadefoot Toad digging himself into soft earth to escape the desiccating rays of the sun.
▽ Green Tree Frog on loquat fruit in Florida, USA. The dark streak in the eye is said to aid the frog in protective camouflage.

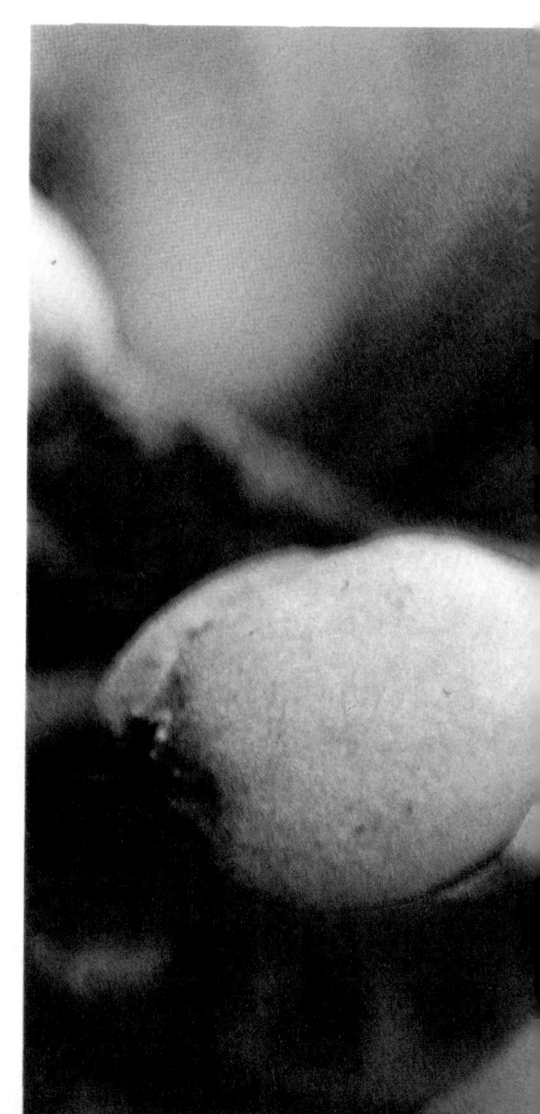

△ Common Toads pairing; the male clasps the female under her armpits while she produces her eggs. The eggs can be seen in strings of jelly, wound round nearby vegetation.

Amphibian numbers

Frogs are attractive morsels to many predators and large numbers are eaten daily. It is therefore not surprising that a clump of spawn laid by a single female may contain two thousand or more eggs, nearly all of which will hatch—and she may lay more than one batch of these during her lifetime. In order for any species to survive, it is only necessary, on average, for each breeding pair to produce two individuals of the next breeding generation. More than this and the world would be swamped with frogs; fewer, and the species would dwindle into extinction. There must therefore be a very exact mechanism to ensure that of the 2,000 eggs, only two survive to adulthood. Tadpoles and froglets are eaten by fish, birds, snakes and mammals: some are desiccated, trodden on, run over by vehicles. Eventually, the requisite number of frogs survive.

This is the situation in natural conditions. However, nowadays there are other factors weighted against the amphibian. Ponds are vanishing, watercourses are being polluted—and frogs take in both water and toxins through their skins. Although there are still Anurans to be found, the threat of extinction through pollution increases.

Unusual modes of development

The oviducts of amphibians contain glands which secrete an outer coat round the eggs as they are laid. However, in a number of species, the secretion comes out of the cloaca as liquid and is beaten into a foam by the feet of the male. The foam encloses the eggs and then hardens. In many cases the foam nest is built in foliage which hangs over a suitable pond. When the tadpoles hatch, the foam liquefies and they fall into the water. In some other cases the foam nest is made in a burrow near the pond.

Tadpoles of certain species develop within the eggs which are laid on moist ground. The very primitive frogs of the genus *Leiopelma* do this, as do those of a number of species of *Leptodactylus*. The eggs of these are large and carry a store of yolk, while oxygen diffuses in through the membranes of the tail which enfolds the tadpole. The eggs finally hatch as froglets.

One or two Treefrogs build private mud pools in streams, within which the eggs are laid and the tadpoles develop, away from both predators and harmful insects. The Frogs of the little-known genus *Nectophrynoides*, parallel the mammals, with the tadpoles developing inside the mother's oviducts.

REPTILES

Like fish and amphibians, reptiles are often said to be 'cold-blooded', a misleading terms since under some circumstances their blood may be quite warm. In fact, their body temperature tends to vary with that of their surroundings (a condition known as poikilothermic), instead of remaining relatively constant as in the 'warm-blooded' or homoiothermic birds and mammals.

Under natural conditions, however, reptiles can control their temperature to a considerable extent, mainly by means of their behaviour. In order to raise the temperature of their bodies to a level suitable for activity they bask in the sun or place themselves where they can absorb heat from the sun-warmed ground. Different kinds of reptiles may prefer to keep their body temperatures within different ranges; these are high (35–40°C.), in diurnal desert Lizards, and lower (25–35°C.), in many Snakes, chelonians Tortoises, etc., and in nocturnal, burrowing or aquatic forms. Since reptiles are so dependent upon external sources of heat they are often called ectothermic, as opposed to the endothermic birds and mammals, which generate the heat they need within their bodies.

Some reptiles have supplementary methods of thermoregulation. Certain Lizards, for example, can change colour and consequently their propensity to absorb or reflect heat. Others, such as Monitor Lizards, incubating Pythons and the leathery Sea Turtle, can make various physiological adjustments, particularly of the blood system, which render them partly endothermic; some extinct forms such as flying reptiles and Dinosaurs may have enjoyed similar advantages.

In general, however, reptiles, lacking an insulating coat of fur or feathers, are unable to maintain indefinitely a body temperature independent from that of their environment, and are much more dependent on climate than birds or mammals. In prolonged cold they are forced into hibernation, and must also shelter from excessive heat beneath vegetation or underground. Since cold rather than heat sets the greatest limit on their activity it is not surprising that they are most abundant in hot countries and are scarce or absent in the colder regions.

Most reptiles lay eggs. The young of chelonians, (Tortoises, etc.), crocodilians and the Tuatara are equipped with a special horny knob on the tip of the snout like that of birds which enables them to break out of the egg when they hatch; in Lizards and Snakes a special egg-tooth serves a similar purpose. The eggs are usually deposited in holes dug in the ground or in some other sheltered position; a few forms, such as certain crocodilians actually construct a nest out of vegetation. The eggs are generally abandoned by the mother and incubated by the warmth of the surroundings. A few reptiles such as Pythons, however, brood or incubate their eggs in their coils, while others, notably Crocodiles, show varying degrees of maternal care towards the eggs or newly hatched young.

A substantial number of Lizards and Snakes bear their young alive and some forms possess a placenta comparable with though less elaborate than that of mammals.

A Galapagos Land Iguana, found only on the Galapagos Islands where the naturalist Charles Darwin made his important observations in the last century.

The Evolution of Reptiles

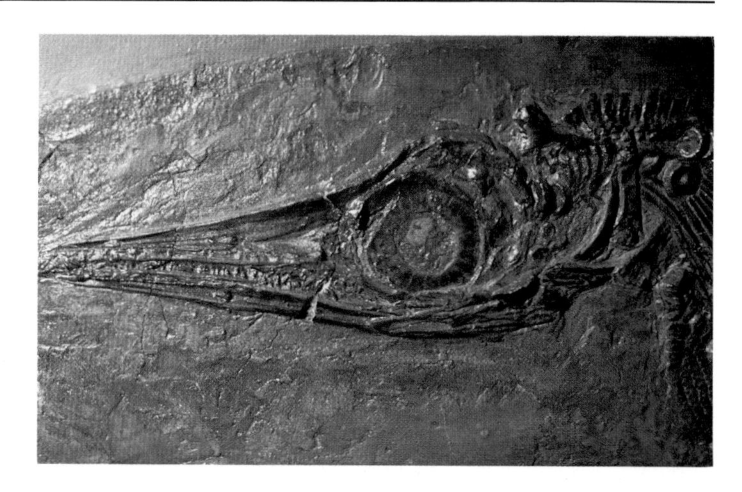

Reptiles form one of the main groups of vertebrates. Existing orders are the Tortoises and Turtles *(Chelonia)*, the Lizards and Snakes *(Squamata)*, the lizard-like Tuatara of New Zealand *(Rhynchocephalia)*, and the Crocodiles, Alligators and their relatives *(Crocodilia)*. These four orders together contain about 6,000 living species; this is only a fraction of the diversity of Reptiles which lived in the past, such as the Dinosaurs, and the flying Reptiles.

In the scale of evolution the Reptiles are intermediate between the fish and amphibians on the one hand and the birds and mammals on the other. Although some Reptiles, such as Turtles, are aquatic, they are as a group much better adapted basically for life on land than Amphibians and possess many characters essential for a fully terrestrial life.

The reptilian skin is modified to form a covering of tough, horny scales or plates which is comparatively waterproof. The male fertilises the eggs inside the reproductive tract of the female, usually by means of a special intromittent organ. The eggs are always laid on land, even if the parents are mainly aquatic; they have tough or hard shells and contain a plentiful food-supply for the embryo in the form of yolk. The embryos possess special membranes, the amnion and allantois (present also in birds and mammals), which play a vital part in the physiology of terrestrial development. The young resemble their parents in most respects other than size and do not pass through an aquatic tadpole stage. Although the majority of Reptiles lay eggs, a substantial number bear their young alive. Some of these characters are paralleled in certain amphibians, but this does not affect the general concept of reptilian organisation.

Like fish and amphibians, reptiles are said to be 'cold-blooded'. Their body temperature tends to vary with that of their surroundings, instead of remaining relatively constant, as in the 'warm-blooded' birds and mammals. They do, however, control their temperature to a considerable extent by their behaviour, basking or alternatively sheltering from the heat, and are thus able to spend a good deal of their time within an optimal temperature range. In prolonged cold, however, they are forced into hibernation, and are much more dependent on climate than birds or mammals. Thus, they are most abundant in hot countries and are scarce or absent in the colder regions.

Reptile beginnings

The study of fossils indicates that the Reptiles were descended from a group of amphibians, now extinct, called Labyrinthodonts. The earliest types appear around the middle of the Upper Carboniferous period, perhaps 300 million years ago. They were small, lightly built animals which seem to have been terrestrial; it is, of course, impossible to tell whether they laid shelled eggs on land in characteristic reptilian fashion. These and other primitive forms are placed in a group known as the Cotylosauria; in more popular terminology they are called 'Stem-Reptiles' since they represent the basal stem from which the more advanced groups of reptiles have descended.

Reptiles did not become numerous until the end of the Carboniferous, when larger and more specialised forms appeared. Some of these seem already to have reverted to semi-aquatic habits. Even at this early date the Reptiles had already split into two principal groups, one containing the ancestors of most, if not all, of the later types such as Dinosaurs, Crocodiles and Lizards and the other leading to the ancestors of mammals.

The Age of Reptiles

The Age of Reptiles lasted throughout the Mesozoic era, from the beginning of the Triassic period, over 200 million years ago, to the end of the Cretaceous, nearly 70 million years ago. By Triassic times the amphibians were on the decline and the Reptiles had firmly established themselves as the dominant forms of vertebrate life, at least on land. Towards the end of the Triassic many important new groups arose: the first mammals, whose reptilian ancestors promptly died out, the Chelonia (Tortoises, etc.), the Crocodiles, the Dinosaurs and the Lizards. Two groups, the fish-like Ichthyosaurs and the long-necked Plesiosaurs, became immensely successful in the sea, filling the role of the whales today; a little later the Reptiles invaded the air, giving rise to the Pterosaurs ('Pterodactyls') and the earliest birds. The mammals remained relatively inconspicuous throughout the rest of the Mesozoic, but dynasties of Reptiles flourished and produced many forms of spectacular size and appearance, exploiting almost every kind of habitat which the world could offer. By the beginning of the Tertiary era the majority of these remarkable creatures had become extinct and the Age of Reptiles had ended, much more abruptly than it began.

◁△Fossilised skull of Ichthyosaur, a reptile which lived over 100 million years ago.
▽ Skeleton of a Boa; vestiges of the hind limbs can be seen at the base of the tail on the right of the picture.
▷△The Crocodiles and their relatives are the sole survivors of the great group of reptiles, the Archosauria, that included the Dinosaurs.
△ Land Iguana, the product of isolated evolution on the Galápagos Islands.

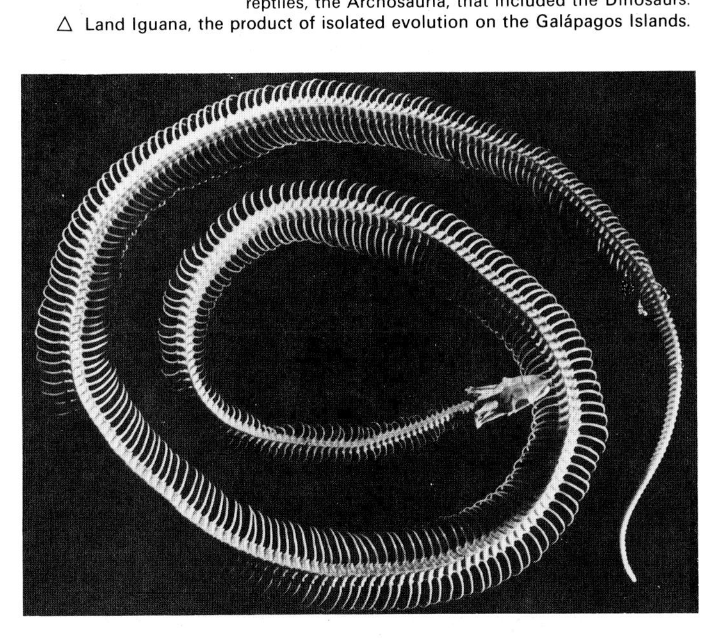

The Dinosaurs were perhaps the most remarkable of all extinct Reptiles. Some were quite small, others gigantic, and they showed an extraordinary variety in appearance and habits. There were two quite distinct groups of them and both flourished throughout the greater part of the Mesozoic, for more than 120 million years.

Many types of Dinosaurs had died out before the end of the Age of Reptiles, but the abrupt extinction of

the remainder at the start of the Tertiary period is one of the greatest mysteries of the past. It has been suggested that the bigger Dinosaurs were able to become more or less warm-blooded under the tropical conditions which prevailed throughout much of the Mesozoic. The Tertiary seems to have begun with a cold spell, and it is possible that these animals, accustomed to a high and constant body temperature but lacking the insulation of fur or feathers, were unable to withstand it. Geological upheaval, loss of habitat and changes in vegetation may have contributed to their downfall. Nevertheless, the extinction of the Dinosaurs is not easy to explain, especially when we consider that so many other types of Reptiles, great and small, terrestrial and aquatic, became extinct at the same time.

Archaeopteryx lived in late Jurassic times, about 150 million years ago, and was the size of a crow. In most respects it was reptilian. It had a long tail, teeth in both jaws, and three clawed fingers on each wing. Its bones were devoid of air-spaces. Its breast-bone (sternum) was small and unlike that of most modern birds, had no keel for the attachment of the flight muscles. On the other hand, its clavicles were fused to form a wish-bone (furcula), and most important of all, some of the fossils show clear impressions of feathers. For this reason alone it can be classified as a bird, for none of the Pterosaurs or flying Reptiles were feathery, though some of them may have had hair.

Archaeopteryx is thought to have been a forest-dweller, able to climb but with only feeble powers of flight. All the five known specimens have come from lagoon deposits of lithographic limestone in Bavaria, suggesting that the creatures had been blown out to sea.

The ancestors of *Archaeopteryx* and of later birds have not been identified with certainty. It is clear, however, that they must have belonged to the Archosauria, the great group of 'Ruling Reptiles' which also contained the Crocodiles, Dinosaurs and Pterosaurs; it is possible that they were actually small primitive Dinosaurs which ran on their hind legs and converted their front legs into wings.

Turtles and Tortoises

Tortoise and Turtle are terms which sometimes cause confusion. In Britain, for instance, land-living chelonians are called Tortoises and the marine species, Turtles. In the USA, Turtle is the term generally used for all chelonians.

The order Chelonia contains about 230 living species of Turtles, Tortoises etc. These are grouped into two sub-orders, the Cryptodira which withdraw their necks in a vertical S-bend into the shell, and the Pleurodira, a much smaller group, in which the neck is withdrawn sideways. In England, the marine forms are called Turtles and have the limbs modified into paddles; the web-footed, amphibious fresh water types are mostly called Terrapins, while the terrestrial ones are known as Tortoises. In America however, the name 'Turtle' is often used collectively for the whole chelonian group, and English writers apply it to some of the larger fresh water species.

Turtles and their relatives are so familiar that one is inclined to forget the remarkable nature of their bodily organisation, which has hardly changed since the end of the Triassic, and which has become profoundly modified with the evolution of the shell. This consists of two pieces, the carapace above and the plastron below, usually joined in the middle by a bridge on each side. The outer layer of the shell is made up of a number of thin horny plates or scutes which correspond with the horny layer of the scales over the head, neck and limbs. In some species these plates show growth rings which may give a very rough idea of the age of the animal. Beneath the horny plates is a layer of living skin cells, and beneath this again is a layer of bony plates. The edges of the horny and bony plates overlap somewhat, but the general arrangement of both is similar; there is a row down the middle of the carapace, a row of larger ones on each side of this, and a series of small ones round the margins of the shell. The ribs and the upper parts of the vertebrae of the trunk (reduced to about 10 in number) are more or less fused with the inside of the bony carapace, and the limb girdles lie inside the ribs instead of outside them

Snapping Turtles

(Chelydridae)

There are two species of **Snappers,** both American. Logically they could be called Terrapins, but even in England the name 'Turtle' is often applied to some of the bigger freshwater Chelonians. The **Common Snapper** *(Chelydra serpentina)*, is found throughout much of the United States of America; some races extend south into South America. It has a big head which cannot be withdrawn into the shell, a powerful hooked beak, a long tail and a small plastron. It reaches a shell length of 15 ins (38 cm) and a weight of 50 lbs. Snappers are ferocious creatures, al ways ready to strike; they feed on fish, waterfowl, invertebrates, etc., but also take some plant food. They like muddy rivers and ponds.

The **Alligator Snapper** *(Macroclemys temminckii)*, is a larger edition of the Common Snapper and is one of the biggest fresh water Chelonians, occasionally reaching a shell length of 3 ft and a weight of 200 lbs. It is found mainly in the south-eastern parts of the United States of America. Its shell is rough, with three longitudinal ridges. Although able to inflict severe injuries to man it is less aggressive than the Common Snapper. This highly aquatic turtle has a forked movable appendage on its tongue which is probably used as a lure to attract fish into its gaping jaws.

as in most other land vertebrates. The neck and tail vertebrae are free and mobile.

A Turtle or Tortoise cannot breathe in the usual reptilian way by moving its ribs; instead, certain abdominal muscles alternately squeeze and expand the body cavity; in expiration the abdominal organs are pressed against the lungs which lie just beneath the carapace. Breathing is assisted by movements of the limbs. As a further peculiarity, Chelonians have lost their teeth, replacing them by a horny beak like that of a bird. All chelonians lay eggs, which are buried in a nest dug in soil or sand.

Armour has served the Turtles and their allies well, despite the restrictions it imposes on their way of life. They are quite a successful group with a surprisingly wide range of habits, and are found in nearly all the warmer parts of the world.

◁ Although Giant Tortoises are protected on the Galapagos islands, the young tortoises are sometimes killed by Wild Pigs and Dogs. On some of the islands, Goats compete with the Tortoise for food.
△ Red-eared Terrapins. When not actually swimming, Terrapins float, with only their snouts showing above the surface of the water, the hindfeet moving slowly to keep their bodies steady.
△ △ Tortoises and Turtles cannot breathe by using their ribs but instead the pumping action is provided by belly muscles. The arched upper shell is called the carapace and the flat under shell is called the plastron.

The Mud, Musk and Big-headed Turtles or Terrapins

(Kinosternidae; Dermatemydidae; Platysternidae).

These are three family groups of small amphibious Terrapins. The **Mud** and **Musk Terrapins** (Kinosternidae), from North and Southern America are rather like small Snappers in some ways. The **Mud Terrapins** *(Kinosternon)*, have two transverse hinges across the plastron which allow the shell to be very completely closed when the extremities are retracted. The **Musk Terrapins** *(Sternotherus)*, are renowned for the strong unpleasant smell they produce from the secretions of glands in the flanks.

The Dermatemydidae contains a single Central American species. The Platysternidae also has only a single species, from south-east Asia, with a very large head and long tail. It lives in mountain streams and feeds on worms and molluscs.

Freshwater Terrapins

(Emydidae)

The big family Emydidae contains many species of typical, flat-shelled Terrapins, some quite big but mostly of small or moderate size. They are found in America, Europe, North Africa and Southern Asia.

Chrysemys picta, the **Painted Terrapin,** and *Pseudemys scripta elegans* are common North American species with beautiful yellow and red markings. Babies of the **Red-eared Terrapin** are often sold as pets in this country, but their rearing requires special care. This species has an elaborate courtship, the male swimming backwards in front of the much larger female and stroking her

face with his long front claws. Other well-known genera are *Malaclemys*, the **Diamond-back Terrapins,** much esteemed as food in the United States of America and *Emys*, which includes the European Pond-Terrapin, *E. orbicularis.* The North American **Box 'Tortoises'** *(Terrapene)*, also belong to this family; they have fairly high shells, a hinged plastron with which they can tightly box themselves in, and live on land.

△Tortoises and Turtles are known for their slow movement and long life-span.

The Land Tortoises

(Testudinidae)

The typical Tortoises have dome-like shells, live on land and feed mainly on plants though they should occasionally be offered meat in captivity. The genus *Testudo* occurs around the Mediterranean and includes the **Common** or **'Greek' Tortoise** *(T. graeca)* which lives in Spain and North Africa but not actually in Greece. It is often sold in pet shops and this trade may threaten its survival in its natural habitats. It can be distinguished from its relatives by the presence of a large spur on the inside of each thigh which can be $\frac{1}{4}$ in (8 mm) long. Such tortoises can be kept outdoors during the English summer but seldom lay fertile eggs and should be allowed to hibernate in a box filled with straw and kept in a cool dry place. Their eggs have hard shells.

The genus *Geochelone* contains the true island giants and also some other quite large species such as *Geochelone radiata* from Madagascar which reaches a weight of 15 lbs and has a shell with beautiful yellow markings. Other genera include *Kinixys*, unique in having a hinged carapace, the American **Desert Tortoises,** *(Gopherus)*, and the African *Malacochersus* with a flat flexible shell, enabling it to squeeze into rocky crevices.

The Giant Tortoises

(Geochelone gigantea; Geochelone elephantopus)

Today, **Giant Tortoises** *(Geochelone gigantea* and *Geochelone elephantopus)*, are found in the truly wild state only in two isolated island habitats; respectively Aldabra in the Indian Ocean and on some of the Galapagos Islands in the Pacific, 600 miles off Ecuador. They formerly had a wider distribution but during the 18th and 19th centuries their numbers were much reduced by mariners who carried them off for food and exterminated many of their races. Aldabra now has a vast tortoise population of perhaps 100,000 but there are only about 10,000 on the Galapagos, where they are highly protected. The males are bigger than the females and on the Galapagos may reach a carapace-length of more than 5 ft (1.6 m) (over the curve), and a weight of over 500 lbs (230 kg).

As in Tortoises generally, the male has a longer tail than the female and his plastron is markedly convex to accommodate the carapace of the female when he mounts her. These animals can make a variety of sounds, the males groaning loudly during mating. Giant Tortoises, like some of the smaller species can live for 100 years, perhaps a good deal longer.

The Leatherback Sea Turtle

(Dermochelys coriacea)

The unique leathery Turtle, **Leatherback** or **Luth** *(Dermochelys coriacea)*, is placed in a family of its own; it is the largest of modern Chelonians, having an average carapace-length of 5 ft (1.7 m); and reaching a weight of over 1,000 lbs. It has no horny shell plates, its body being covered with leathery skin, and no typical bony shell of large plates; instead both carapace and plastron are mainly composed of enormous numbers of small bones which fit together in a mosaic. This is raised up to form seven longitudinal ridges along the back, accentuating the creature's streamlined appearance. The vertebral column and ribs are quite free from the carapace instead of being fused with it as in chelonians generally. The Leatherback ranges around the warmer oceans of the world and occasionally strays to Britain. It is an immensely fast swimmer, and may be partly warm-blooded. Its general habits resemble those of other sea turtles.

The Typical Sea Turtles

(Cheloniidae)

Apart from the Leatherback, the **Sea Turtles** all belong to the family Cheloniidae and comprise the **Green** and (rare) **Flatback Turtles** *(Chelonia mydas* and *Chelonia depressa)*, the **Hawksbill** *(Eretmochelys imbricata)*, the **Loggerhead** *(Caretta caretta)*, and the **Ridley Turtles** (genus *Lepidochelys).* Their limbs are paddle-like and their bony shells somewhat reduced; except for nesting purposes they seldom come

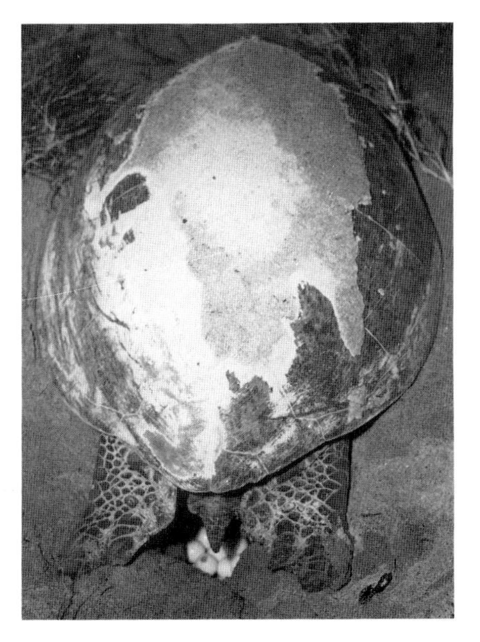

◁ △ From left to right: The female Green Turtle crawls up the beach during the night to nest. Using her back flippers, she digs first a shallow pit to take her body and then a deeper pit under her tail to take the eggs. A clutch of eggs can number up to 150.

After laying the eggs, the female Turtle covers the eggs with sand to incubate them and returns to the sea.
The hatched turtles return to the sea, guided to the water by such visual stimuli as moonlight over the horizon. Only a few young Turtles reach the sea, many of them falling prey to birds and crabs.

▽ Breeding map of the Green Turtle, showing major and minor nesting beaches and principle feeding grounds. The Green Turtle is migratory, swimming hundreds of miles to reach favoured nesting grounds to lay a clutch of 100–200 eggs in two or three visits. Former nesting beaches can be seen marked with open circles.

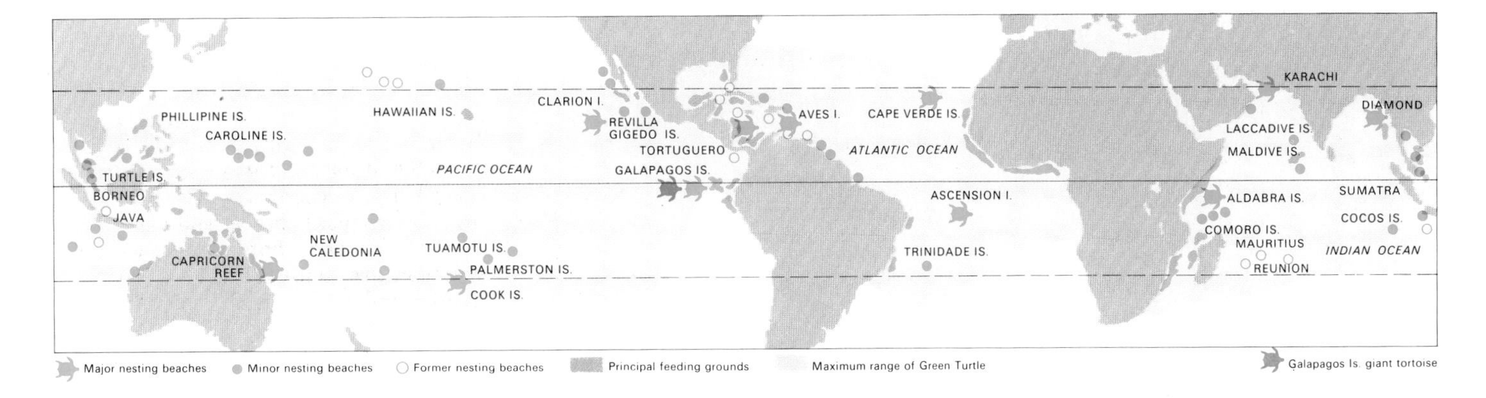

on land. All species lay large clutches of fairly soft shelled eggs on tropical beaches, burying them in the sand. The hatchlings are guided down to the water by visual stimuli such as moonlight over the broad horizon of the sea. During this short journey they are preyed on by many animals such as birds and crabs and the infantile mortality rate is very high. Tagging experiments have shown that the Green Turtle is migratory, swimming for hundreds of miles to favoured nesting grounds. The Green Turtle mainly eats seaweed; the others take more animal food.

The eggs of Turtles, and the flesh of the adults, especially of the Green Turtle, are eaten by man, while the horny plates of the Hawksbill provide the decorative material known as tortoiseshell. Owing to their nesting habits, Turtles are highly vulnerable and most species require urgent conservation. Controlled farming, with liberation of a quota of young, is advocated by some conservationists and could render these animals a valuable and lasting asset to man.

Soft-shelled and Plateless Turtles

(Trionychidae; Carettochelidae).

The soft-shelled Turtles (Trionychidae), live in lakes and rivers of North America, Africa and south-east Asia. They are highly aquatic, having very flat, pancake-like shells which blend with the river bottom, and nostrils at the end of a long proboscis through which they can breathe when almost submerged. They also have filaments in the throat which act as gills. Like the Leatherback, they have lost the horny shell plates, and the bony plates are reduced. They are carnivorous, catching fish by strikes of the long neck; some species reach a shell length of 1 m.

Carettochelys, the only member of its family, is a large and remarkable species from New Guinea; it resembles the Trionychidae in having a covering of soft skin instead of horny plates and is the only freshwater Turtle with paddle-like limbs.

The Side-necked Turtles

(Pelomedusidae; Chelidae)

The families Pelomedusidae (sometimes called **Hidden-necked Turtles)** and Chelidae belong to the sub-order Pleurodira; they differ from other modern chelonians (Cryptodira) in withdrawing the neck sideways instead of vertically into the shell. They are virtually confined to the southern hemisphere, being found in fresh waters of South America, southern Africa and Australasia. *Podocnemis expansa* is a very large, herbivorous pelomedusid from the Amazon; its eggs are used as a source of oil by the Indians, who also eat its flesh. The Chelidae contains the **Snake-necked Turtles** of Australia and New Guinea *(Chelodina)* and the curious **Matamata** *(Chelus fimbriatus)* of Brazil, a big Turtle with a knobby shell, a long proboscis and fleshy filaments round the mouth which may serve as a lure to attract little fish.

The Great Ruling Reptiles

Surprisingly the Crocodiles' nearest living relatives are the Birds, which now form part of their diet. Large Crocodiles will attack animals the size of a deer, dragging their prey into the water and tearing it apart with their fearsome teeth.

The order *Crocodilia* belongs to the great group called the 'Ruling Reptiles' (subclass *Archosauria*), which includes the Dinosaurs and extinct flying Reptiles—and also the ancestors of Birds, which are the Crocodiles' nearest living relatives. Crocodiles appeared in the later Triassic and became numerous in the Jurassic period, when there lived some marine types with paddle-like limbs and finned tails. Forms essentially like those of today are found in the Cretaceous period, including the gigantic *Phobosuchus*, 36 to 45 ft (12 to 15 m) long, which probably ate Dinosaurs.

The existing Crocodilians comprise about 22 species of Crocodiles, Alligators, Caimans and Gharials. These animals are basically very similar and differ from each other mainly in trivial characteristics, such as the proportions of the snout and the appearance of the teeth. The Crocodilian heart differs from that of other reptiles in that the right and left ventricles are completely separate.

All Crocodilians show many adaptations to their amphibious, predatory mode of life. The eyes and the valvular nostrils are placed on top of the head and can be kept above the surface when the rest of the animal is submerged. The palate, separating the nose from the mouth, is extremely long and the internal nostrils open far back, into the throat. These can be shut off from the mouth by valvular flaps so that a Crocodile can drown its prey without taking water into its own windpipe. The feet are partly webbed, but the tail, flattened from side to side, is the main swimming instrument; it can also be used as a weapon

Crocodilians can move quite well on land, often slinking along with the legs extended and the belly held well off the ground. Their hearing is excellent and their ears are protected by scaly flaps. They are partly nocturnal, but spend much of the day basking, gaping their jaws to cool themselves by evaporation of water through the lining of the mouth. Like most other reptiles, they can replace their teeth throughout their lives and a Crocodile 12 ft (4 m) long has had about 45 generations of teeth. These conical teeth are for seizing, not cutting or chewing. A Crocodile has to pull large prey to pieces before it can eat it, unless, as some believe, the carcass is hidden in a hole until it rots. Some Crocodilians, at least are territorial, the males fighting in the breeding season and roaring loudly, perhaps as a challenge to rivals. The female makes a nest of some kind for her hard-shelled eggs, laid in clutches of 20–80, and shows a considerable degree of maternal care, before and after the young are hatched.

Giants in distress

Today most species of Crocodilians are in danger. Their skins make excellent leather and they have been wiped out of many of their haunts in African and South American countries. Even in nature reserves they are liable to poaching and can easily be shot at night since they are attracted to the light of a lamp; they are also caught by means of hooks and snares.

The International Union for the Conservation of Nature and Natural Resources has established a special group for the study of crocodilian biology and conservation; similar groups deal with other endangered animals such as turtles and spotted cats. Legal protection offers some hope and seems to have given the American Alligator a partial reprieve.

▽▽ Crocodiles, like other reptiles, bask in the sun to raise their body temperature to levels suitable for activity.

▽ Salt Water or Estuarine Crocodiles grow up to 20 ft long and are dangerous to man.

Crocodiles

(Crocodilinae)

The true Crocodiles belong to the sub-family Crocodilinae. The best known species are the **Nile Crocodile** *(Crocodylus niloticus),* found throughout much of Africa, the **Salt Water** or **Estuarine Crocodile** *(Crocodylus porosus),* of south-east Asia and Australasia, the **American Crocodile** *(Crocodylus acutus),* which ranges as far north as the tip of Florida, and the **Indian Marsh Crocodile** or **Mugger** *(Crocodylus palustris).* The first two species, (and probably some others), may exceptionally grow to 18 ft (6 m) in length and are regarded as particularly dangerous. Most Crocodiles have fairly long, pointed snouts; in a few, such as the West African *Crocodylus cataphractus,* which feed mainly on fish, the snout is very long and slender, while in certain others it is very broad and short. These stumpy-nosed Crocodiles *(Osteolaemus),* only grow to between 3 and 6 ft (1 and 2 m) long.

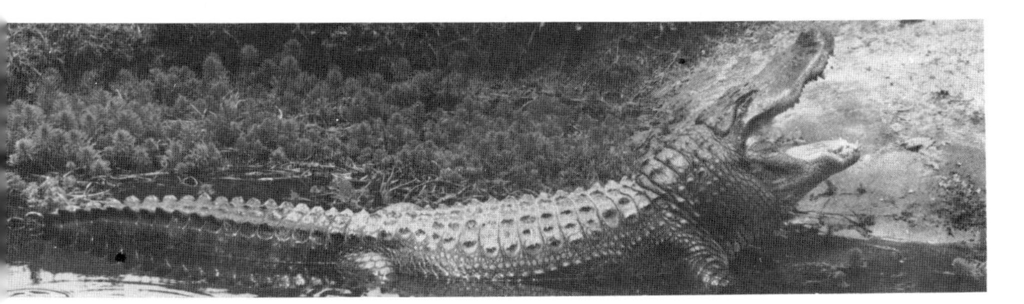

△ A female Crocodile lays about 60 eggs in a nest and covers them up. The young hatch about 2–3 months later, fully formed and equipped with sharp teeth.

▽ The Crocodile has a projecting fourth lower tooth while the Alligator's fits into a pit.

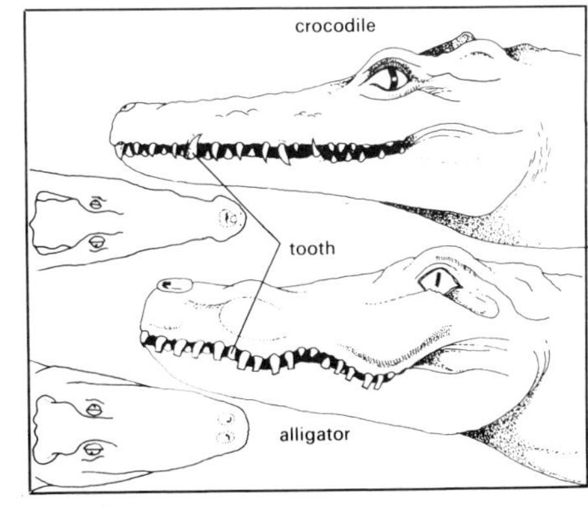

The Nile Crocodile has been the most fully studied species and is renowned for its maternal care. The mother guards her eggs which are buried in a sand-bank, helps the young to escape from the nest and may even escort them to and in the water. Basking Nile Crocodiles are attended by certain birds such as the Spur-Winged Plover which warn them of danger and remove parasites from their gaping jaws.

Alligators and Caimans

(Alligatorinae)

Alligators and their allies, the **Caimans,** belong to the subfamily *Alligatorinae* and differ from Crocodiles in the fact that the large fourth tooth on each side of the lower jaw fits into a pit instead of a notch in the upper jaw and so does not project visibly when the mouth is closed. Although this feature is not always reliable, the Alligators can be distinguished from most Crocodiles by their broad, bluntly rounded snouts. In Caimans, the proportions of the snout are variable; these reptiles differ from most other Crocodilians in having bony plates beneath the horny scales of both the belly and the back, instead of on the back only.

Caimans, (e.g. genera *Caiman* and *Melanosuchus*), are found in Central and South America.

The **American Alligator** *(Alligator mississipiensis),* lives in the swamplands of the southern United States of America. It grows to a length of 6 ft (2 m) within 6 years of hatching; the males become bigger than the females and may eventually reach 4 metres or more. The female builds a nest-mound out of vegetation and guards her eggs. Alligators feed mainly on fish but may attack mammals and, very occasionally, man. The other species *(Alligator sinensis),* is found in China.

Gavials

(Gavialis)

Gavials or **Gharials** are Crocodilians with exceptionally long, slender snouts and feed predominantly on fish which they catch with rapid sideways snaps. The **Indian Gharial** *(Gavialis gangeticus),* reaches a length of more than 18 ft (6 m) but much bigger fossil forms are known. When mature, it grows a curious excrescence of unknown function near the tip of its snout.

The **Malayan** or **False Gharial** *(Tomistoma schlegeli),* is found further east, including Borneo in its range. Superficially, it looks rather like the Indian species but is not closely related to it and is classified with the true Crocodiles in the Crocodilinae.

The Versatile Lizards

Surprisingly some Lizards are able to glide through the air, others are swimmers, runners or burrowers. Some resemble snakes, others have flattened bodies and five-toed limbs. They range in size from 2 inches to the 10 foot long Komodo Dragon.

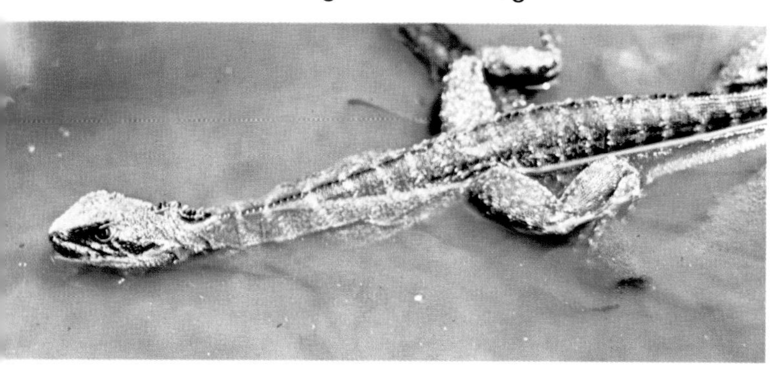

Lizards are remarkable for their versatility. They are found in all warm, and not-so-warm countries, from the Arctic Circle to the tip of South America. Species have become specialised for running, climbing, burrowing and swimming; and a few can even glide through the air!

Existing Lizards number around 3,000 species, and together with their relatives the Snakes, are by far the most successful and abundant of modern reptiles. They are placed in the suborder *Sauria* or *Lacertilia* of the order *Squamata* (which also contains the Snakes), and appeared towards the end of the Triassic period.

The Lizards and Snakes share many interesting features, some of which may have been responsible for their success. Above the front of the palate on each side is a saccular structure called the Organ of Jacobson which opens into the mouth through a narrow duct. This is an important sense organ which smells particles picked up by the tip of the protrusible and often forked tongue; with its aid a Lizard or

Tuatara

(Sphenodon punctatus)

The **Tuatara** looks like a clumsy Lizard, growing up to 2 ft long, with a big head and a crest of spines down its back. It is related to the Lizards but, because of certain features of skull structure and other peculiarities, such as the absence of a copulatory organ in the male, it is placed in an order of its own, the Rhynchocephalia or 'beak-heads'. It is something of a living fossil, for creatures very similar to it existed about 200 million years ago, in the Late Triassic period.

The Tuatara is now restricted to a number of small, off-shore islands of New Zealand, though formerly it occurred on the mainland. It lives in burrows, often taking over those made by sea-birds which share its habitat, and eats insects, Snails and other small animals. It is partly nocturnal and remarkably tolerant of cold, sometimes foraging when its body temperature is as low as 10°C. It lays parchment-shelled eggs in a hole in the ground which take 14 months to hatch. It is protected by the New Zealand government.

Like many Lizards, the Tuatara can regenerate its tail, and has a well-developed parietal (or pineal), eye at the top of its head. Experiments on Lizards suggest that this organ acts as a kind of register of solar radiation, telling the animal when it is time to stop basking and retreat from the sun.

△ Eastern Water Dragon which is found in the Far East and in Northern Australia, in the tropical rain forests. This is a large lizard, growing up to 3 ft in length.

▷ One of the Australian Agamids, known as 'Dragons'; many species live amongst trees or rocks.

Snake is able to track its food or a mate. The Organs of Jacobson are less well developed or absent in other groups of reptiles. The upper jaw of most Lizards is able to move somewhat in relation to the brain-case, increasing the gape and helping the reptile to pull prey into its mouth. Many types of Lizards show a tendency to lengthen their bodies and lose their limbs, moving by means of serpentine undulations. Male Lizards and Snakes have a pair of copulatory organs known as hemipenes, instead of the single penis of most other reptiles; only one organ is used at a time, however.

In general, Lizards differ from Snakes in the following important respects. They nearly always retain vestiges of both limb girdles, even though all external traces of limbs may be lost. Most Lizards possess movable eyelids instead of a transparent eye-covering called a spectacle. Some Lizards have small plates of bone called osteoderms beneath their horny scales. Many Lizards have a special mechanism for throwing off their tails when they are attacked, and can subsequently grow new ones.

Most Lizards and Snakes lay eggs, usually with leathery or parchment-like shells; a few species show rudimentary forms of maternal care. A considerable number, however, have become viviparous (or ovoviviparous), retaining the young within their bodies till they are ready to be born. This type of reproduction is characteristic of, (though not exclusive to), forms which live in cold places, and is found in four out of the six British species of reptiles.

▽ Most Geckos produce only one or two eggs at a time although they can breed most of the year. The Gecko shown has specialised pads on its feet enabling it to climb smooth surfaces.

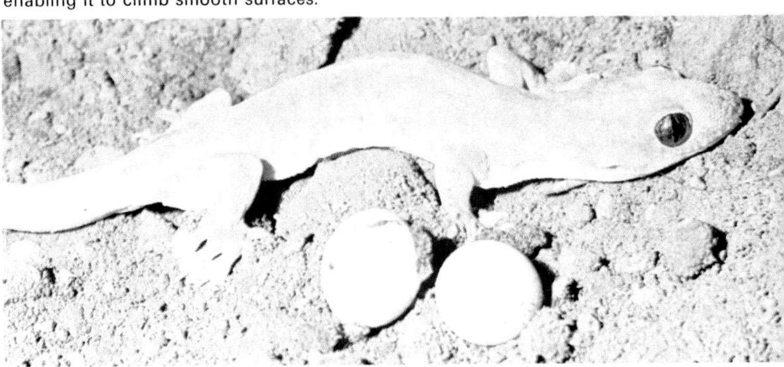

The Geckos

(Geckonidae)

The **Geckos** are one of the largest of the 18 or so families of lizards and are almost world-wide. They are mostly small, although the **Tokay** *(Gekko gecko)*, is the largest 12 ins (30 cm) in length. Most of them have an immovable, transparent eye-covering called the spectacle instead of eyelids, and are more or less nocturnal. They make cheeping or clucking sounds—hence the name 'Gecko'—which may play a part in their social behaviour.

Many Geckos have toes modified for climbing; the rows of large scales on their undersides bear thousands of tiny bristles which help the Lizard to cling to smooth surfaces or even to run upside-down over the ceilings of houses, which some species frequent. The *Ptychozoon,* from south-east Asia, has scaly fringes along its flanks and tail which enable it to parachute through the air. Other species live on the ground and may have fringed or webbed toes which support them on the desert sand. The scientific names of many genera end in 'dactylus' meaning a finger, (e.g. *Hemidactylus, Phyllodactylus),* indicating the importance of these structures in this group. Most Geckos lay eggs, with hard shells.

The Geckos have some unlikely-looking relatives, notably the *Pygopodidae*, a family of snake-like Lizards found only in Australasia.

▷ The Leaf-tailed Gecko which lives in Australian rain forests. At rest, among leaves, the Gecko can camouflage itself extremely well, the large flattened tail blending with the surroundings. Here it is lying across a rock and yet it would still be difficult to spot.

The Agamids

(Agamidae)

The *Agamidae* is a large family of Lizards found in southern Europe, Africa, Asia and Australasia, where they are popularly known as **'Dragons'.** They are related to the **Chameleons** and the **Iguanids** and vary greatly in appearance and habits; some, (species of *Agama* for instance), live in rocky

places and deserts, and others, such as *Calotes*, in trees. A few, such as the **Water Dragon** *(Physignathus)*, of Australia are amphibious, while the **Flying Lizards** *(Draco)* of the East can glide through the air on wing-like membranes supported by elongated ribs.

Agamids are active, sun-loving, diurnal creatures. They are highly territorial and the larger and more brightly coloured males have elaborate courtship and challenge rituals during the breeding season; such display is accompanied by bobbing gestures of the head and fore-quarters, erection of the crests and dewlaps with which many species are adorned, and often by colour change. A few Agamids, such as the *Uromastyx*, are herbivorous, unlike most Lizards which feed on insects.

△ Boyd's Forest Dragon of Australia. Many Dragons have well-developed crests.

Beards and Frills

Among the many Agamids found in Australia are the **Bearded Lizard,** (or **Dragon**) *(Amphibolurus barbatus)*, and the **Frilled Lizard** *(Chlamydosaurus kingi)*. The Bearded Lizard grows to nearly 2 ft (60 cm) and has a semicircular 'beard' or ruff beneath its throat which is erected when the animal is threatened. The Frilled Lizard grows to 3 ft (1 m) long and has a large frill which is normally folded back along the neck. When alarmed this is erected by movement of the hyoid apparatus (a series of bones, muscles etc., associated with the throat and tongue), and stands out like a fan, sometimes nearly 12 ins (30 cm) across. At the same time, the jaws are gaped, showing the yellow mouth-lining, and producing quite a frightening effect.

Frilled Lizards can run fast on their hind legs, the front part of the body being raised off the ground and the long tail acting as a counterpoise. Like other Agamids, they lay eggs.

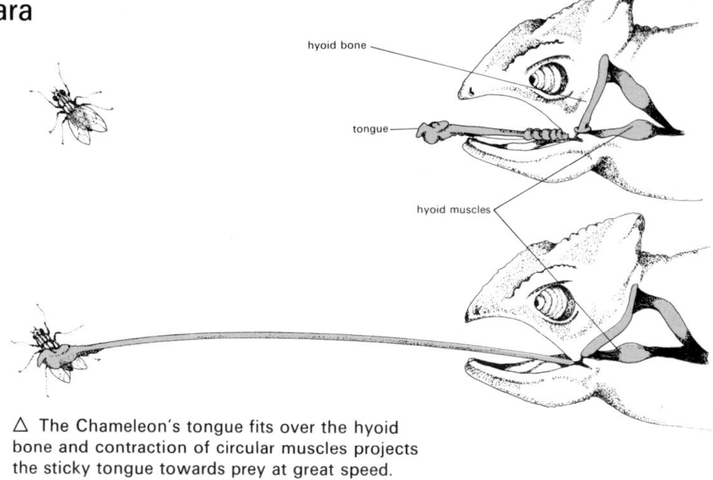

△ The Chameleon's tongue fits over the hyoid bone and contraction of circular muscles projects the sticky tongue towards prey at great speed.

The Moloch

(Moloch horridus)

The **Moloch Lizard** or **Thorny Devil** *(Moloch horridus)*, is a peculiar kind of Agamid found in the deserts of Australia. It is only about 6 ins (15 cm) long and is covered with spines; it has a particularly large pair situated on a kind of hump behind its small head. It has the curious ability of attracting water on to its skin which must be useful during occasional rain. The water does not pass through the skin but flows in minute channels along it until it reaches the mouth, when it is swallowed. This harmless little creature feeds entirely on small black ants, catching them with its short sticky tongue and consuming several thousand at a sitting.

The Chameleons

(Chamaeleonidae)

The **Chameleons** are highly specialised **Tree Lizards** found in southern Europe and Asia, but mainly in Africa and Madagascar. Some are only a few centimetres long but others can extend to 60 centimetres in length. Their heads are often adorned with casques and flaps and some have one, two or three horns or curious projections arising from the nose. The feet are highly adapted for grasping twigs and the toes are arranged in two opposable groups; on the front feet there are two toes on the outside and three on the inside, while on the hind feet this arrangement is reversed. The tail is also prehensile. The eyes are set in turret-like projections which can be swivelled in all directions independently of each other. When a Chameleon sees an insect it brings it into binocular focus and stalks it until it is within range of its extremely long tongue. This is then shot out by a remarkable muscular mechanism and drawn back with the prey adhering to its sticky tip.

Chameleons have considerable powers of colour change which are influenced by factors such as illumination and excitement. They do not necessarily match their backgrounds, though quite often they blend with them. Most Chameleons lay eggs in the soil beneath their bushes but others, which live in cooler regions, are viviparous.

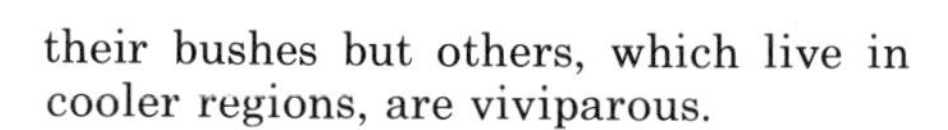

The Iguanids

(Iguanidae)

The *Iguanidae* is one of the biggest families of Lizards, containing some 700 species. They show many resemblances to the Agamids but differ from them in certain dental features and in occurring mainly in the New World. The green **Iguana** of tropical America grows to almost 6 ft (2 m) in length and has a conspicuous dewlap and a spiny crest down its back; it lives in trees but takes readily to water. Ground-living relatives include the **Rhinoceros Iguana** *(Cyclura cornuta)*, of Haiti which has three short, horn-like scales on its snout. A few species of Iguanas, as the bigger members of this family are mostly called, are found outside the New World in Madagascar and Fiji. The **Basilisks** *(Basiliscus)*, are biggish crested **Tree Lizards** which can run on their hind legs, even dashing for a short distance over the surface of water. The family also includes many small and medium-sized forms such as the **Spiny Lizards** *(Sceloporus)*, and the **Collared Lizards** *(Crotaphytus)*, found in the United States of America. Some of the bigger Iguanids are more or less herbivorous.

The Horned Toads

(Phrynosoma)

These grotesque little creatures are the counterparts among the family Iguanidae of the **Thorny Devils** of the Agamid family. The **Horned Toads** live in the desert regions of North America and burrow in the sand, often lying with only their heads exposed. They have short tails, squat, toad-like bodies and a number of spines growing out from the rear of the head which may afford some protection against predators. They also have another, more unusual means of defence, being able to spray drops of blood from their eyes for a short distance; this blood is said to have an irritant effect if it reaches the eyes of an enemy.

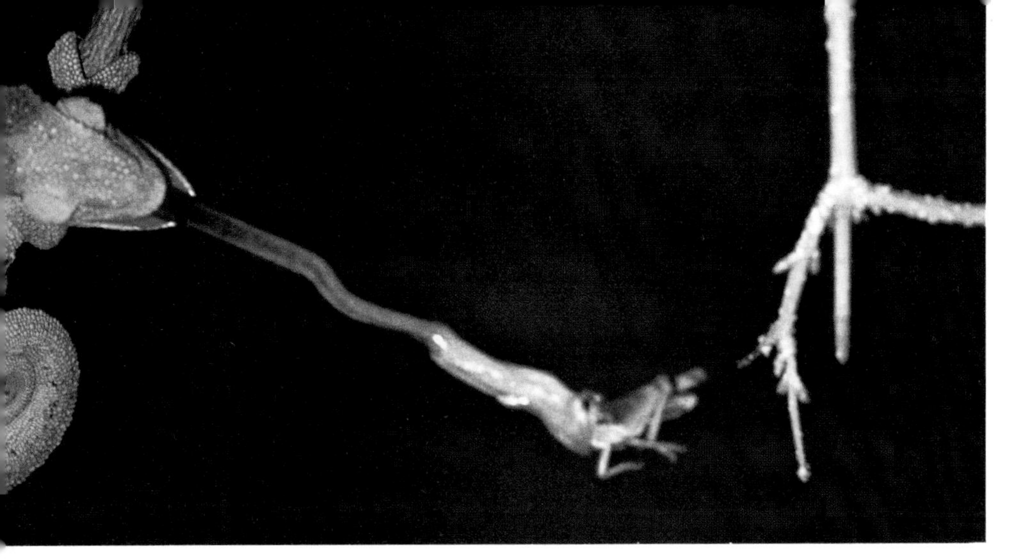

△ Chameleon eyes move independently but can also have binocular vision, giving the lizard very precise aim. The tongue, when fully projected, measures half the length of the Chameleon itself.
▽ A Frilled Lizard, erecting its large conspicuous frill and opening its mouth wide as a warning gesture.

The Anoles

(Anolis)

The **Anoles** form a large group of arboreal Iguanids; about a quarter of the species of the huge family Iguanidae belong to this genus. They are found in North, Central and parts of South America and are well represented in the West Indies. Most of them are quite small, but the **Knight Anole** *(Anolis equestris)*, grows to 18 ins (45 cm). These Lizards have long, triangular heads and well developed dewlaps or throat-fans. When these structures are suddenly distended, as in the challenge display of rival males, the bright coloured skin, (red in some species), at the hinges of the stretched scales suddenly becomes visible, making the animal very conspicuous. Anoles can also change colour readily by the movement of pigment in their skin cells and are sometimes miscalled 'Chameleons'. It is uncertain how far such colour change serves the purpose of concealment.

The Teiid Lizards

(Teiidae)

The **Teiid Lizards** *(Teiidae)*, are restricted to the New World, many species being found in Central and South America. They vary greatly in appearance, and for the most part their habits are poorly known. Some, such as the **Racerunners** *(Cnemidophorus)*, and **Ameivas** are typical, small, agile Lizards. Others, such as the **False Monitor** *(Callopistes)*, of Peru and the massively-built **Tegus** *(Tupinambis)*, reach a metre in length and are powerful predators, able to kill small vertebrates. The larger **Caiman Lizard** *(Dracaena)*, which grows up to 4 ft (120 cm) long, is one of the few amphibious Teiids and feeds on snails, which it crushes with its large flattened teeth. Certain Teiids, such as *Cnemidophorus*, are able to run bipedally when they are in a hurry, like some Agamids and Iguanids.

A number of small Teiids have followed a very different line of specialisation, becoming more or less limbless and creeping on the ground or burrowing beneath it.

The Lacertids: typical Lizards

(Lacertidae)

The family *Lacertidae* consists of typical generalised Lizards found throughout the Old World from Europe to China, though not in Australia. A few show minor specialisations, such as scaly fringes along the toes for running over sand. The main genus, *Lacerta*, includes two species which occur in Britain, the **Common Lizard** *(Lacerta vivipara)*, which differs from its relatives in bearing its young alive, and the slightly bigger **Sand Lizard** *(Lacerta agilis)*. Male Sand Lizards are more brightly coloured than the brownish females, being green, with blue spots on the back and flanks. This species has a restricted distribution in parts of south-western England and of Lancashire, and is in need of conservation. The **Wall Lizard** *(Lacerta muralis)*, and the **Green Lizard** *(Lacerta viridis)*, are common European species, while the handsome **Eyed Lizard** *(Lacerta lepida)*, which reaches 60 cm in length is found in southern Europe and North Africa.

The African family *Cordylidae* is broadly related to the Lacertids, Skinks etc. It includes the spiny **Girdle-tailed Lizards** *(Cordylus* or *Zonurus)*, and the **Plated Lizards** *(Gerrhosaurus)*, as well as some snake-like, nearly limbless forms.

The Anguid Lizards

(Anguidae)

Like certain other groups of Lizards, the family *Anguidae* includes reptiles of very different appearance. The **Alligator Lizards** *(Gerrhonotus)*, of North America are suggestive of miniature Crocodilians with their long bodies, short but well developed legs and broad scales; the biggest species reaches 50 centimetres in length. The **Galliwasps** *(Diploglossus)*, found in Central America and the West Indies show varying degrees of limb reduction, while in the **Slow-worm,** all external traces of limbs have disappeared. The Slow-worm *(Anguis fragilis)*, is found over much of Europe including Britain and grows to about 15 ins (38 cm) long. It burrows well and is seen mainly on banks and under stones; it is viviparous, and feeds mainly on slugs. Its tail is easily broken off, but the new one is much shorter than the old. Like other members of its family, the Slow-worm has a well developed armour of osteoderms.

The Skinks

(Scincidae)

The **Skinks** *(Scincidae)*, are a very large, cosmopolitan group containing some 700 species. A typical Skink, such as a member of the genera *Scincus* or *Mabuya*, is under 12 ins (30 cm) in length with a rather elongated body, fairly short limbs and smooth, shiny scales; such forms usually live close to the ground, under soil or sand, or among the leaves of the forest floor. Some of these forms, such as *Ablepharus*, have a spectacle over the eye like a snake or a transparent window in the lower eyelid. A few species grow to 12 ins (30 cm) or more, like the **Shingleback** *(Trachysaurus rugosus)*, of Australia; this untypical Skink has rough scales and a very short, stumpy tail.

One of the most interesting features of the Skinks is the tendency to lose the limbs which can be seen in about two-thirds of the genera, for example *Scelotes* and *Acontias*. In some forms, the limbs are tiny, often with reduced number of toes, and in others there are no external limbs at all. Limb reduction is often correlated with burrowing habits and with reduction of the eyes and ears as well. Many Skinks are viviparous and some have a well developed placenta, an organ characteristic of mammals, which allows oxygen, food etc. to pass from the mother to the embryo.

The Marine Iguana

(Amblyrhynchus)

Although a few other species of Lizards may live along the shore and sometimes enter the sea, the **Marine Iguana** *(Amblyrhynchus cristatus)*, is the only existing Lizard which has really taken to marine life. It is found only on the Galápagos Islands where it lives on the lava rocks along the shore, swimming out to sea to feed on seaweed. It grows to over 3 ft (1 m) in length and has a spiny crest down its back. Most Marine Iguanas are sooty black in colour but on Hood Island the breeding males are marked with crimson on the flanks and have green crests. The males of this species practise a curious form of ritual combat, two rivals clashing their heavily scaled heads together until one retreats or assumes a posture of submission. These Lizards have special glands in the nose which help them to get rid of excessive salt by exhaling it as a spray.

A large **Ground Iguana** *(Conolophus subcristatus)*, is also found on the

Life on isolated islands

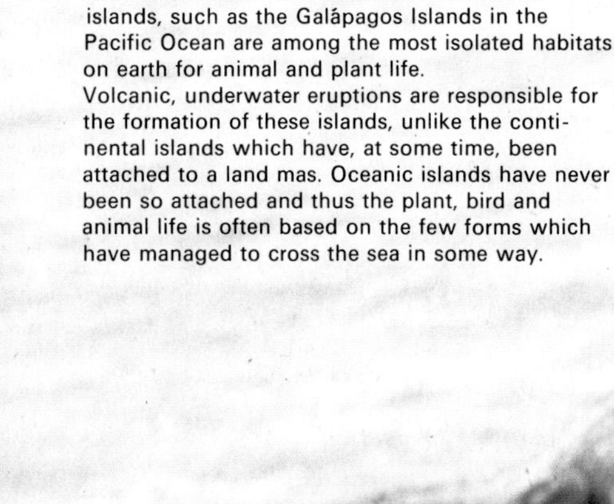

The sea is the greatest barrier to the movement of land animals and for this reason, the oceanic islands, such as the Galâpagos Islands in the Pacific Ocean are among the most isolated habitats on earth for animal and plant life.
Volcanic, underwater eruptions are responsible for the formation of these islands, unlike the continental islands which have, at some time, been attached to a land mas. Oceanic islands have never been so attached and thus the plant, bird and animal life is often based on the few forms which have managed to cross the sea in some way. Because the weather surrounding such islands is usually rather inclement, being wetter and cloudier than that on the mainland, plants may become stunted in their growth and not develop their flowers or fruits. This in turn affects the insects who depend on the food plants and some groups may therefore be missing. Birds too, may find their particular food difficult or impossible to obtain. But those plants and animals which do survive can be outstandingly successful. The absence of mammals grazing on the herbage can mean the development of certain grasses. Birds and small mammals flourish in the absence of predators. A successful species, cut off from breeding with its parent stock and in the absence of competition, radiates and new species can evolve. Indeed, one such species can produce several different species, quite endemic to the island.

▽ Galâpagos Marine Iguanas.

Galapagos. It lives on land, and like its marine relatives is almost entirely herbivorous. Like most members of their family, both these Iguanas lay eggs.

△ The Slow-worm is a limbless lizard, common to many parts of Europe. It is sometimes mistaken for a snake but can be identified by its moveable eyelids and partly fused lower jaw.
▷ Sand Goanna: this lizard is common throughout Australia, growing to lengths of over 5 ft. It is carnivorous, feeding on other lizards and small mammals, such as rodents.
▽▷The Water Skink lives besides streams and takes refuge in the water when frightened or disturbed.

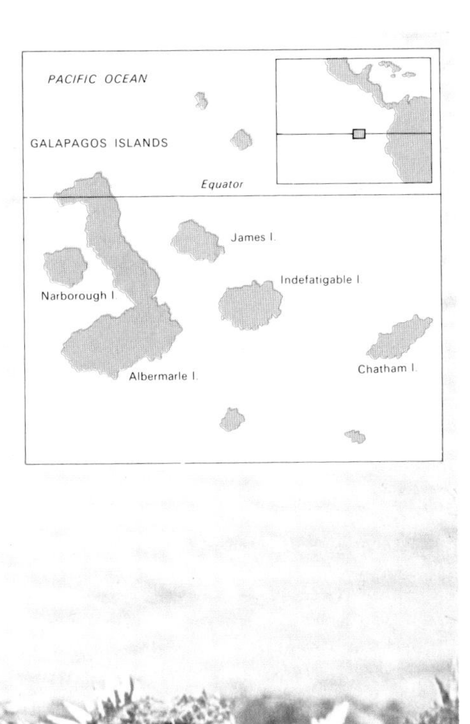

The Gila Monster

(Heloderma suspectum)

The **Gila Monster** from the arid regions of the southern United States and its Mexican relative, *Heloderma horridum* are the only poisonous Lizards and belong to the family *Helodermatidae*. Their poison-glands and fangs are in the lower jaw, not in the upper as in venomous Snakes; their bite, maintained with great tenacity, has occasionally been fatal to man. The Gila Monster is a stoutly built Lizard reaching nearly 2 ft (60 cm), with a big blunt head and a thick tail which contains reserves of fat. Its pink and black colouring is often conspicuous in captivity but may help to conceal it in its natural, semi-desert habitat. The **Mexican Heloderm** is somewhat larger and has a slightly different colouring. These Lizards probably feed on small mammals, young birds and birds' eggs.

The non-poisonous **Earless Monitor** *(Lanthanotus)*, found on the island of Borneo is a relative of the Heloderms and Monitors, though it is placed in a family of its own. It is thought by some to be related to the ancestors of Snakes, but this view is controversial.

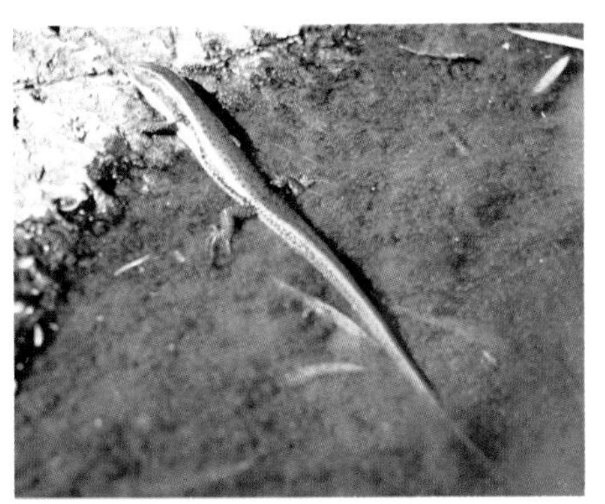

The Monitor Lizards

(Varanidae)

The **Monitor Lizards** *(Varanidae)* include some of the largest living Lizards, and are absent from the New World. Several species such as the **Nile Monitor** *(Varanus niloticus)*, the **Water Monitor** *(Varanus salvator)* of south-east Asia and the **Perentie** *(Varanus giganteus)* of Australia (where the majority of species are found) grow to almost 6 ft (2 m) in length. The **Komodo Dragon** *(Varanus komodoensis)* reaches about 10 ft (3 m) and is restricted to a few small islands in Indonesia. These Lizards are swift, alert creatures which feed on any animals they can overcome, as well as carrion. The Komodo Dragon, named after the island of Komodo, is a formidable predator and can kill large mammals such as pigs and even buffaloes after biting through the tendons in their legs and bringing them to the ground.

All Monitor Lizards have well developed limbs and sinuous bodies, devoid of crests and generally of spines. Their tongues are long and protrusible like those of Snakes. Some live in dry country, others combine arboreal with amphibious habits. All lay eggs, sometimes in the nests of termites which act as natural incubators. Some remarkable relatives of the Monitors called **Mosasaurs** flourished in the late Cretaceous period; these were predatory **Sea Lizards** with paddle-like limbs and which grew 29 ft (9 m) long.

Worm Lizards: Subterranean Squamata

(Amphisbaenia)

The *Amphisbaenia* or **Worm-Lizards** *(Amphisbaena, Trogonophis)* are among the most remarkable of all reptiles. They are highly specialised for subterranean life and like certain other burrowing **Squamata** have tiny, degenerate eyes and have lost their external ear openings. Their limbless worm-like bodies are covered with rings of scales and they are able to progress backwards as well as forwards. One form is exceptional in having a pair of well developed front legs, though the hind pair is absent. Some Amphisbaenians have short, blunt tails which resemble their almost eyeless heads so that in medieval bestiaries they are figured as having a head at both ends. Their skulls, used for ramming the earth, are compact and strongly ossified. These curious creatures, mostly quite small, are found mainly in Africa and the warmer parts of America. Some modern zoologists believe that they should be classified in a separate group from the Lizards.

Snakes–Sophisticated Killers

Snakes are well-equipped to capture their prey. Some species can kill, paralyse or momentarily blind a victim with their venom: others kill by constricting.

The Snakes belong to the suborder *Serpentes* or *Ophidia* of the order Squamata. They seem to have been the last group of reptiles to appear and are not known as fossils until the Cretaceous. The group expanded enormously during the Tertiary era and today, with about 10 families and 2,700 species, has become as successful and diverse as the Lizards. Snakes have clearly descended from Lizards of some kind although almost certainly not from any of the existing limbless, saurian types. Detailed study of comparative anatomy, however, especially of the eye, suggests that the ancestors of Snakes were adapted for burrowing life.

The Snakes have carried to extreme many of the evolutionary trends which are evident among Lizards. Their eyes are covered by a transparent spectacle instead of eyelids, they have no outer ears or eardrums and their forked, protrusible tongues, acting in conjunction with the Organs of Jacobson, are very important as sense organs. Their bodies have become very elongated and the vertebral column usually consisting of 200 to 400 vertebrae, is strengthened by the presence of additional joints. All traces of the shoulder-girdle have disappeared although some primitive Snakes retain a vestigal pelvis and hind limb. Some of the internal organs have become modified so that space can be saved in the long slender body. Thus, the left lung is reduced, (as in some snake-like Lizards), and the kidneys set at different levels in the body. Snakes usually moult several times a year, the horny layer of the old skin often being shed almost entire, instead of in large flakes as in many Lizards.

Typical snakes show many complex adaptations which enable them to swallow large prey. The brain-case is hard and compact but the mobility of both upper and lower jaws is enormous. The jaw bones, armed with sharp, recurved teeth, can move independently on the two sides of the head; the jaw muscles are correspondingly elaborate and the salivary glands, necessary for lubricating the prey, are very well developed. Moreover, many snakes have evolved sophisticated methods of killing their victims, either by constriction or by poison, secreted by modified salivary glands.

Giant Constricting Snakes

(Boidae)

The **Boas** and their allies the **Pythons** are non-poisonous snakes which kill by constriction and are placed together in the family *Boidae*—one of the first snake families to appear in the fossil record. Although their mobile jaws are well adapted for eating large prey, they show certain primitive characters, such as vestiges of the pelvis and hind limbs. In many species, these limbs project as spurs on either side of the vent; they are often larger in the male and are used for stimulating the female during coition.

The majority of Boas, including the **Boa Constrictor** *(Constrictor constrictor)*, are found in the warmer parts of the world. This snake occasionally grows to 19 ft (6 m). It spends much of its time up in trees but some of its relatives, such as the **Emerald Boa** *(Boa canina)*, are much more highly specialised for arboreal life. Certain others, such as the **Sand Boas** *(Eryx)*, of Asia and North Africa are quite small snakes which burrow in sand.

The Anaconda

(Eunectes murinus)

The **Anaconda** is a huge Boa found in the forest regions of tropical South America and in Trinidad. Though specimens in zoos and museums are usually under 20 ft (6.5 m), much bigger ones have been reported and there are apparently reliable records of an Anaconda 37 ft (12 m) long, from Colombia. This would make this thick-bodied species easily the world's largest snake. The Anaconda is dark olive green in colour with black oval spots. It spends most of its time lying in water where it is able to ambush its prey; mammals such as large rodents, fish and even Caimans. Like the big Pythons, the Anaconda very occasionally attacks Man. The Anaconda is viviparous, producing between 28 and 42 young.

Pythons

(Python reticulatus)

The **Pythons** include the world's largest Snakes, apart from the Anaconda. The biggest species are the **Reticulated Python** *(Python reticulatus)*, from south-east Asia, which grows to about 33 ft (10 m), the **African Rock Python** *(Python sebae)*, growing to about 30 ft (9 m), the **Indian Python** *(Python molurus)* which reaches 20 ft (6.5 m) and the **Amethystine Python** *(Liasis amethystinus)* of Australia and New Guinea which reaches a length of 22 ft (7 m). These dimensions are only reached exceptionally however and there are a number of smaller species, almost all of which are found in the Old World and, in greater variety, in Australia.

A large Python can kill and swallow an animal the size of a medium-sized deer but generally takes a smaller prey. Many pythons have a row of pits in the scales along the upper lip; these are heat-sensitive organs and may help the Snake to locate warm-blooded prey. Pythons lay large clutches of eggs, the females of some species coiling round them and brooding them.

The Indian Python may actually incubate her eggs for her body temperature rises while she is brooding.

Primitive burrowing Snakes

These are small, harmless, unfamiliar snakes which possess certain primitive lizard-like characters. For example, their jaws are less movable than those of typical snakes and they feed on small, often invertebrate prey. Many of them retain vestiges of the pelvis and hind limbs and they mostly lack the broad ventral scales which, in typical snakes, extend across the full width of the belly and perhaps assist in locomotion. These snakes are all more or less adapted to burrowing and the condition of their eyes ranges from small to highly degenerate. It is possible that they represent survivors from the earliest phases of ophidian history.

Five families of these snakes are usually recognised: **Thread Snakes** *(Leptotyphlopidae)*, and **Blind or Worm-snakes** *(Typhlopidae)*, from the warmer parts of both the New and Old Worlds; the **Pipe-snakes** *(Anilidae)*, from South America and south-east Asia and the *Uropeltidae* and *Xenopeltidae* from southern and south-eastern Asia. The Blind and Thread Snakes are not closely related to the rest.

Racers and Whip Snakes

(Colubridae)

The family Colubridae is an enormous group containing about two-thirds of all the known types of living Snakes and divided into a number of sub-families. Many of these Snakes are quite harmless but others secrete a mildly toxic venom and have grooved fangs at the back of the jaw. The group includes many familiar fangless species such as the **Grass Snake** *(Natrix natrix)*, and **Smooth Snake** *(Coronella austriaca)*, found in England; the **Aesculapian Snake** *(Elaphe longissima)*, of southern Europe; the **Rat Snakes** *(Ptyas)*, of Asia, the **Whip Snakes** *(Coluber)* of America and Europe and the **Racers** *(Coluber, Masticophis)*, of America. The latter are long, slender snakes which appear to move with tremendous speed among rocks or vegetation—yet the fastest speed actually recorded over an appreciable distance is under 4 mph.

◁ Female Carpet Python brooding her eggs. It is uncertain whether the body temperature of this species rises during brooding, as is the case with the Indian Python. The Carpet Python is a close relative of the Diamond Python.

▽ Grass snakes are common in Britain and many parts of Europe and are distinguishable from Adders by a yellow collar. The snake is feigning death as protection from a supposed enemy.

▽▽ A Grass Snake swimming.

Egg-eating Snakes

Many snakes like birds' eggs and a few colubrid species of the African genus *Dasypeltis* and the Indian *Elachistodon* are highly adapted for dealing with them and appear to eat little else. The harmless *Dasypeltis*, 3 ft (1 m), long and only an inch thick, can eat a hen's egg, $1\frac{1}{2}$ in (40 mm), in diameter. The skin of the snake's throat is highly distensible and its anterior vertebrae are equipped with hard, bony projections. Some of these are long and pointed and penetrate the gullet lining; they pierce the egg-shell and weaken it so that it can be crushed against blunter projections from the other vertebrae. The egg contents pass back into the stomach and the fragments of shell are regurgitated from the mouth. The **Indian Egg-eater** shows similar specialisations and these also occur in rudimentary form in a few oriental **Rat Snakes.**

The King Snakes

(Lampropeltis)

The **King Snakes** of America are moderate-sized terrestial colubrids, non-poisonous but quite efficient as constrictors. They feed on Frogs, small mammals and other reptiles, particularly snakes. Although they are not the enemies of **Rattlesnakes,** as folklore suggests, they do eat Rattlers on occasion and have some immunity to their venom. In fact, Rattlesnakes have evolved a special means of defence which they use solely against King Snakes; instead of biting, the Rattler tries to slap at the King Snake with a raised loop of its body.

King Snakes show interesting colour variations. The common species, *(Lampropeltis getulus)*, is speckled, striped or banded in yellow in different parts of its range. Some sub-species of *Lampropeltis doliata*, (which are also called **Milk Snakes** because they are erroneously believed to suck milk from cows), are beautifully ringed with red, black and yellow. These, and certain other harmless species are sometimes called **False Coral Snakes** because of their resemblance to the poisonous **Coral Snakes** of the family *Elapidae*.

◁ Egg-eating snakes have very few teeth but have downward projections from the neck vertebrae which penetrate the eggshell. The snake can dislocate its jaw to swallow the egg which is larger than its mouth.
△ Brown Tree Snake of north Australia. This back-fanged snake is not supposed to be dangerous.
▷ The Tiger Snake is large and grows to 6 ft. It is the most poisonous snake in Australia, belonging to the same family as the Cobras and Mambas.

Tree Snakes

Many snakes can climb and some members of every family, except for the primitive burrowers and the **Sea Snakes,** are highly adapted for arboreal life. They are often green in colour to match their surroundings and some forms, such as the **Emerald Tree Boa** *(Boa canina)*, and arboreal **Pit Vipers,** (*Trimeresurus* and *Bothrops* species), have strongly prehensile tails. The front teeth tend to be extremely long in Tree Boas, perhaps an adaptation for catching birds. Colubrid and elapid Tree Snakes are often long and thin; some forms, such as the **Vine Snakes** *(Oxybelis)*, of southern America and the oriental **Whip Snakes** *(Ahaetulla)*, may be as thin as a pipe stem.

These Snakes also have long, pointed snouts, sometimes with grooves in front of the eyes to allow good binocular vision. They look like vines and move with great agility, spanning several branches with their attentuated bodies. The so-called **Flying Snake** *(Chrysopelea ornata)*, is a back-fanged colubrid found in south-east Asia, marked with black and green-yellow.

It is exceptionally agile and can spring short distances by rapidly uncoiling. It can also parachute obliquely downwards from a branch, drawing in its belly to form a concavity and thus increasing its resistance to the air.

Locomotion

Snakes move in several different ways. The commonest method is known as horizontal undulatory, or 'serpentine' locomotion. The body of the snake is thrown into lateral waves which begin at the head end; its sides are thrust against irregularities on the surface of the ground, such as stones and plant stems, and propel the creature forwards. This type of movement serves also for swimming. It is obvious that when a snake is moving across very smooth surfaces without any irregularities, its sides can get no purchase at all and the reptile is thus unable to progress.

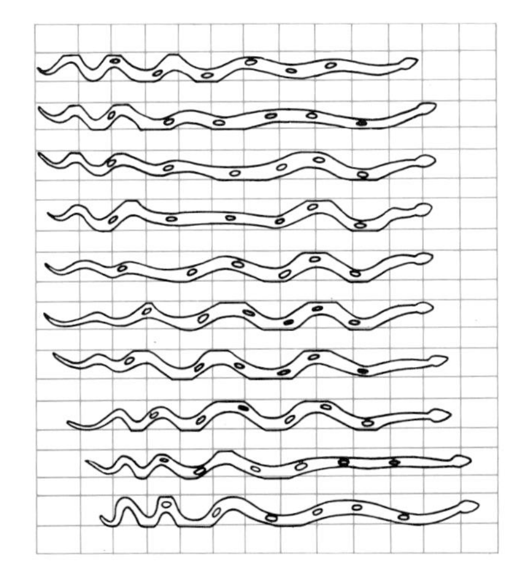

Concertina: This movement is used by a snake when it is moving through a crevice or a tube. The movement involves bunching up and then straightening out different parts of the body, the snake pushing against the walls which surround it to provide the necessary purchase or friction.

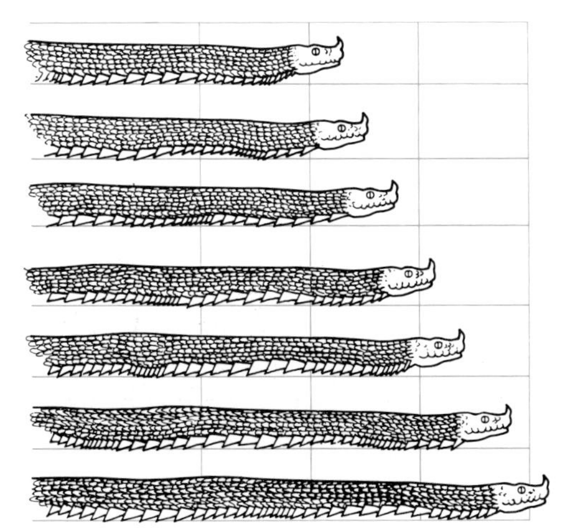

Rectilinear locomotion: Some thick snakes, such as Boas, Pythons, the big Vipers and Rattlesnakes can move forward in almost a straight line by using the broad scales on the underside of the body. The contraction of the scale muscles is from the front to the back.

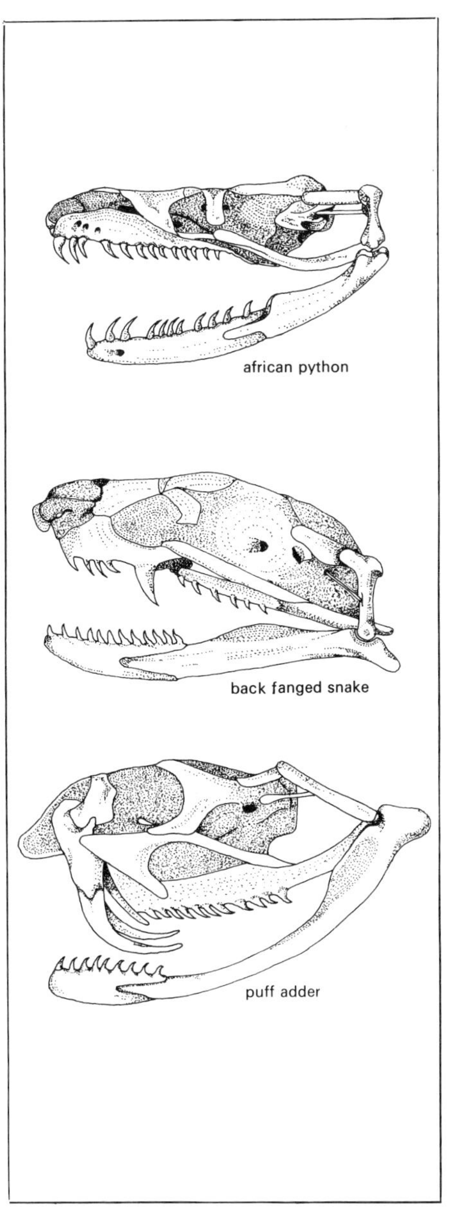

△ The drawings above show skulls of snakes illustrating the dental characteristics of the constrictor, back-fanged and viperine species. Teeth point backwards to move prey in the direction of the throat.

Back-fanged Snakes

In many colubrid snakes, the saliva is not poisonous and the teeth are all of simple thorn-like shape. In a considerable number however, a part of the salivary gland on each side of the upper jaw, (Duvernoy's gland), has become specialised for secreting venom, and a few of the teeth at the back of each maxillary bone have developed grooves for conducting the venom into the wound.

The position of these fangs renders them somewhat inefficient and the snake has to hang on and chew in order to bring them into action. Moreover, the venom itself is usually rather feeble in strength and these back-fanged snakes, which feed mainly on lizards and small animals, are seldom dangerous to Man. A few species however, such as the **Boomslang** *(Dispholidus)*, a big African Tree Snake, which inflates the front part of its body in threat, have exceptionally effective venom.

Cobras and Mambas

(Elapidae)

The family *Elapidae* contains some of the world's most deadly Snakes such as the **Cobras** (*Naja*, etc.), **Kraits** *(Bungarus)*, **Mambas** *(Dendroaspis)*, and **Coral Snakes** *(Micrurus)*. Nearly all the poisonous snakes of Australia, for example the **Taipan** *(Oxyuranus)*, and the **Tiger Snake** *(Notechis)*, also belong to this group.

The fangs are situated at the front of the maxilla and are canalised like a hypodermic syringe for injection of the venom. The venoms of many elapids predominantly affect the heart and

Sidewinding: This movement is used by desert snakes, such as Asps and some Rattlesnakes, to move across sand. Muscles contract and loops of the body are lifted, the snake travelling a path at an angle to the mark its body makes.

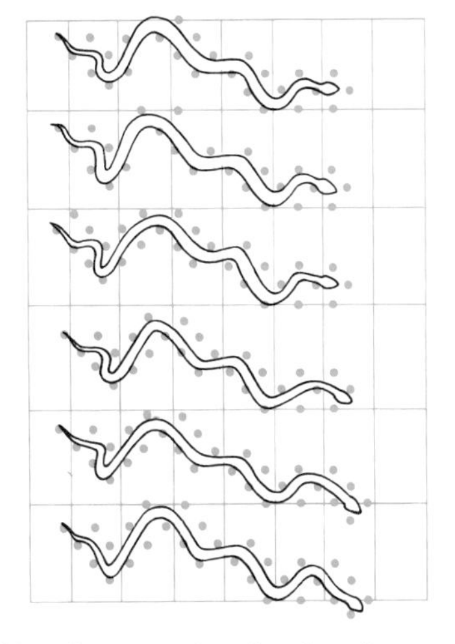

Serpentine: Muscular contractions flow from the head of the snake towards its tail, its sides pushing against any projections which there may be on the surface of the ground. This pushes the snake forward and a similar movement is made when the reptile is swimming.

Another type of locomotion known as 'concertina movement', is used when a snake is creeping down a narrow tunnel, or sometimes when climbing. This involves alternate bunching up and straightening out of parts of the body. Desert snakes, such as Asps, use a method called sidewinding, loops of the body being raised and thrown sideways; this is very effective over sand. Finally, some snakes are able to creep along slowly with their bodies almost straight by muscular movements of their ventral scales. This is known as 'rectilinear locomotion' and is often practised by heavy-bodied snakes such as Boas, Pythons, big Vipers and Rattlesnakes.

nervous system and may cause death by producing cardiac or respiratory failure.

The Cobras, characterised by possessing a hood which can be erected by movement of the elongated neck ribs, occur widely in Africa and Asia. The common 'Indian' species *Naja naja*, ranges from India eastward to southern China; some specimens have distinctive spectacle-like markings on the back of the hood. The **King Cobra** *(Ophiophagus hannah)*, of south-east Asia is the longest poisonous Snake, exceptionally reaching 18 ft (5.5 m). Some Cobras such as the African **Ringhals** *(Hemachatus)*, are able to spit venom at the eyes of an aggressor, causing pain and temporary blindness.

Coral Snakes

The name Coral Snake is given to several genera of elapid snakes found in various parts of the world. Notably, one of these genera is *Micrurus* and its allies in America, which are marked with alternate rings of black, yellow and red. Opinion differs as to how far this coloration has a warning function and on the significance of the similar coloration found in certain harmless snakes which have been regarded as Coral Snake mimics.

Coral Snakes are mostly fairly small, secretive and often nocturnal. They do not bite readily, although some species have the habit, (paralleled in other snakes), of hiding their heads and waving their tails in the air in a threatening fashion.

Sea Snakes

The **Sea Snakes** *(Hydrophiidae)*, are close relatives of the *Elapidae* but have become adapted for marine life. Their eyes and their valvular nostrils are placed towards the top of the head and their tails are deep and compressed laterally for swimming. Most of them lack the broad ventral scales of typical land snakes and can hardly progress out of the water. The majority are viviparous and hence do not need to come ashore to lay eggs. They are all venomous with fangs at the front of the maxilla, as in Cobras and their allies. They feed on fish and are not, as a rule, aggressive.

Nevertheless, their venom is exceptionally powerful, and fishermen and bathers have been killed by their bites. Some species appear to be gregarious. The majority of Sea Snake species are found in the far eastern and north Australian waters and the animals generally remain fairly near to the coast.

The **Black and Yellow Sea Snake** *(Pelamis platurus)*, however, is truly oceanic and has reached the western coast of America and the eastern coast of Africa, as far south as the Cape.

Courtship and rivalry in Snakes

In temperate countries, snakes generally mate in the spring, though individual females may only breed every other year. The male locates the female mainly by smell and Jacobson's Organ sense, rather than by vision, as in many lizards. It is perhaps significant that whereas in lizards the male is often larger and more conspicuous than the female, in snakes the two sexes generally look much the same. Courtship procedure varies in different species of snakes, but in many, the male crawls alongside the female, often rubbing his head along her back while convulsive waves pass along his body from the tail.

Although snakes do not obviously defend territories, the males of some species have peculiar forms of ritual combat. Male Rattlesnakes and Vipers, for instance, rear up, sometimes with the front parts of their bodies entwined, and thrust against each other until one is exhausted and retires. As a rule no serious injuries are inflicted. Such exhibitions have often been erroneously regarded as courtship dances involving members of the opposite sexes.

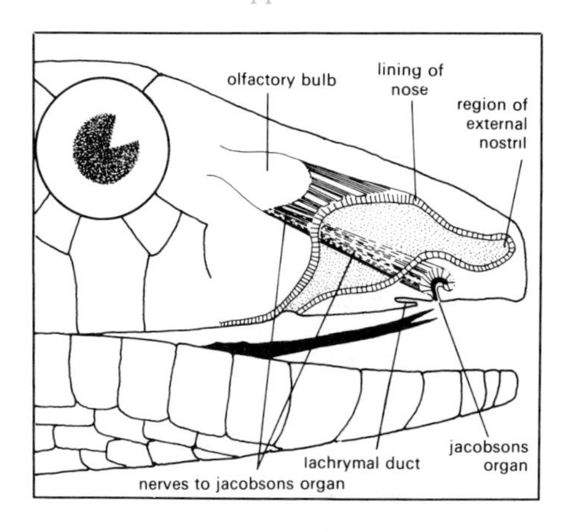

△ Jacobson's Organ is an important sense organ and is used by the snake to smell particles picked up by the tongue. It can be seen lying above the front of the palate and opens into the mouth through a narrow duct.

Hog-nosed Snakes

(Heterodon)

The **Hog-nosed Snakes** are a group of small, harmless North American colubrids with burrowing proclivities.

They have a pig-like appearance, owing to the presence of a prominent, upturned scale on the tip of the snout, thick bodies and triangular, rather viper-like heads. They feed mainly on toads. These snakes are renowned for their defensive behaviour. When molested, they flatten their heads and inflate the front part of their bodies; at the same time, they hiss loudly and strike, usually without actually biting. If this intimidating behavior fails to deter an aggressor, the snake shams dead, rolling over limply onto its back—a position to which it will return even if righted. When left alone, the Snake soon turns over and beats its retreat.

Rattlesnakes

The **Rattlesnakes** are distinguished from other Pit Vipers by their caudal appendage and mostly belong to the genus *Crotalus*. They are viviparous. The rattle is a remarkable structure consisting of a number of horny cups or segments which partly interlock in such a way that they are jointed movably together. A new segment is formed at each moult, several times a year, but when an appreciable number of segments have been accumulated, the end usually breaks off. Consequently, Rattles with more than about 10 segments are rare, except perhaps in captive snakes, living peaceful lives. When vibrated, the rattle makes a dry, sinister sound and serves as a warning to potential enemies and is not a call to other members of the species.

The majority of Rattlesnakes are found in the United States of America, although the group extends northwards into Canada and southwards into South America. The **Tropical Rattler** *(Cro-*

◁ The European Adder or Viper is the only poisonous snake found in the British Isles. It is distinctive for the dark zig-zag along its length and the V or X mark behind the head. Vipers generally are thick-bodied and rather sluggish snakes and live mainly on the ground.

▽ Russell's Viper which is found in both India and Ceylon is said to kill more people in each year than any other snake. It sometimes hides inside houses and can grow to lengths of 4 ft.

talus durissus), is a very dangerous species and differs from most of its relatives in producing a strongly neurotoxic venom. Common Rattlesnakes of America are the **Timber Rattlesnake** *(Crotalus horridus)*, the eastern **Diamond-back** *(Crotalus adamanteus)*, and the western **Diamond-back** *(Crotalus atrox)*. The last two are heavy, big snakes, reaching a length of just over a metre.

Pit Vipers

The **Pit Vipers** *(Crotalinae)*, differ from the typical Vipers in possessing a remarkable sense organ, a pit containing a membrane between eye and nostril on each side of the head. This is able to register minute changes in the surrounding temperature. The pits found along the upper lip scales of Pythons are sense organs of similar type although they are less specialised. Many of these snakes are partly nocturnal and it is likely that the pit organs help them to locate warm-blooded prey in the dark.

The Pit Vipers have a wide distribution and are particularly numerous in the New World. Apart from the Rattlesnakes they include such species as the **Copperhead** *(Agkistrodon contortrix)*, and **Water Moccasin** *(Agkistrodon piscivorus)*, as well as the deadly **Fer-de-Lance** *(Bothrops atrox)*, and the **Bushmaster** *(Lachesis muta)* of southern America and certain islands in the West Indies.

The Old World Pit Vipers (usually placed in the genus *Trimeresurus*), range from extreme eastern Europe across Asia to Siberia, China and Japan, and southwards to Malaya. They are mostly of moderate size and a number of species are arboreal.

The Vipers

The family *Viperidae* is divided into two groups, the typical **Vipers** of the Old World *(Viperinae)*, and the **Pit Vipers** including the **Rattlesnakes** *(Crotalinae)*. All of these snakes are poisonous, producing a venom which in general acts on the blood system and also causes a great damage to the tissues in the vicinity of the bite. The Viper venom apparatus has been more highly perfected than in any other Snakes the fangs being long and completely canalised. Moreover, the maxillary bones to which they, (and the reserve fangs which replace them), are attached are able to pivot on the skull so that the fangs can be folded back in sheaths along the roof of the mouth when not in use.

The typical Vipers include the **Adder** *(Vipera berus)*, which is common in Britain, and the related European species; the **Asps** *(Cerastes)*, of north Africa which may have horns on the nose and such formidable snakes as the Asiatic **Russell's Viper** *(Vipera russelli)*, and the African **Puff Adder** *(Bitis arietans)*, and the **Gaboon Viper** *(Bitis gabonica)*. The last three species, which can grow to about 4 ft (1.5m) in length, are highly dangerous to Man. With some exceptions, Vipers tend to be thick-bodied, rather sluggish and to live mainly on the ground.

▽ Largest of the Rattlesnakes, this Western Diamond back shows its warning device. The rattle is made up of horny cups which partly interlock and a new segment is formed at each moult. When a number of segments have formed, the end breaks off.

Birds comprise the Class Aves, one of the eight major groups of living vertebrate animals, and one of only three such Classes which are entirely air-breathing, the others being the reptiles and mammals. Both the birds and mammals are believed to have evolved from so-called 'cold-blooded' reptilian ancestors and both have evolved warm-bloodedness. This has been one of their most significant developments, enabling them to become relatively independent of their environment, to remain active even at low temperatures and in inhospitable climates, and to survive in a wider variety of habitats than 'cold-blooded' creatures. The advantage of conserving heat was probably instrumental in giving rise to the plumage of birds. The uniqueness and great adaptability of this plumage was the key to the development of the power of flight, for being very strong and light and reasonably rigid, the feather was the ideal structure for the aerofoils necessary for flight.

Feathers thus serve as insulation; the extent of which can be varied according to the ambient temperature by flattening or erecting the feathers over the skin. The ability to fly has enabled birds to exploit the earth's surface to a very high degree, though they are still restricted during the breeding season to the proximity of land areas where their eggs can be laid. They breed from the Antarctic continent to well

BIRDS

inside the Arctic Circle, on oceanic islands, from lowlands to mountains and in virtually every habitat from desert to tropical rain-forest.

Although the exploitation of a wide range of habitats and feeding niches has evolved along with a great number of structural modifications, those of the bills and feet being particularly obvious, the power of flight is in the majority of species the dominant factor affecting their unique anatomy. The trunk consists of a box-like cage of bone, much more enclosed and rigid than that of other vertebrates, while the breastbone is large, broad and flat with a deep keel for the attachment of the relatively enormous pectoral muscles necessary for moving the wings against the resistance of the air. Many of the bones are hollow with internal cross-struts to give maximum strength with minimum weight. The lungs have supplementary membranous air-sacs which serve to keep the air moving through the lungs, avoiding the residual 'stale' air remaining in the lungs after expiration. They are therefore extremely well adapted to make maximum use of the inspired air, an essential to a creature with the inevitable high metabolic rate associated with the strenuous muscular activity of flight.

The unforgettable sight of Flamingos in flight over Lake Nakuru in Kenya. Many thousands of birds will live in a single flock.

The Evolution of Birds

Birds evolved from reptilian ancestors at least 150 million years ago, in the Mesozoic era. The reptile ancestors were probably small Dinosaurs that ran on their hind legs, looking like scaly Lizards and called Thecodonts. Some of these may have climbed into trees and, with the aid of their strong legs and long balancing tails, they leapt amongst the branches. Some of their arm scales may have developed and become feather-like. The earliest known fossil bird is the *Archaeopteryx*.

Ancestors of Waders

The next fossil birds date from 70–100 million years ago and were all waterbirds from the Cretaceous period. *Hesperornis* was a large flightless bird that still possessed teeth and probably looked rather like a present-day Loon. *Icthyornis* was a flying bird which looked rather like a Gull and probably fed on fish. These birds were probably the ancestors of the present day Waders, Gulls and Auks. Other fossil birds known in the Cretaceous were probably the ancestors of the Grebes, Pelicans, Herons and Flamingoes.

70 million years ago, some of the large oceans were formed and the Cretaceous seas became land. The giant reptiles died out and the birds became more common in this, the Eocene period. Fossil Cormorants, Rails, Loons, Penguins, Pheasants and some perching birds are known from this time. Several large flightless birds evolved and then became extinct. *Diatryma* is known from fossils in Europe and North America. *Phororhacos* was about 5 ft (1.5 m) tall and had long legs, a large head and a huge hooked bill.

The Golden Age of Birds

The greatest number of bird fossils have been found in the Pliocene and Pleistocene deposits which are both less than 10 million years old. All of the birds now living probably existed in those periods, in much their present forms. Fourteen Pliocene species still live, the oldest of which is probably the Sandhill Crane. The Golden Age of birds was probably about half a million years ago when there may have been about 12,000 species in existence. Bird species have slowly declined since that time and today there are about 8,600 in existence. Thus, the reptilian Thecodont ancestors have given rise to a very diverse number of warm-blooded birds.

Another group of animals that descended were the Crocodiles, large Dinosaurs and the Pterodactyls. The Pterodactyls or Flying Reptiles, were a very successful group of animals for a while but they became extinct about 65 million years ago. They had a wing span of up to 20 ft (6 m) and were probably the largest animals to have ever flown.

Archaeopteryx

Archaeopteryx is the earliest known fossil bird and lived about 150 million years ago in the Jurassic period. Three fossils are known and were found in a deposit of limestone in Bavaria. The creature shows a curious mixture of reptile and bird features. Like many reptiles, it had a long tail with twenty vertebrae, a backbone with simple vertebrae, three clawed fingers, free hand bones, teeth present in both jaws and small eyes and brain. The bird features present include feathers like those of modern birds. It had primary feathers attached to the hand and wrist and secondary feathers attached to the forearm. The two collar bones were joined to form a wish-bone. It had a very small keel or breast bone, and was, at best, a weak flier. *Archaeopteryx* probably lived in trees and glided from one branch to another. One toe is placed opposite the other three and the bird could probably perch in the branches. The fact that it possessed feathers suggest that it was warm blooded. *Archaeopteryx* was about the size of a Magpie and probably fed on flying insects.

Adaptive radiation

After the primitive birds evolved from their reptile stock and had become established, they were ready to spread out over the world and adapt themselves to different modes of life. Most species of birds have retained the power of flight but others have lost it; Ostriches and Rheas for instance have adapted themselves for a ground living existence and have become fast runners. Penguins have taken to a life of swimming. The flying birds colonised forests, grasslands, tundras, deserts and oceans. In order to live in these areas they adapted to different diets, and developed special ways of dealing with different food groups. This usually concerns the beak being variously modified. This is well shown by the Galápagos Finches. A general Finch probably colonised the Galápagos islands and, in the absence of other bird competitors, evolved into several groups, which in time became separate species. The seed-eaters developed the largest, stoutest bills and the

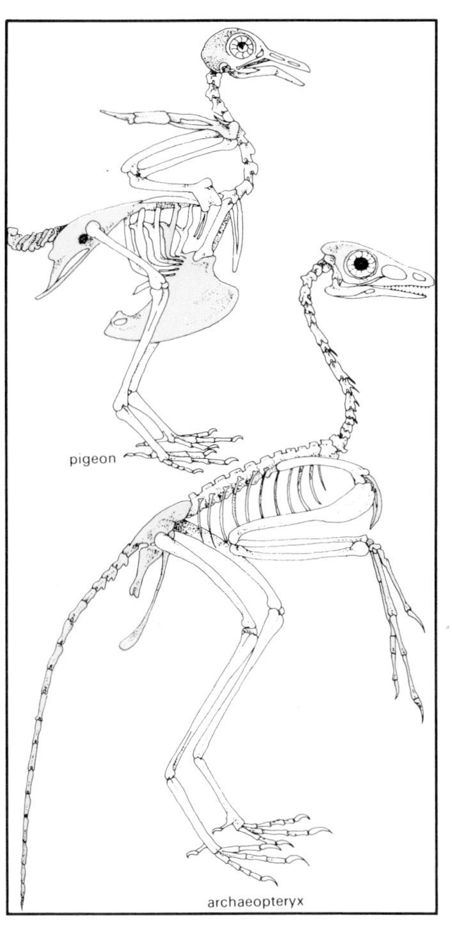

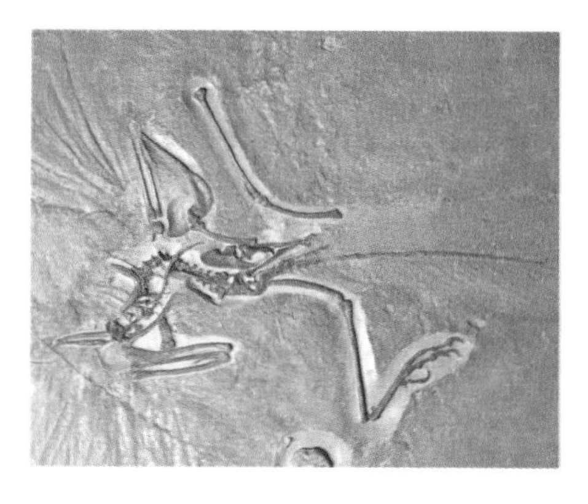

△ Fossil of *Archaeopteryx* in limestone, dating from the upper Jurassic period. Only a few fossil specimens are known, all from Bavaria.

◁ The pigeon has developed the deep breast bone for more efficient flight and as a firm anchorage for the wing muscles. *The Archaeopteryx* skeleton shows a short skull with teeth and the tail also shows reptilian similarities.

◁ ◁ A Crowned Crane, one of the 15 species of Cranes which include the Sandhill Crane, probably the oldest of the 14 species of birds now living that existed in the Pliocene period.

◁ ◁ ◁ Moorhen hatching: The 'eggtooth' can be seen, on the end of the red bill and this is shed in a few days. Moorhens are one of the birds which have feathered young on hatching.

▽ ◁ Museum reconstruction of *Archaeopteryx*. The bird seems to have been about the size of a Magpie and pigmentation tests suggest that the plumage was brown. The flight feathers were similar to those of modern flying birds. The claws on the wings were used for climbing.

insect-eaters developed the smallest, slimmest bills. Other species evolved with bills suitable for fruit eating or cactus eating. Some species live on the ground, and some in trees and other vegetation of differing heights.

What it takes to fly

In order to fly, birds need a body that is suitably adapted to travelling through the air and organs capable of generating enough energy to get them into, and keep them in, the air. A bird's body is usually long and streamlined; the contour feathers press closely against the body and help in the streamlining. The body is light and many of the bones are hollow, containing large air spaces. There are strong shoulder and hip girdles. The first is important because it acts as a base for the wing attachments and the second because it has to stand the stresses imposed on the legs when the bird lands. The movement of the wings enables the bird to fly. The secondary feathers are responsible for giving the bird lift and keeping it in the air, whilst the primary feathers propel it along.

The power to move the wings comes from the large pair of breast muscles which are firmly anchored to the large keel or breast bone. These large breast muscles can be regarded as the bird's engines and they burn up a lot of energy. In order to provide enough energy for these muscles, birds eat a great deal of food, have a rapid digestion rate and a very quick heart beat and blood circulation to convey the necessary fuel to the breast muscles.

Flightless Birds

As European civilisation spread through the world Man found that flightless birds were an easy source of food; and so did the dogs, cats and rats that he brought with him. Since the 17th century a number of flightless birds, like the Dodo, have become extinct and others are in danger of extinction.

All the members of two major groups of birds are unable to fly, but there are flightless species in other groups. The largest flightless group is the Penguins (Spheniscidae), with 17 members. The other group is the Ratites, or Running Birds, which covers Ostriches (Struthionidae, 1 species), Rheas (Rheidae, 2 species), Cassowaries (Casuariidae, 3 species), Emus (Dromiceiidae, 2 species), and Kiwis (Apterygidae, 3 species).

Despite their inability to fly, Penguins have clearly defined wings that are used as flippers when they swim. These flippers are similar to those of Dolphins, which evolved from forelegs. The Ratites have very small wings which may sometimes be used as sails when they are running. Their well-developed legs enable them to run well and provide a good weapon.

Penguins are not confined to the Antarctic, but they are confined to the Southern Hemisphere. The Emperor Penguin *(Aptenodytes fosteri)* breeds only on the Antarctic continent, while the Galapagos Penguin *(Spheniscus mendiculus)* lives almost on the Equator. Each is highly specialised to live in its particular environment.

The Ratites, generally, share the same niche in their environment as the large grass-eating mammals, relying on their speed to avoid predators.

Other flightless species tend to be confined to islands, where there are no native mammals to prey on them. For instance, New Zealand, apart from its Kiwis and Penguins has three other species that are totally flightless and two Wattlebirds that can barely fly.

When a bird cannot fly, its feet take on special importance. Penguins use their wings to propel themselves underwater, but their feet act as rudders. In the case of the Emperor Penguin, feet are also used to carry the egg. The Ratites, on the other hand, rely on their legs and feet for locomotion. All have strong, well-developed legs and feet, which some use for defence as well as for running.

The feet of the Ostrich have an interesting development. Instead of the four toes on each foot that most modern birds have, the Ostrich has only two—and one of these is considerably smaller than the other. It is possible that the Ostrich could be following the evolutionary course of the horse, which has developed from a five-toed animal to one-hoofed animal. The other Ratites have 3 toes on each foot. The Emu's three toes enable it to retain its balance in rocky terrain.

◁ King Penguins, related to the Emperor but smaller.

Penguins

(Spheniscidae)

All the species of **Penguin** are found only in the southern hemisphere. They are the most primitive group of living birds, over 60 million years old. They are well adapted to existence in the sea. Although their flipper-like wings have no flight feathers, the structure of their bodies would suggest that they have evolved from ancestors that could fly. Their wings move only at the shoulder joint; in other birds, there is a flexible joint at what would be the elbow on a human arm.

The short legs make movement on land awkward and clumsy; set at the back of the body, they force the Penguins to walk upright, but in the water they act as a rudder. In the sea it is important that the Penguins should be able to move fast and efficiently to catch their food (usually fish) and to avoid the large marine mammals which prey on them. On land the main predators are other sea birds, which concentrate on taking young Penguins or eggs, and the adults have little need to move fast. They sometimes slither forwards on their bellies on snow and ice, using their flippers as oars.

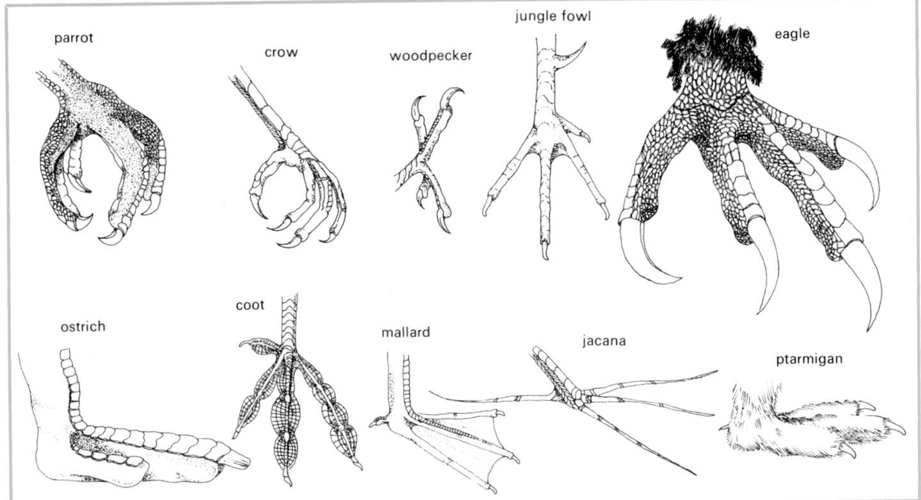

△ The feet of birds show adaptation to their habits: Parrot and Crow for grasping; Woodpecker for gripping trees; Jungle Fowl—males only—have spurs for fighting; Eagle for seizing prey; Ostrich for running; Coot for paddling; Mallard for swimming; Jacana for walking on lily pads and Ptarmigan on snow.

△ △ Penguins are sociable birds and large colonies are characteristic although the birds tend to be quarrelsome about territorial rights.

▽ Penguins are specialised for a marine life, their webbed toes and strong flippers providing considerable swimming and diving power.

Emperor Penguins

The **Emperor Penguin** *(Aptenodytes forsteri)* is the largest of the family standing 3 feet (91 cm). It breeds in large colonies on the mainland of Antarctica. When the sea-ice forms in autumn the eggs are laid. Only one egg is laid per pair. For 64 days the male takes care of the egg, resting it on his feet, covered by a fold of skin on his belly. Meanwhile the females disperse, travelling many miles to reach the open sea and food. At the end of incubation the female returns to look after the chick and allow the male to break his fast. The young are reared by the parents until the beginning of the short summer in October or November, when they gather in groups at the edge of the ice. As the ice breaks up, they use the ice-floes as rafts on which to drift northwards. The young birds can only survive in the water when they have completed their moult and, therefore, if the raft melts before this has taken place, they will drown.

King Penguins *(Aptenodytes patagonica)* look similar to Emperors but are smaller and weigh from 30 to 40 lbs (approximately 15 kg). They are found on the coasts of the Antarctic and sub-Antarctic islands.

Crested Penguins

There are six species with yellow crests. They are members of the same genus and are similar, except for their crests and slight differences in size. Each has a black head and face, except the Royal which has a black cap and white face, with a red bill and yellow crest. Because Penguins spend much of their life in the water, it is necessary for the difference between species to be apparent on that part that protrudes above the water. Hence the differences in the crests. The crested Penguins are the **Erect-crested** *(Eudyptes sclateri)*, **Rock Hopper** *(Eudyptes crestatus)*, **Macaroni** *(Eudyptes chrysolophus)*, **Fiordland Crested** *(Eudyptes pachyrhynchus)*, **Royal** *(Eudyptes schlegeli)* and **Snares Island** *(Eudyptes robustus)*. The genus is circumpolar, various species breeding on various islands and shores, some as far north as Australia and New Zealand.

Jackass and Adelie Penguins

The **Jackass Penguin** *(Spheniscus demersus)* is confined to South African waters. It has well-defined black and white plumage and breeds in large, noisy colonies. The nest is made from twigs, roots and stones in burrows or rock crevices. Its main source of food is small fish, squid and crustaceans. Because it takes pilchards, the Jackass Penguin numbers are controlled, but it is conserved, because of the commercial value of its guano and eggs.

The **Adelie Penguin** *(Pygoscelia adeliae)* is the most widely distributed Antarctic Penguin and has the classic black and white 'evening dress' appearance. To reach their breeding colonies when they return in October they may have to cross up to 60 miles of ice. Nests are built from stones. About four weeks after the young have hatched, they congregate in crèches of up to 100 young birds, while the parents go in search of food. These crèches serve to protect the young from predators and work on the principle of safety in numbers.

Cassowaries

(Casuariidae)

Living in the dense forests of New Guinea, Northern Australia and the neighbouring islands requires special adaptations in a flightless bird. The three species of **Cassowaries** have developed long, drooping, harsh feathers to protect them from the thorny underbrush; large, wiry, vaneless wing-feathers which curve along their bodies to give further antithorn protection, and the large, bony helmet-like structure on the head to push aside branches as they run head-down through the brush.

The **Australian Cassowary** *(Casuarius casuarius)* and the **One-wattled Cassowary** *(Casuarius unappendiculatus)* have decorative wattles hanging at their throats. The third species is **Bennett's Cassowary** *(Casuarius bennetti)*.

Cassowaries are hard to see in the wild; in spite of their size, the Australian species standing some 6 to 7 feet (2 to 2.5 metres) high, and striking colours, they are shy and tend to be nocturnal. In defence of their territories, they are very fierce and it has been recorded that they will kill humans.

Ostrich

(Struthionidae)

The adult male **Ostrich** *(Struthio camelus)*, is the largest living bird, with a height of about 8 ft (2.5 metres), and weight of up to 345 lbs (approximately

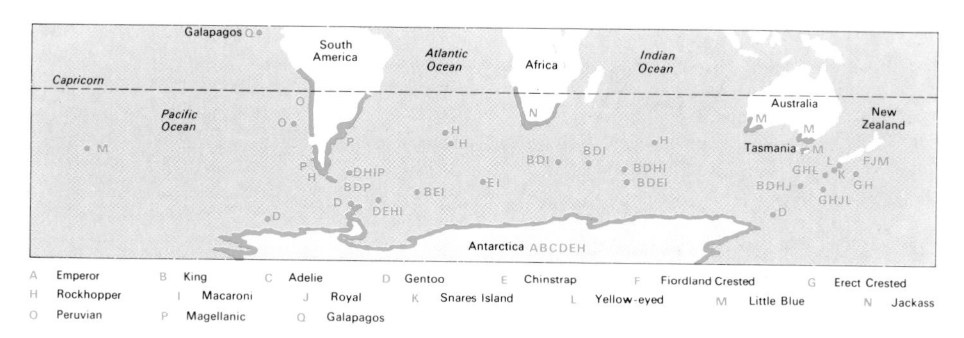

△ Breeding grounds of Penguins. These birds are confined to the southern hemisphere.

◁ The Cassowary is a large flightless bird of Australia and New Guinea, standing between 3 ft and 4 ft tall. The distinctive horny casque on the head is especially well developed in older birds. The head and neck are brightly coloured but the rest of the body is brown or black.

▽ The Common Emu stands over 6 ft tall and is the second largest living bird. Emus are found on the desert plains of Australia and are nomadic, moving about in search of food and water.

▷ Kiwis have degenerate wings and feathers. Eyesight is weak but unlike most birds, Kiwis have a good sense of smell. They are nocturnal, feeding off insects, earthworms, leaves and soft fruits.

170 kg). It is black with white plumes on the wings and tail. The long neck and legs are featherless and pink. The female is smaller and grey-brown.

Ostriches favour dry open country, which they roam in small flocks. Often they are to be seen with herds of grazing mammals, like zebras and gnu. They seem to have a mutually beneficial relationship with these mammals; the Ostriches, being tall, make good sentries and the herds of grazing mammals disturb seeds and insects, on which the Ostrich can feed.

When threatened with danger Ostriches, especially the young, will sometimes stretch their necks flat on the ground. This behaviour probably gave us the myth about them burying their heads in the sand.

Rheas

(Rheidae)

The only Ratites found in the New World are two species of Rhea. Rheas are not as large as the Ostrich, being only 5 feet (1.5 metres) in height and weighing only 44 to 55 lbs (20 to 25 kg). Once, the **Common Rhea** *(Rhea americana)* roamed the pampas of Argentina and Brazil in huge flocks, but the conversion of this habitat to agriculture has reduced numbers. Added to this, the Rhea has been hunted strenuously by the cowboys of the pampas. The other species, **Darwin's Rhea** *(Pterocnemia pennata),* is smaller and more brown with white-tipped wing feathers. It is found in the foothills of the Andes southwards from Peru and Bolivia. Like Ostriches and Emus, the Rheas rely on running speed to avoid pursuers.

Emus

(Dromiceiidae)

Of four species of **Emus** known a century ago only one, the **Common Emu** *(Dromiceius novae-hollandiae),* is not extinct. It lives in the open, arid country of Australia. At almost 6 feet (2 metres) tall, it is the second largest living bird. There is no plumage variation between the sexes, but the male is usually larger than the female. Like the other Ratites of open country, it runs fast—up to 30 miles per hour; it also swims well.

The bulk of the Emu's diet is vegetable, but they will eat almost anything; during grasshopper plagues they gorge on the insects. In wheat-growing areas they can become pests. In 1932 a flock of 22,600 ravaged cornfields in Western Australia to the extent that the army was called in to exterminate them with machine-guns. Despite a month in the field, the soldiers were unable to make any impact, because the birds split into smaller flocks making the use of the guns impractical. Since then, control has been through destruction of eggs.

Kiwis

(Apterygidae)

Kiwis are such peculiar birds that it is hardly surprising that New Zealand should have taken a Kiwi as its national symbol. In doing so it has ensured the birds' conservation. Kiwis are sedentary and live in the humid forests of New Zealand. These forests are dense and Kiwis nocturnal, so that despite their celebrity, they are rarely seen in the wild by any but the most devoted bird-watchers.

Kiwis' wings are so very small that they have the appearance of being wingless. Because there are no indigenous predatory mammals in New Zealand, the power of flight was not important to these ground-living birds. In many ways they have features usually associated with nocturnal mammals; the feathers seem more like Badger hair, the sense of smell is well-developed, they have whiskery feathers on the face and strong, clawed feet and they live in holes in the ground. They feed on insects and worms using their long decurved bills to probe the earth.

The three species are the **Common Kiwi** *(Apteryx australis),* the **Great Spotted Kiwi** *(Apteryx haastii)* and the **Little Spotted Kiwi** *(Apteryx owenii).*

Eggs and Young

Often large birds that live a long time lay few eggs. This is certainly true of the largest Penguin, the Emperor, which lays one egg a year. To maintain the continuity of a species it is necessary that there should be as many young produced to breeding age as are lost by mortality. In other words, at its death, an individual should have produced at least one offspring that has reached healthy breeding maturity. Penguins live a long time; a King Penguin has lived to over 28 years in captivity and Yellowed-eyed Penguins *(Megadyptes antipodes)* reach at least 22 years in the wild. Most Penguins reach breeding maturity in their second or third years and the greatest mortaility occurs before this stage is reached. Those species that make nests do so in comparatively exposed positions and the eggs and young are at considerable risk to predatory sea birds. In the case of young Emperors, many are lost on ice-floes that melt. If they survive to reach the open sea after their moult, they then face the risks of being caught by a marine mammal, such as a killer whale or leopard seal.

Because of their size, one might expect the large Ratites to lay few eggs, but they lay very large clutches. Ostriches will lay between 10 and 25 eggs—larger clutches are the product of more than one female (the Ostrich is polygamous)—Rheas between 20 and 30, and Emus between 7 and 12. In most Ratites the male incubates the eggs, but if the male is polygamous, often the case with Rheas, he has no hope of incubating all the eggs successfully. The Ostrich male incubates the eggs by night and the female takes over by day. An additional hazard is the fact that the eggs are laid on the ground and are consequently prey to Vultures, other birds and mammals. When the young hatch, they are vulnerable still, even though they emerge from the egg able to stand and to run. The large number of eggs not only ensures the continuation of the species but, in terms of the total ecosystem, also provides food for other animals.

Several groups of birds live on or near water. Some live and feed in freshwater while others are exclusively marine. Many species, which habitually dive and swim, have very streamlined bodies, and legs placed far back on the body. Their feet are often webbed. Those birds that wade have developed very long legs, often without webbed feet.

There is great specialisation in beak shapes and structure in those groups which are associated with different diets. Some birds spear fish, others grasp them in toothed mandibles; some birds have developed filtering beaks for straining out small water animals and plants. Many swimmers and waders particularly the Ocean Wanderers breed in dense colonies.

Swimmers and Waders

The Grebes

In appearance, **Grebes** resemble Loons. Many species have long bodies and long, sharp bills. They are highly adapted for their aquatic life and their partially webbed or lobed toes help them to swim underwater where they catch fish, crustaceans, insects and other small water animals. The Grebe family (Podicipedidae), is composed of 20 species which are distributed virtually over the whole world. They are solitary or only slightly gregarious creatures and in the breeding season they frequent freshwater lakes and ponds which have reedy margins. Grebes indulge in 'mutual' courtship—where the actions are either similar or can be performed by either sex. The displays include preening, weed holding, head flagging and some species perform 'dances' on the water. Bulky nests are made in shallow water or floating nests in deep water. Three to nine dull white eggs are laid and are incubated by both parents.

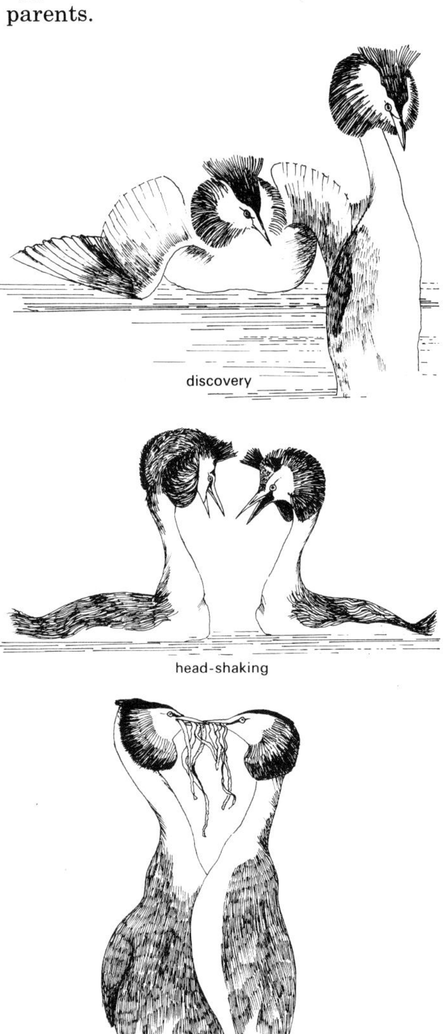

△ Mutual courtship dance of Grebes.
◁ Part of a nesting colony of Herons and Spoonbills in a Cork Oak. Water birds of different species can live in close proximity as here.

Loons and Tinamous

Divers or **Loons,** of which there are four species, are large aquatic birds found in the higher latitudes of the northern hemisphere. They have long bodies, thick necks, small pointed wings, straight, sharply pointed bills and webbed feet. The feet are set at the back of the body and are well positioned for swimming. They can dive to great depths and feed mainly on fish. However, on land the birds are clumsy and cannot hold their bodies high enough off the ground to walk properly.

These migratory birds generally remain at sea in the winter, sometimes in small loosely-formed flocks. They pair off before the breeding season and build nests on islands or on the shores of lakes and ponds. Two dark, mottled eggs are usually laid and are incubated by both parents.

The 45 species of Tinamous are found in Central and Southern America. They are primitive ground birds which live in forests and grasslands and feed mainly on fruits and seeds together with some insects. The females take the lead in courtship and the males incubate the eggs and rear the young.

Wandering Albatrosses

(Diomedeidae)

Albatrosses are ocean-living birds which are found mainly in the southern hemisphere, as far as the Antarctic although some species are found in the North Pacific ocean. They are large birds and some species are about 53 inches (135 cm) long. A stout body supports a pair of very long, narrow wings which are adapted for gliding flight. The **Wandering Albatross** *(Diomedea exulans)* which is the largest of all living flying birds, has a wingspan of over 11 feet (3.5 metres).

Most species of Albatross are migratory and wander considerable distances from their breeding colonies. They have a habit of following ships and many superstitions and legends surround them; their presence has been associated with high winds and bad weather and this is probably true because these are ideal conditions for the birds to glide in. Albatross colonies are usually found on remote islands and they probably remain paired for life. They lay a single egg sometimes on bare or vegetated ground and both sexes share in the incubation which lasts for two to three months. The adults feed their young for a considerable time; this lasts for about 5 months for the smaller species and up to a year for the Wandering Albatross. Feeding the young Albatross is obviously so demanding that the parents breed only once in two years.

Petrels and Kin

Shearwaters, Fulmars and **Petrels** are very closely related to the Albatrosses. All of them have nostrils that extend onto the bill in short tubes. **Storm Petrels** (Hydrobatidae) spend most of their lives at sea and are distributed over all the oceans of the world. They nest colonially, on offshore islands, in burrows and rock crevices. The single white egg is incubated for about five to seven weeks. Storm Petrels pick their food, usually small animals and bits of offal, from the surface of the sea. The **Diving Petrels** (Pelecanoididae) catch their prey underwater and use their wings to help them swim.

The Shearwaters and Fulmars also wander great distances over the oceans of the world. Shearwaters get their name because they glide so close to the sea where they skim the surface for food. Fulmars have similar feeding habits but also settle on the sea to feed on carrion and offal. Shearwaters nest in burrows and Fulmars in the open or in rock crevices. Single eggs are laid and the young are fed on half digested plankton. The young birds put on weight very quickly and are finally abandoned by the parents. After about a week in their burrows with no food, the hungry chicks emerge at night, fly out to sea and do not return to the breeding colony for about three years.

Tropic Birds

(Phaëthontidae)

The three species of **Tropic Birds** wander great distances over the tropical and sub-tropical oceans of the world. Outside the breeding season they travel either singly or in pairs. They fly with quick wing beats and plunge into the sea to catch fish and squid. Basically the plumage is white. The long central tail feathers sometimes exceed the length of the body. The legs are short and are set far back on the body so that the bird walks awkwardly. They usually nest in cavities on steep cliffs from which they find no difficulty in taking off.

There is a lot of fighting between rival pairs at the nest sites. The single egg is incubated by both parents in stints of two to five days for a total of about six weeks. The newly hatched chicks are downy and they leave the nest after about 14 weeks. The small chicks are often left alone and are sometimes killed and eaten by other Tropic Birds.

Pelicans

Pelicans are large, mainly white birds with large wings and well-developed webbed feet, the web joining all four toes. The most remarkable feature is the pouch which is a large bag of distendible skin attached to the throat. The pouch is also attached to the undersurface of the long straight bill and is used as a sort of 'net' for catching fish. Pelicans tend to fish together in flocks. They drive a school of fish into shallower waters, swim closely together and dip their beaks into the water simultaneously, scooping up the fish. The **Brown Pelican,** however, which lives in coastal areas of the Americas, catches its prey by diving into the sea, often from a considerable height. Pelicans nest colonially in trees or amongst ground vegetation. One to four white eggs are laid and are incubated by both parents.

Cormorants and Snakebirds

Cormorants, of which there are 30 species, are found on most of the coasts, larger rivers and lakes of the world. Their plumage is predominantly black and they have long bodies and necks, short wings and legs and large webbed feet. Like the Pelicans, they have all four toes joined by the web. All the species of Cormorant dive from the surface of the water to chase and catch their prey which is mainly fish. The largest species is the **Common Cormorant** *(Phalacrocorax carbo)*, which breeds around the coasts of the British Isles. It nests colonially on rocky islets, or

cliff ledges. Nests may be constructed of seaweed or other vegetation. Usually two to four eggs are laid and are incubated by both parents. The newly hatched chicks are naked and helpless, and at first their parents feed them with little pieces of half-digested food.

Closely related to the Cormorants are the two species of **Snakebirds** or **Anhingas.** Their bodies are long and extremely narrow. These birds inhabit freshwaters in sub-tropical and tropical areas, swimming with their bodies submerged and their long snake-like head and neck emerging from the water. When a Snake Bird catches a fish or a Frog, it impales the prey on its bill, beats and then kills it before swallowing the animal down.

Gannets or Boobies

Gannets are large stoutly built birds which are typically predominantly white. They have large, conical, pointed bills. There are 9 species, which are widely distributed in the tropical and temperate seas. Gannets are gregarious and feed by diving into the sea from the air, often from as high as 30 feet. They pursue for a short distance and catch their fish prey underwater. They nest close together in large colonies which are often on the tops and gentle slopes of rather steep-sided islands. Both sexes incubate the one to three eggs and care for the young.

Herons and Bitterns

The **Heron** family contains about 60 species which are distributed practically over the whole world. They are mostly large birds with long bills and necks and narrow bodies. The feet are not webbed and the toes are long and slender. Plumage varies greatly and some species have plumes on the head, neck or back, especially in the breeding season. Herons and **Bitterns** are adapted for wading and are usually found on rivers, pools and lakes. They feed on fish, amphibians and other water animals. Characteristically, a Heron will stand motionless in the water and when the prey is within striking distance will

dart out its long beak and neck and grab it. Although they are solitary for most of the year, Herons breed in colonies. Three to six eggs are usually laid in nests that are built in the tops of tall trees, among reeds or, sometimes, on the ground.

The **Boat-billed Heron** *(Cochlearius cochlearius),* which lives in the mangrove swamps of Central and South America has developed a large scoop-like bill for catching its animal prey. The **Hammerhead** *(Scopas umbretta)* also has an unusually wide flat bill, whilst the boat-like bill of the **Whalehead** *(Balaeniceps rex)* is adapted to catch lungfish and other animals in the swamps of East Africa. Each of these species is classified in a family of its own.

Storks and Ibises

There are 17 species of **Storks** and 28 species of **Ibises** which are found in the warmer parts of the world. The northerly species are migratory. They are usually gregarious and fly strongly with their long necks extended, unlike Herons which retract the neck. The wings and legs are long. The bills are usually long and either straight or decurved.

Storks and Ibises live near shores and marshes and feed on fish, amphibians, reptiles and a variety of invertebrates. Storks nest in trees, on cliffs and on buildings and are well-known for their large bulky nests. They return to

◁ Gannets with their chicks partially covered in thick down. Gannets nest close together in large colonies.
△ ◁ The Double-crested Cormorant is common to North America and breeds in large colonies.
△ A nocturnal bird, the Little Bittern sits quite still during the day with its neck outstretched. Thus, its black, brown, yellow and white plumage makes it difficult to see against reeds and stems. Little Bitterns are about 15 inches long.
◁ ◁ Pelicans in flight at Lake St Lucia, Zululand. The distendible throat pouch is used for catching fish, being trailed through the water rather like a net.

the same nesting place every year. Ibises and **Spoonbills** make their nests in trees, on cliffs or on the ground in marshy places. The 2 to 5 eggs are incubated for three to four weeks and the young are cared for by both parents. The **Scarlet Ibis** *(Eudocimus ruber)* is a very beautiful bird and in the past has been heavily hunted for its feathers and meat by the natives of South America.

Flamingos

Six definite species can be identified and are mainly found in shallow water lakes and lagoons of warm climates. Typically the plumage is pink. These large birds have extremely long legs and necks and have a particularly specialised bill which has special filtering structures which strain out food particles from the water that is pumped through it by the action of the tongue. The bill is placed into the water in an upside down position when the birds are feeding. The filtered foods include snails, planktonic animals and plants and algae. **Flamingos** nest colonially, often in very large numbers. Some colonies have been estimated to contain almost a million pairs. The nest is a cone of mud with a shallow depression scooped out from the top of it. The one or two eggs are incubated for about four weeks. The young are cared for and fed by both parents for about 10 weeks until their filtering mechanisms have developed.

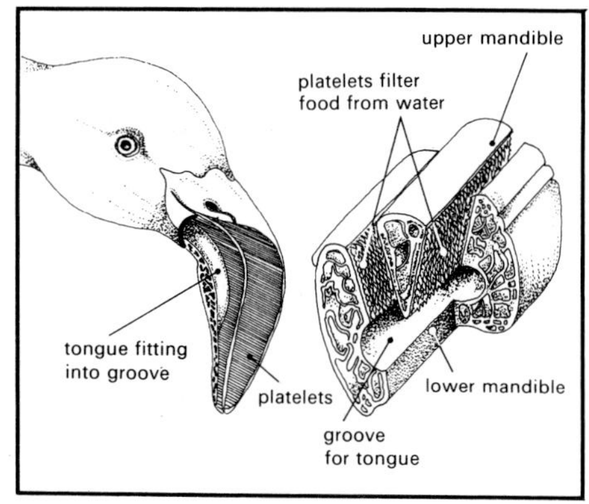

△ △ In the early part of the year, the male Stork returns to the old nest and awaits the return of his mate from migration. Once she arrives, the pair rebuild their nest with new material. Over a few years, a Stork nest can become very large, as much as 8 ft in diameter and 8 ft deep.

△ The specialised bill of the Flamingo with a cross section detail showing filtering structure.

◁ The Sacred Ibis (shown here with Crowned Cranes), has white plumage, black head and neck and partly black wings.

▷ Flamingos feeding with their characteristic beak-upside-down posture. The bill of the Flamingo has special filtering structures to strain out food particles when water is pumped through by the tongue.

The White Swans

The **White Swans** are the largest of all the waterfowl. The **Mute** *(Cygnus olor)*, **Whooper** *(Cygnus cygnus)*, **Bewick's** *(Cygnus bewickii)*, **Trumpeter** *(Cygnus buccinator)*, and **Whistling Swans** *(Cygnus columbianus)*, all breed in the higher latitudes of the Northern hemisphere. Outside the breeding season some species migrate south to more temperate areas. All Swans have long necks and feed on aquatic plants and small animal forms, particularly on lakes and marshes. The nests are very bulky and are built in reedbeds or actually in the water. The eggs are incubated about five weeks and the newly hatched downy cygnets leave the nest soon after they hatch.

In Britain the Mute Swan was kept in a semi-domesticated state for hundreds of years and was an important source of food. They are commonly found on town and park ponds at all times of the year. The yellow and black-billed Bewick's and Whooper Swans are winter visitors to Britain from the tundras and are confined to the wilder, more remote watery places.

The Black Swan

The **Black-necked Swan** *(Cygnus melanocoryphus)* is found in the southern parts of South America and the Falkland Islands and the **Black Swan** *(Cygnus atratus)* is native to Australia and Tasmania. The latter has curly black feathers and a bright red bill. It has been successfully introduced into several parts of the world. In some areas it has been domesticated and in other places semi-domesticated flocks occur.

Grey Geese and White Geese

The five species of **Grey Geese** are confined to the northern hemisphere and breed in Arctic and sub-Arctic areas. Outside the breeding season they migrate to more temperate, southerly countries. They are highly gregarious and breed in widely-spaced colonies and feed and fly in flocks. They nest on the ground and incubate their eggs for about four weeks. The newly hatched goslings are covered in down and leave the nest soon after hatching. They feed on young plant shoots and small animals. Adult Geese feed largely on vegetation and the winter populations in Britain also eat waste grain, potatoes and turnips.

The **Greylag Goose** *(Anser anser)*, is the ancestor of the most familiar domestic Goose and several strains have been developed. The **Bar-headed Goose** *(Anser indicus)* flies at considerable heights when it migrates from its breeding grounds in Tibet southwards to the plains of India.

The three species of white Geese are found in North America. The **Snow Goose** *(Anser caerulescens)* can either be white or blue-grey. **Ross's Goose** *(Anser rossii)* breeds near the Arctic coast of Canada.

The Canada Goose

The **Canada Goose** *(Branta canadensis)* consists of 12 races which are native to Canada and the United States. The races vary in size but have the same basic plumage details. All of them have black heads and necks with white check patches, whilst the body colours vary from light to dark brown. The beaks, legs and feet are black. Canada Geese have been introduced into several places in Britain and Europe and several large feral populations are in existence.

The **Brent Goose** *(Branta bernicla)* is smaller than the Canada Goose and the four races breed in areas of the high Arctic. They winter in the intertidal waters of temperate coasts and feed mainly on sea grasses. Unlike Canada Geese, they are rarely found in freshwater areas.

Dabbling Ducks

This group of Ducks contains the familiar **Mallard** *(Anas platyrhynchos)*, **Pintail** *(Anas acuta)*, **Wigeon** *(Anas penelope)*, **Shoveler** *(Anas clypeata)* and **Teal** *(Anas crecca)*. The females are usually brown and the males have gaudy plumages, especially in the breeding season. Dabbling Ducks are found all over the world and frequent both coastal and inland waters. They feed on a variety of small animals, plants and seeds which they find in or near water. Typically, Dabbling Ducks feed in shallow water and grub or filter out their foods and there are many adaptations to differing diets. The Wigeon has a stout short bill for plucking grass. The bill of the **Blue or Mountain Duck** *(Hymenolaimus malacorhynchos)* is developed as a sucker for sucking the green algae off stones. The Shoveler has many long combes in its spatulate bill which enables it to strain out microscopic animals and plants from the water. Most Dabbling Ducks nest on the ground near water in tussocks of vegetation. Up to 16 eggs are laid and are incubated by the female. Most male Ducks abandon the female and ducklings and gather together in post-breeding flocks. The ducks that breed in the Arctic and sub-Arctic areas migrate to water bodies in the more southerly temperate countries after the breeding season.

Diving Ducks

This group of birds differ from the Dabbling Ducks in that they have larger webbed feet, smaller wings, and legs set further back on the body. Although ungainly on land they are well adapted for swimming and diving for their food. They also nest closer to water. Dabbling Ducks take off from the water by 'jumping' out but Diving Ducks have to 'patter' or run along the surface of the water before they take off. Familiar examples include the **Tufted Duck** *(Aythya fuligula)* and the **Pochard** *(Aythya ferina)* which mainly frequent freshwater areas feeding on seeds and small animals. Other species such as the **Scoters** live at sea and feed on marine animals. **Mergansers** and **Goosanders** have sharp, tooth-like serrations lining the bill so that they can easily grasp their prey—usually fish.

Other Ducks

Shelducks are the most goose-like of the Ducks and mainly live on coastal areas. The sexes are both brightly coloured and usually similar and have handsome black, white, grey and chestnut plumage together with black, grey, pink or red bills. The most brilliantly coloured Ducks are probably some of the Wood Ducks—the **Mandarin** *(Aix galericulata)* of Eastern Asia and the **Carolina** *(Aix sponsa)* of North America. The **Eider Ducks** are marine and nest in fairly dense colonies. The female lines her nest with down plucked from her breast. This is harvested in areas such as Iceland and the feathers used to make warm 'eiderdowns'.

The **Stiff-Tails** are nocturnal freshwater Ducks with long, stiff tail feathers. One species, the **Black-headed Duck** *(Heteronetta atricapilla)* of South America, is a brood parasite—it lays its eggs in the nests of other waterbirds and relinquishes the parental duties of incubating the eggs and rearing the young.

Birds of Prey

Birds of prey are swift and efficient killers of, amongst other victims, fish, small animals and other birds, but their killing is not wanton. Prey is sought and killed only for food not for sport and some birds of prey, including Eagles, will feed on carrion.

The order Falconiformes covers an extremely diverse collection of what are generally called 'birds of prey' and, depending on the classification adopted, includes between 270 and 290 species, which between them fulfil the role of avian predator over practically all of the earth's land surface.

Their size range alone—from the tiny Falconets of the tropics to the immense New World Condors and Old World Vultures—indicates the considerable variation of adaptations within the order. All have certain things in common, notably strongly hooked bills for tearing and dismembering prey and, in most species, powerful, sharp talons used in killing. But according to the size and type of prey eaten, and its mode of capture, here too there is a great deal of variation. Birds of prey occur in virtually every type of land habitat, (and a number of species hunt in either marine or freshwater habitats), and avail themselves of an enormous variety of vertebrate and invertebrate prey. There are bird and mammal killers, large and small; insectivorous birds; specialists in fish, snakes, snails, and other types of prey; and many which utilise carrion as a major source of food. These include not only the best-known types, the Vultures of the Old and New Worlds, but also a number of Eagles, Kites, Caracaras and others. All are superb fliers in one way or another, according to their mode of hunting, and all have highly efficient powers of vision.

Except in certain, special circumstances, birds of prey do not control the numbers of their prey; in fact the reverse is the case. This is a complex subject but in simple terms predators crop a surplus from a prey population and are dependent on this surplus for their own numbers.

New World Vultures

(Vulturidae)

The **Vultures** of the Americas closely resemble those of the Old World, being birds adapted to carrion feeding with powerful bills, feet designed for walking and perching rather than grasping prey, naked heads and necks and long, broad wings for soaring flight in search of food. The family includes the wide-ranging **Black** and **Turkey Vultures** of North and South America and the colourful **King Vulture** of Central and South America.

The most famous bird is the huge **Andean Condor,** with a wing span of up to 9½ ft, a bird of Western South America and occurring from sea level to the highest points of the Andes. The **Californian Condor** is almost as large and, with only 40 or so left in the mountains of California, is one of the rarest birds in the world.

Hawks

This large group includes a very large variety of birds of prey, ranging in size from small **Sparrowhawks** to huge **Eagles** and including true **Hawks, Buzzards, Serpent-Eagles, Harriers, Harrier-Hawks,** true **Kites, Fish Eagles, Honey-Buzzards, White-tailed Kites, Bat-Hawks, Ospreys** and the Old World Vultures. About 200 species are included in this general grouping of the family *Accipitridae.*

The **European Sparrowhawk** and **Goshawk** are typical of a cosmopolitan group—small to medium-sized Hawks—with relatively short wings, long tails and long, slender legs. These are essentially woodland birds, hunting by stealth and surprise; the Sparrowhawk is mainly a killer of small birds, while the Goshawk is well capable of taking prey as large as Rabbits and Pheasants. The Buzzards, or Hawks as they are called in North America, are basically medium to large in size, with broad wings and soaring flight, living in both open and wooded country and taking, for the most part, mammals and reptiles caught on the ground. The Buzzard found in Britain is typical of this group, as is the **Rough-legged Buzzard,** one of the small number which are true migrants.

Serpent or **Harrier-Eagles** are also large soaring birds, feeding mainly on Snakes and other reptiles. Their feet are specially adapted for this prey, with short, extremely strong toes. Harriers are slender, long-winged birds of open country—even nesting on the ground, unlike most birds of prey—and spending much time on the wing in search of their varied prey. Three species occur in Britain. Kites are a widespread and varied group, including the highly migratory **Black Kite,** the **Red Kite**—still present in Wales, but now very rare, the **Everglades Kite,** a Snail-eater with a specially adapted bill for extracting these from their shells—and many more.

A variety of other medium to large birds form the **Honey Buzzard** group, insect specialists which include the **Eurasian Honey Buzzard** which feeds largely on Bees and Wasps' grubs and honey. **White-tailed Kites** are rather small, Falcon-like Hawks, while the unusual **Bat-Hawk** is also Falcon-like and catches Bats, Martins and Swallows.

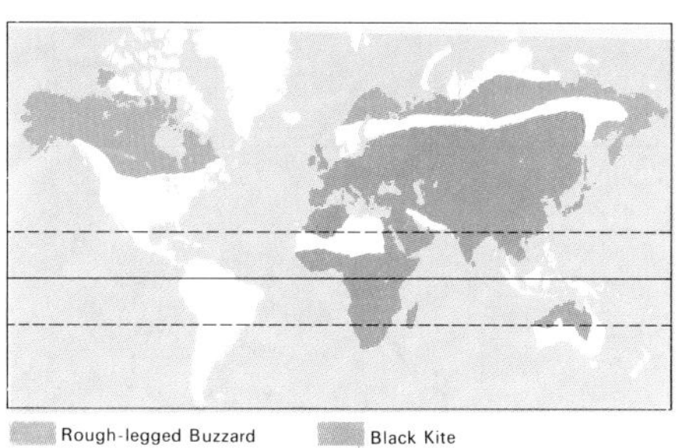

◁ Distribution of Buzzards and Black Kites.
△ Golden Eagle: the nest of the Golden Eagle has a foundation of tree branches, grass, heather and tree bark and initially is about 10 in deep. Over the breeding seasons, the nest is added to and can finally be as much as 4 ft deep.
▽ Galápagos Hawks feeding on Goat kid. These birds were on the list of rare and threatened species then a law was passed in 1959 protecting them. In 1962, the numbers were down to 200 birds.
◁◁ The food of Fish Eagles consists of fish and small birds. Some of their prey is taken on land and they are thus not as well adapted for fishing as Ospreys.

Eagles

The true **Eagles** are found all over the world and are almost as variable as the other large bird of prey groups, varying in size from the small, highly active **Hawk-Eagles** to the huge tropical South American **Harpy Eagle** and the very rare **Monkey-eating Eagle** from the Philippines. All are active, powerful birds which prey on a wide range of mammals, birds and reptiles, of all sizes. Some are carrion eaters, including the **Golden Eagle,** while others rob

other Hawks of their prey or even eat birds' eggs. The Golden Eagle is perhaps the best known; there are still between 250 and 300 pairs in Scotland and in recent years, a pair has recolonised the Lake District of England. It is a large Eagle, almost 3 ft long with a wingspan of around 6½ ft.

The **Fish** and **Sea-Eagles** are large, or very large birds of prey, more closely related to the Kite group than to the true Eagles. Among them are the famous **Bald Eagle,** the emblem of the United States of America, the noisy and beautifully coloured **Fish Eagle** of tropical Africa and the huge **White-tailed Eagle** of Europe—the last being a British breeding bird until the early part of this century.

Old World Vultures

The Vultures of the Old World differ slightly from their New World counterparts in structure. All are large, with 4 ft to 8 ft wingspan and all have powerful hooked beaks and feet adapted for walking rather than grasping. They soar for long periods in search of carrion, on which most feed exclusively, over the open or cultivated terrain in the warmer parts of Europe, in Africa and in Asia where they are found. Several species are gregarious, and large numbers gather at carrion in a surprisingly short space of time. The larger, more powerful species, such as the **Black, Griffon** and **Lappet-faced Vultures,** take precedence over the smaller ones, such as **Hooded** and **Egyptian Vultures,** different species tackling different parts of their prey according to size and strength.

The uncommon **Lammergeier** lacks the typical naked head of the other species and is altogether a narrower-winged, longer-tailed bird, though still of great size. Its most notable habit is that of dropping bones from a height, splitting them so as to obtain their marrow. The other odd-man-out is the **Palm-Nut Vulture,** a smallish Vulture, also with a feathered head, which feeds on the fruit of the oil palm and will also catch small fish.

Secretary Bird

(Sagittarius serpentarius)

This odd, terrestrial bird of prey suggests a **Hawk-Faced Crane** more than anything, with its long legs and long-striding gait. The **Secretary Bird** stands over 3 ft high and has a wingspan of 8 ft. Confined to Africa, it is a bird of open grasslands and bush, where it preys on a wide variety of ground-living prey, being especially well-known for its ability to kill Snakes by pounding them to death with its powerful feet, but also taking small mammals, birds and insects.

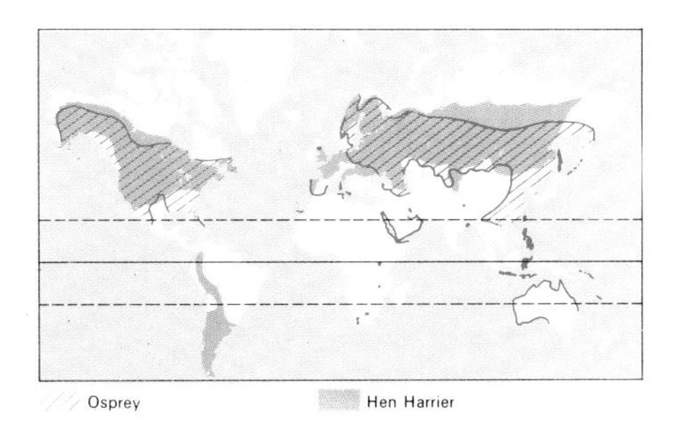

△ Distribution of Osprey and Hen Harrier.
▽ ▽ An Australian Falcon. Falcons feed on smaller birds which are taken in flight—sometimes birds as large as pigeons. It has very pointed wings to give greater swooping speed. Being a predator at the top of the food chain, pollution has depleted its numbers.
◁ Ospreys, sometimes called Fishing Eagles, fly at about 100 ft looking for fish. Upon spotting one, the bird drops fast to water level and sometimes under the water, to seize the fish in its talons. The talons have tiny spikelets under them to assist in holding the fish fast.
▽ Vultures feeding on carrion. A Marabou stork can be seen on the left. While not a bird of prey, a Marabou stork will feed on carrion.

Osprey

(Pandion haliaetus)

The **Osprey** is a specialist feeder on fish, which it catches in its talons, plunging feet-first into the water, often from a considerable height. Its feet have specially adapted, rough, spiny pads for grasping its slippery prey. It is a relatively large bird with a wingspan of about 6 ft.

Ospreys occur almost throughout the world, except South America. They are essentially birds of lakes, rivers and the sea coast, often nesting in trees and, where common, in loose colonies.

In Britain, the Osprey was much persecuted and became extinct as a breeding bird in the early part of this century. However, breeding began again in the Scottish Highlands in the mid-1950's, and following intensive protection, especially by the Royal Society for the Protection of Birds, is now well established, with about 10 pairs present by the early 1970's.

Falcons and Caracaras

Although a few rather hawk-like forest Falcons from South America belong to this group, the majority are 'true Falcons' ranging from the very small **Falconets** and **Pygmy Falcons** of the tropics to the stocky, long-winged members of the large genus *Falco.*

Falcons are exceptionally swift and agile fliers, taking their prey—mostly birds, but a variety of insects in some species—on the wing; the **Kestrels** are an exception to this general rule, taking most prey from the ground. All Kestrel species, including the very common one found in Britain, habitually hover while hunting.

Most are birds of open or relatively open country. Among the best known in Britain are the **Peregrine,** a Pigeon-sized Falcon of mountains and cliffs, a flier perhaps without equal in the bird world; the tiny and dashing **Merlin,** a moorland bird which preys on small passerines; and the handsome **Hobby,** a southern species not only adept at taking large flying insects such as Dragonflies, but fully capable of outflying and catching even Swifts, Swallows and Martins.

Caracaras are large, long-legged birds, found in open country, savannah and forests in Central and South America. They are more or less omnivorous and are partial to carrion; although agile on the ground, they are generally rather sluggish birds.

A bill for any diet

One of the most extraordinary features of birds is the variety in size and shape of their bills—adaptations for feeding in every case.

The birds of prey, Owls and a few other species such as **Shrikes** have strongly hooked bills for tearing prey. Species from those groups which prey on fish, have many adaptations, including the dagger-shaped bills of **Herons, Bitterns** and **Kingfishers,** for example, or the serrated mandibles of the **Sawbill Ducks.** Sieves are not uncommon—many wildfowl and **Flamingoes** are among the birds able to separate small food items from the water in which they feed in this way. Wading birds show an incredible diversity in bill shape and length for picking and probing for invertebrates, both in and beneath mud.

Small, fine bills of many shapes serve insectivorous birds, and stouter bills those which eat fruit and seeds; **Crossbills** even have crossed mandible tips enabling them to extract the seeds from fir cone scales. **Woodpeckers** have strong, chisel-shaped bills, used both for feeding and for boring nest-holes, while a number of other birds (**Swifts** and **Nightjars** for instance), have insignificant bills but huge gapes, the better to enable them to engulf flying insects. **Humming Birds** and some other families, have long, slender bills, used for extracting tiny insects from inside flowers as well as for obtaining nectar.

▽ Drawings showing different types of bird' bills sizes and shapes having become adapted for specialised feeding. The birds of prey have strongly hooked bills for tearing while wading birds show diversity in both shape and length for picking invertebrates out of mud.

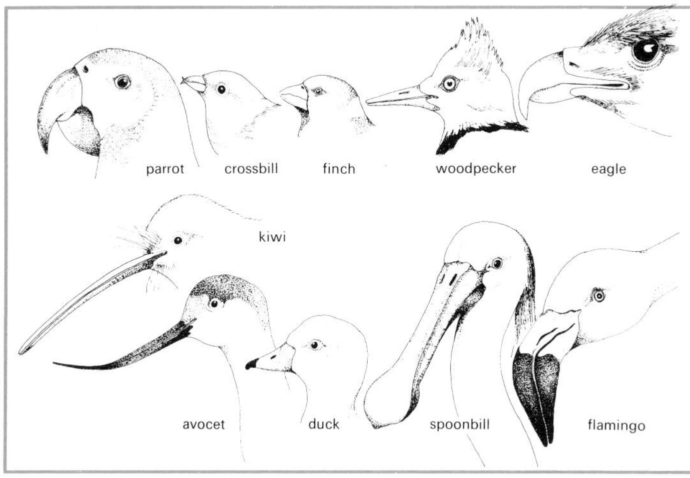

The Domesticated and the Hunted Fowls

The Fowl-like birds or Galliformes are a large order widely distributed throughout the world, except in Polynesia and Antarctica. They are described as fowl-like, because from their ranks comes the ancestor of our domestic chicken, the Jungle Fowl *(Gallus gallus)* of south-east Asia. The order contains 24 species in seven families.

Characteristically, they are all basically ground birds; their large feet have three forward-pointing toes and a small hind toe, giving them the ability to run well. Their short, rounded wings allow them to fly strongly for short distances, often in close cover; because they are mainly non-migratory, they do not need longer wings adapted for sustained flight. Only four species can be regarded as truly migratory and they tend, as in the Quail *(Coturnix coturnix)*, to have more slender wings.

All members of the order have short, stout, decurved bills which they use, often in conjunction with feet, to scratch for food. They are vegetable and invertebrate feeders. Like domestic Fowls, many have wattled or bare faces.

No other group of birds, except perhaps the Ducks and Geese, has become so closely associated with Man.

The Jungle Fowl, the Common Turkey *(Meleagris gallopavo)* and Guineafowls (Numididae) have been domesticated; the Grouse (Tetraonidae), Pheasants, Quail and Partridges (Phasianidae) are widely shot and are even bred for shooting; and others because of their startlingly beautiful plumages have been captured and reared for ornament. Because of this close relationship, some species have become established far from their native habitats. The classic example is the Common Pheasant *(Phasianus colchicus)* which is native to Asia, but which has now been introduced almost everywhere within the temperate zones.

The seven families that make up the order are—the Megapodes, which are found in Australia, Malaya and various islands between Samoa and Nicobar; the Curassows, Guans and Chachalacas which are found in forests in the New World; Grouse; Pheasants, Partridges and Quails; Guineafowls; Turkeys (Meleagridae); and Hoatzins (Opisthocomidae).

Grouse

There are 18 different species in the **Grouse** family, found across North America, Europe and Asia. All have at least partially feathered legs and some have feathered feet and covered nostrils and all have some form of wattle on the head. Like other Galliformes, they tend to be dramatically aggressive when displaying to attract mates or defend territories. The largest, the **Capercaillie** *(Tetrao urogallus)* fans its tail and throws its head back in display, while a smaller European species, the **Black Grouse** *(Lyrurus tetrix)*, takes part in communal displays with a number of other males, each fanning its lyre-shaped tail, puffing up its neck, raising the red wattles above its eyes and threatening others.

The North American **Sage Grouse** *(Centrocercus urophasianus)* and **Prairie Chicken** *(Tympanuchus cupido)*, perform similar but even more dramatic rituals, involving inflating the air sacs on the sides of their necks to impress their females.

Pheasants, Quails and Partridges

With 165 species, the family Phasianidae is the largest within the order. The range of sizes is dramatic from **Indian Peacock** *(Pavo cristatus)* at 90 inches (2.30m), in length to the tiny Painted Quail *(Coturnix chinensis)* at only 5 inches (127 mm). Even more breathtaking is the range of colours that occur within the family.

Any division into natural groups is difficult, but basically there are New World **Quails,** Old World Quails and **Partridges** and true **Pheasants** and **Peafowls.** Typical of the New World Quails is the **Bobwhite** *(Colinus virginianus)* which is a favourite quarry of sportsmen because it flies well and tastes good. It has been introduced to parts of Europe and New Zealand.

Old World Quails and Partridges, of which there are 95 species widespread throughout Europe, Asia and Africa, have amongst them the four migratory gallinaceous birds, all of them Quails.

The Pheasants and Peafowl includes Jungle Fowl, Tragopans, Peafowl and Pheasants. Where the Quails and

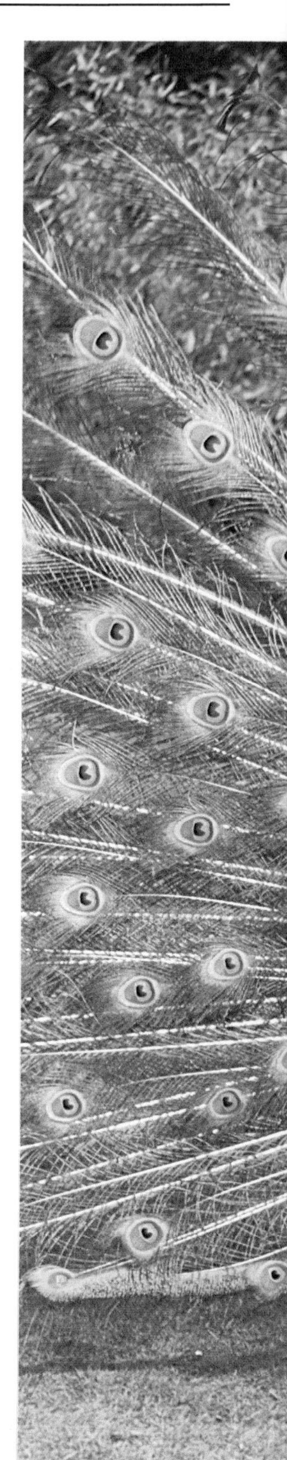

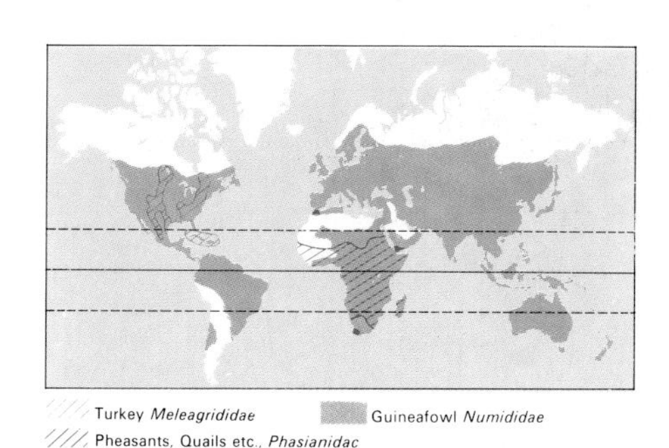

△ To attract the peahen, the Peacock displays its train in a brilliant fan, shaking it so that the iridescent feathers shimmer. The peahen is without a train and her colouring is a dull brown.
◁ Distribution map of Quails, Pheasants and Partridges.
▷ Helmeted Guinea Fowls are found mainly in tropical Africa. Guinea Fowls have very few feathers on the neck or head but some species have a tufted crest while others have a bony casque or helmet.

Partridges favour open country, the Pheasants prefer woods and forests, although some individuals will nest in open country and reed-beds. All nest on the ground, except the Tragopans which build in trees.

Guineafowl, Turkeys and Hoatzin

Guineafowl are restricted to Africa south of the Sahara and to Madagascar. They have large bodies, long legs and apparently small heads. All seven species are very similar in appearance. One of the largest, the **Vulturine Guineafowl** *(Acryllium vulturinum)* is about 26 inches (66 cm) in length and is so-named because its head resembles that of a Vulture.

There are two species of wild Turkeys—**Ocellated Turkey** *(Agriocharis ocellata)* and **Common Turkey** *(Meleagris gallopavo)*. Both are New World species. The Common Turkey is found in open woodland southwards from the eastern United States to southern Mexico. In the Yucatan, Guatemala and British Honduras it is replaced by the Ocellated Turkey.

The **Hoatzin** *(Opisthocomus hoazin)* is unique. Superficially it is not unlike a thin Pheasant, but it has a number of extraordinary features. Its crop extends over more than half of its breast-bone, which is itself keeled at the base and covered with a leathery callous; the bird sometimes uses its breast-bone as a third foot on which it rests. The young birds are even stranger; within two days of hatching, covered with a thin down, they leave the nest, using their well-developed four-clawed feet, their thick, slightly hooked bills, and the claws on their wings to clamber through the trees. The wing claws of the young Hoatzin eventually disappear. Found only in the basin of the Amazon, the Hoatzin, although in many ways primitive, seems well-adapted to its life. There is, however, still much to be discovered about its behaviour and the reasons for its apparent strangeness.

Tragopan Tragedy

Extinction threatens 14 of the world's 48 living Pheasants. Three of these are **Tragopans,** of which there are only five species anyway. They are quite large and the males have brightly coloured 'horns' and a large, brightly coloured bib, which becomes extended during courtship. Tragopans are forest birds. Their range extends through the Himalayas from Kashmir across Burma to Central China and Northern China.

Sedentary, rather solitary birds, Tragopans depend on forests for cover, food and nest-sites. The probable cause of the extreme rarity of three species is felling of their forest habitat, combined with trapping by the local humans.

The **Western Tragopan** *(Tragopan melanocephalus)* has not been seen since 1965. Its range is limited to the Western Himalayas and its future is doubtful. None exist in captivity, so that, unlike other rare Pheasants, it is impossible to breed a captive stock for release among the wild population. Similarly, there are no breeding stocks in captivity of **Blyth's Tragopan** *(Tragopan blythix)* which is extremely scarce in its natural range in north-west India. **Cabot's Tragopan** *(Tragopan caboti)* has been bred in captivity at the Pheasant Trust in Norfolk, England. The main threat to the wild population is forest clearance in its native mountains of south-east China. It may soon disappear completely as a wild bird.

Cranes, Rails and Relatives

Cranes are shy and wary but gregarious birds. Rails, in contrast, are solitary, more often heard than seen.

The Order Gruiformes comprises twelve families of birds which live, nest and feed on the ground. Many of them seldom, and some of them never, fly. Among them there are aquatic and marsh birds. Of this Order, only the Cranes (Gruidae), and Rails (Rallidae), are found world-wide, although Bustards (Otididae), are widespread in the Old World.

Cranes

There are 15 species of **Cranes** in the world and they are found in every continent except South America and Antarctica. This family has the distinction of including three of the rarest bird species in the world, and is famous for its ceremonial dances.

Cranes are tall, elegant birds, some species being as tall as 5 ft. Some Cranes wear red bands on their heads, some have crests, and their history is a long one, dating back to the Pliocene epoch.

▽ ▽ ◁ The Sunbittern, *Eurypyga helias*, is the only species in its family and is found in central and south American tropical forests near to water, hunting for insects and small fish. Sunbitterns live either alone or with a mate. Here the Sunbittern is performing a display, showing the superb chestnut markings on the wings.

▽ Crowned Cranes are striking looking birds with their beautifully marked bodies and black heads topped with a bristle-shaped tuft of brilliant feather feathers. It was once known as the Balearic Crane being found as far north as those islands. The courtship dance of this species is highly ritualised and is intended to reduce the natural aggressiveness between the birds. The Crowned Crane is the national emblem of Uganda.

◁ Demoiselle Cranes which are found in south eastern and eastern Europe and Asia, are distinctive for the brilliant white tips on each side of the head and the trailing dark-coloured feathers under the breast. The Demoiselle Crane is smaller than other cranes and is extremely graceful in movement. In winter, the birds migrate to north east Africa and across Asia to China.

All Cranes breeding in the Northern Hemisphere are migratory and watching the departure of migrating Cranes is a springtime ritual with the Japanese. There is a magic about the migration of these birds which is accompanied by their continuous bugling call.

Cranes live to a great age and will mate for life in captivity and probably, in the wild also. They have been persecuted because of the damage they do to wheat crops but besides cereals, Cranes also eat insects, molluscs, Frogs, Snakes, Mice and small birds, roots, acorns and vegetation, which they browse.

Rails, Gallinules and Coots

(Rallidae)
This is a ground dwelling family, often aquatic and living in dense vegetation. There are over 100 species of Rails, varying in length from 5½ ins–20 ins and which are found in all the continents except Antarctica. Typical is the **Water Rail** *(Rallus aquaticus)*, which occurs from Great Britain and Iceland to Eastern Asia, a shy, dark bird with a long red bill. Water Rails can often be detected by their squealing, grunting cries.

The **Corncrake** or **Landrail** *(Crex crex)*, is a grassland bird which has now become rare in Southern Britain but may still be seen in the north and in Ireland. Both these Rails are migratory.

A large species of Rail is the **Purple Gallinule** or **Swamphen** *(Porphyrio porphyrio)*, which ranges from southern Europe to Australia, New Zealand and islands in the Pacific and to Africa and Madagascar.

Of particular interest in this family, is the **Takahe** *(Notornis mantelli)*, of New Zealand. This bird was thought extinct from 1898 to 1948, when 12 pairs were discovered in a remote valley in the South Island. In spite of protection, this bird is on the list of species in danger of extinction.

The commonest member of the Rail family is the familiar **Moorhen** *(Gallinula chloropus)*, a ubiquitous breeding bird in parks and countryside. It occurs in all continents except Australia and Antarctica.

The **Coot** *(Fulica atra)*, has a similar wide distribution. Distinguishing features of the Coot are its lobed feet, (a fringe of web round each separate toe) and white bill.

Bustards and Sun-bitterns

The **Bustards** are an Old World family of stout bodied, long necked birds, varying in size from 1 ft 3 ins to 4 ft 4 ins in length, which live in deserts or grassy plains. There are 22 species of which a few are Eurasian, one Australian and the rest African.

The **Great Bustard** *(Otis tarda)*, used to breed in Britain but does so no longer. An interesting experiment is currently taking place to reintroduce birds taken from the Continent to Salisbury Plain in the hope of establishing a breeding colony.

Bustards are omnivorous, eating animal and vegetable food. The Great Bustard has extraordinary courtship display during which it contorts itself until it resembles a heap of feathers, with the neck and throat inflated like a balloon.

Sun-bitterns *(Eurypyga helias)*, have a family to themselves. They are native to forest regions of America, ranging from Mexico through Central America to northern South America. They resemble Bitterns and **Herons** in being long necked and long billed, but are smaller. Although somewhat soberly coloured, in their display they disclose an area of rich orange surrounded by pale orange on the wings 'like the sun setting in a clouded sky' as one writer described it. This feature gives the birds their popular name. Strangely enough, for a bird which spends its life like a Wader foraging along streams and the edges of swamps, the Sunbittern usually builds its nest in a tree.

Crane crisis

Cranes do not continue to exist without danger of extinction. One of the world's rarest birds is the **Whooping Crane,** an American species. In the mid-1940s, its world population reached the figure of 20 birds, and by 1969, in spite of protection, there were only 68. Sportsmen along its migration route, from its breeding grounds in Canada to the coast of Texas, are largely responsible for its decline. Its wintering area in Texas has been declared a reserve, now called the Aransas Wildlife Refuge, and since its breeding ground was discovered in 1954 in the Wood Buffalo Nature Park, both ends of its journey are carefully guarded. Unfortunately, the Aransas Refuge is not large enough to accommodate a large winter population of these territorial birds, and oil drilling in the area causes disturbance and pollution. The prospects for this species are not encouraging, in spite of efforts to breed them in captivity.

The **Japanese** or **Manchurian Crane** *(Grus japonensis)*, has done better, although it is strange that this species which has played so important a part in Japanese art, should have sunk as low as 20 birds in Japan. It was not uncommon in feudal times, but the damage it does to farmers' fields caused it to be exterminated almost everywhere. However, due to the protection now afforded these noble birds, a count by schoolchildren in Hokkaido in 1968–1969 totalled 171. In Hokkaido there are now local Crane Clubs in schools and the children feed the birds in winter.

The **Sandhill Crane** *(Grus canadensis)*, also of America, survives in precarious numbers in part of its range. The Florida population is declining, but it is the **Cuba Sandhill Crane** which is in most danger, for the entire population does not exceed 150 birds. However, they are now believed to be holding their own although no protective measures appear to be in force. Strangely enough, the clearance of forests which has harmed other species seems to be benefitting these open-country birds, so there seems some hope of their survival.

The Shore Birds

For 75 million years, the shore birds have lived in the marshes, near inland lakes and on sea shores, evolving specialised bills and feet for their watery habitat.

The order of Charadriiformes contains 16 families of birds making a total of 306 species. The order is composed mainly of three groups of birds—waders or shore birds, Gulls and Terns, and the Auks and their allies. The Gulls, Terns and Auks have evolved from the wader group and have some similar body features, such as tufted oil glands and feather

arrangements. The fossil record is 75 million years old and the order seems to be related to the Gruiformes (Rails and Cranes). This is especially so for the Jacanas, Seed Snipes, and the Thick knees.

Typical wading birds feed and nest mainly on the ground and often inhabit open water places, marshes and coastal areas. They are usually gregarious outside of the breeding season and are common on the seashore. Many species undertake long migrations to spend the winter feeding in such areas. Waders that have long bills usually have very long legs, the Curlew for instance. Other species have bills and legs of differing lengths and are adapted for walking in wet and muddy areas and probing into the ground for small animals, such as Sand Worms. In some species the front toes are webbed. The Jacanas have very long toes which helps them to walk steadily on the floating lily leaves of tropical and sub-tropical pools.

▷ ▷ Guillemot colony. Overcrowding on narrow rock ledges is no inconvenience to seabirds as they use the ledges only for nesting.
△ △ Lapwing feeding chicks. Although the Lapwing's eggs are well-camouflaged by their colouration, many are eaten by other birds. A female Lapwing may lay five clutches in a season before being able to raise a brood.
△ ▷ The Oystercatcher lays its eggs in a shallow depression scraped in the sand.
△ The long toes of the Lilytrotter enable it to walk over floating vegetation.

The Gulls and Terns also frequent coastal and inland waters. They have webbed feet and their plumage has a lot of white or grey colour in it. They are very gregarious, feed in flocks and usually nest in compact colonies. They are distributed all over the world except in some deserts and many of them are migratory.

The Auks and their allies are exclusively marine birds, spending the greatest part of the year feeding out in oceans. They come on to islands or mainland cliffs to breed, sometimes in very large colonies. These birds are highly specialised for swimming under water and feed on small fish and other marine life.

Oystercatchers

(Haematopodidae)

The **Oystercatchers** are large shore-birds 15–20 in (381 to 508 mm), whose plumage is either black or black and white. They are also called **Sea-pies**. The most conspicuous feature is the long, blunt, sideways flattened bill which is bright red in colour. The feet are slightly webbed and are also red. There are six generally recognised species and they breed on most of the temperate and tropical sea coasts from North Scandinavia south to New Zealand. They also breed inland in some parts of Europe and Asia. They are gregarious birds and have very noisy, shrill piping calls. They are strong and direct fliers. The powerful bill is used to open up the shells of Oysters, Mussels, Cockles and Clams, so that the soft flesh of these animals can be eaten. Other shorefoods include Limpets and Sandworms which the Oystercatchers catch by probing their bills deeply into the mud or sand. The sexes are similar in appearance. Nests are made on the ground, often in shingle and may be lined with grass or moss. Two to four eggs are laid and are incubated by either sex for about 2½ weeks. The downy chicks are cared for by the parents for about five weeks after they have hatched. The loud, shrill cry and the striking colours of the Oystercatcher make it unmistakable.

Plovers and Lapwings

(Charadriidae)

There are 38 species of true **Plovers** and 25 species of **Lapwings.** They are found all over the world and live in open, bare areas, often near water. Most species migrate. They are compact, plumpish birds with straight, medium length bills. Their plumages have strong colour patterns containing brown, olive, grey, black and white. The sexes are usually similar in appearance. The nest is usually a scrape on the ground with little or no lining. Four eggs are usually laid and incubated by both parents for about three weeks. They are very dutiful parents and often lead predators away from their young by pretending to be injured and attracting attention to themselves. The diet consists mainly of small animals together with a little plant material.

Typical of the true Plovers are the small **Ringed Plovers**. They breed as far north as the Arctic tundra and migrate to southerly beaches for the winter. Here they commonly feed on beaches at the edge of the tide, picking up small animals that have been stranded by receding waves.

Some Lapwings have crests at the back of their heads. They are primarily found inland and in upland areas.

Phalaropes and Thick knees

(Phalaropodidae: Burhinidae)

Two of the three species of **Phalaropes** breed in Arctic and sub-Arctic regions and winter in certain parts of the tropical oceans: **Wilson's Phalarope** *(Phalaropus wilsonii)* breeds in North America and winters inland in South America. They are all small birds and are well adapted for living at sea. The very fine plumage on the belly traps a layer of air which allows the bird to float buoyantly on the sea. Each toe is webbed in a similar fashion to the Grebes and Coots. Female Phalaropes are larger and more brightly coloured than the males and they take the lead in courtship and mating. The males build the nest alongside a small fresh-water pool, incubate the eggs (usually

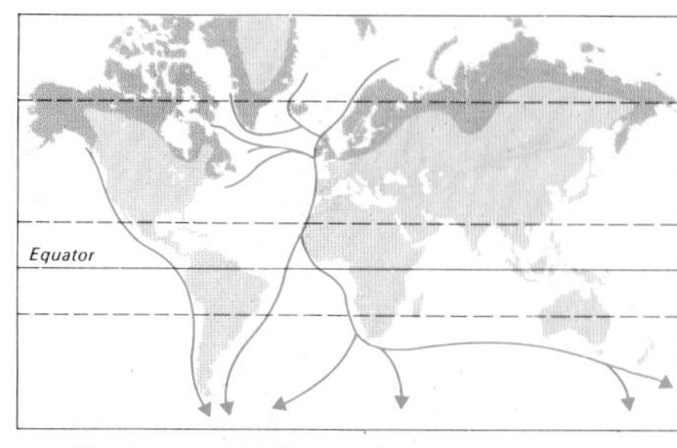

Migration routes of the Arctic tern

△ Migration routes and summer breeding grounds of the Arctic Tern.
◁ Skuas fly superbly but are not as accomplished in swimming and diving. To obtain food easily they harass smaller seabirds, compelling them to disgorge food from the crop.
▽◁ Razorbills are superb swimmers and divers, plunging to depths of 30 ft, and are able to stay underwater for up to 2 minutes to catch fish.
▷ Kittiwake breeding colonies contain thousands of nests, built on high cliff ledges.

four) and rear the young. The foods include crustacea and insects.

The nine species of **Thick Knees** or **Stone Curlews** are characterised by swollen 'knee' joints. They have long legs and large eyes which are essential to nocturnal birds. Most species are birds of dry or semi-desert areas. The toes are slightly webbed indicating their wader ancestry. These noisy birds feed on animals such as Worms and Mice and lay two eggs in a shallow scrape on the ground.

Stilts and Avocets

(Recurvirostridae)

There are seven species of **Stilts** and **Avocets** and they are characterised by their very long bills and legs. Avocets have strongly up-curved bills which they sweep to and fro in shallow water. The partly open bill catches the prey which includes aquatic insects and crustaceans. Stilts have straight bills which they use for probing deeply into muddy areas in search of food.

This group of birds is widely distributed between the temperate and tropical regions of the world frequenting shallow fresh or saltwater lakes, marshes and pools. Most species have webbed feet and are capable of swimming. They nest in loose colonies. The Avocet's nest is usually a scrape in the ground lined with pebbles or grass whilst Stilts build nests of vegetation at the edges of pools. Four eggs, incubated by both sexes, are usually laid.

Coursers, Seed Snipes, Sheathbills

(Glareolidae: Thinocoridae: Chionididae)

The Courser, or pratincole family, comprises 17 species which live mainly in tropical and sub-tropical areas. The habitat is usually stony or sandy and often near water. Pratincoles often hawk for insects in flight whilst **Coursers** live and feed mainly on the ground. All are ground nesting birds and lay two or three eggs.

The four species of **Seed Snipes** live in South America. These small waders are ground living birds and are gregarious. They run rapidly and when disturbed fly away with a zig-zag flight. Although they eat some insects they have adapted themselves to living on seeds and buds in the open country where they live.

The two species of **Sheathbills** have white plumage and are characterised by a horny sheath at the base of the bill. They live on the sea coasts in Antarctica and sub-Antarctic areas. They look very like Pigeons and have a very varied diet including fish, molluscs, birds' eggs and young and offal. They also scavenge at rubbish dumps.

Skuas

(Stercorariidae)

The four species of **Skuas** are closely related to the Gulls but have much darker plumage. They live in Arctic and Antarctic areas and have a wide variety of feeding habits. They can fish and they scavenge along shorelines for carrion. They follow whaling ships to feed on the offal thrown overboard and are well known for chasing other birds at sea and making them disgorge their food. **Great Skuas** are particularly skilful fliers and often catch the disgorged food in the air before it hits the water. They nest in loose colonies and usually lay two eggs in a depression in the vegetation. Both sexes care for the young and obtain a lot of food from nearby colonies of seabirds.

Gulls

(Laridae)

The 43 species of **Gulls** are probably the best-known seabirds. They are found along most of the world's coasts and at most of the inland waters and marshes. Most of them are large, stout birds with long pointed wings and have rather blunt and slightly hooked bills. They are strong fliers and some species are migratory. Some Gulls catch their own fish but many are inter-tidal scavengers. They sometimes scavenge far inland and supplement their diet with berries, grain and offal from rubbish dumps and animal farms. They are highly gregarious—feeding, roosting and breeding in large numbers. The nest may be placed on a cliff in the case of a **Herring Gull** or on floating, marshy vegetation in the case of the **Black-headed Gull.** Two to three eggs are usually laid and both parents incubate the eggs and take care of the young. Gull colonies are very noisy places because there is a lot of calling and displaying between individual birds.

Terns

(Laridae)

Terns are very closely related to the Gulls but are smaller, slimmer and have relatively longer wings and tails. The bills too are usually longer and more pointed. They are more specialised feeders than the gulls. Most species feed on small fish which they catch by plunge-diving into the sea. **Marsh Terns** often catch insects on the wing. Terns often nest in very large colonies, laying two or three eggs on the ground or amongst scant vegetation. Marsh Terns build nests amongst floating water plants. The 39 species have a world-wide distribution and are mainly found along sea coasts. Ringing studies have helped to show the migration routes of Terns and the long distances they travel. The **Arctic Tern** *(Sterna paradisea)* breeds in the Arctic and sub-Arctic areas of the northern hemisphere and in the next 10 months may travel as far as the Antarctic before returning to its breeding colony again. Birds nesting in Europe, Greenland and North America seem to travel down the eastern shores of the Atlantic. Those nesting in Siberia and Alaska fly southwards down the shores of the eastern Pacific ocean. Because these nesting and wintering areas are the places where the sun seldom sets the Arctic Terns probably live in more daylight each year than any other bird.

Auks

(Alcidae)

The 22 species of **Auks** are squat, heavy-bodied birds. They have short legs set far back on the body and when they try and walk on land they look ungainly. The wings are fairly short and pointed and the birds fly very directly with very rapid wing beats. All species swim and dive well. They feed mainly on fish and crustaceans which are usually caught below the surface of the water. Auks propel themselves under water by using their wings seeming to 'fly' under water. They remain at sea all the time except for the short nesting season when they gather in colonies in the coastal areas of the North Atlantic, Pacific and Arctic oceans. Very large numbers breed in the Bering Sea area.

The plumage of most Auks is black, grey or white, but the **Puffin** develops a beautiful multi-coloured beak during the breeding season. The **Horned Puffin** *(Lunda cirrhata)* also develops a pair of long feather crests on the head. Most species nest in dense colonies. Usually one or two eggs are laid. The nest site may be on a cliff ledge in a rock crevice, or a burrow.

Pigeons and Doves

The Columbiformes is one of those orders containing a very large number of species and includes the familiar Pigeon, which has flourished in Man's cities and towns. The order also includes the extinct Dodo, which perished through Man's depredations.

The Columbiformes which includes the Pigeons and Doves (Columbidae), and the Sandgrouse (Pterodidae), is a distinctive group of world-wide distribution except the polar regions. They are land birds with thick heavy plumage, whose feathers are loosely attached in the skin. All of them live mainly on vegetable foods and their method of drinking is unusual in birds—they immerse the bill and suck without raising their heads.

Pigeons and Doves

(Columbidae)

This is a large nearly cosmopolitan family of 289 species. The smaller species are popularly known as **'Doves'** and the larger as **'Pigeons'** but the terms are not consistently used. They vary in size from that of a Lark to that of a small Turkey but most have compact bodies and small heads. Most species are arboreal but a few are cliff-dwelling or terrestrial. Their food is generally seeds, berries and other vegetable matter. The nest is a simple structure woven of twigs or roots and 1–2 eggs are incubated by both sexes. The helpless young are initially fed on 'Pigeon milk', a curd-like substance from the crop.

True Pigeons

This group consists of a large number of primarily seed-eating species of which the **Domestic Pigeon** is a universally known representative. All the many different breeds are descended from the cliff-dwelling **Rock Pigeon** or **Rock Dove** *(Columba livia)*, of Eurasia.

▷American White-winged Dove. In desert regions where nesting habitats are limited birds make use of existing holes which have already been excavated in the giant saguaro cactus.
▷△The Domestic Pigeon demonstrates distinct social habits. Both in the wild and in urban habitats, the birds flock together during the day and roost together at night. In cities, Pigeons are regarded as pests but they were of special interest to Darwin, providing evidence to support his theories.
▷▷The Dodo was a large, heavy bird weighing about 50 lbs. Because it moved slowly it was fairly easy to catch and though once common on the island of Mauritius it became extinct at about 1680. Visiting sailors, who killed them in large numbers for food, contributed to their extinction.

Extinction in birds

'As dead as a Dodo' is a familiar phrase applied to anything totally extinguished. This strange-shaped bird has become the symbol of those species that have entirely ceased to exist within historic times.

The true Dodo of Mauritius disappeared in the late 17th century and all that remains are a few museum specimens and fairly abundant skeletal remains. According to contemporary descriptions the bird was incredibly heavy, weighing up to 50 lb (100 kg). It had a very large head and the bill was strongly hooked at the tip and about 9 in (23 cm) long. The plumage was mainly ashy grey and the short curly tail was placed high on the body. The bird moved slowly and was considered stupid because it had no fear of Man. The depredations of visiting sailors were obviously the main factor contributing to its extinction. The diary of a Dutch ship's captain dated 1602 reveals that they killed at least 50 birds for food in a few weeks. The introduction of Pigs and Monkeys to the island in the 16th century probably hastened its decline.

The most spectacular extinction was that of the **Passenger Pigeon** *(Ectopistes migratorius)* of North America, because it existed in countless millions as late as 1880 but it became extinct early in the 20th century. It was a fairly large Pigeon, about 17 in (43 cm) long, with a bluish-grey head, brownish-grey upper parts, rufous under parts, and white vent and outer tail feathers. It nested in huge colonies in the original hardwood forests of central North America, feeding on many types of seeds, especially beech-mast. It seems that this colonial way of life was necessary for its existence because once civilised Man with his resources became a predator and broke up the colonies, the species was doomed. The birds were killed in incredible numbers for food and because they were a pest in crops. The last wild nest was recorded in 1895 and the last wild sighting was 1899.

The wild form is bluish-grey with black wing-bars and an iridescent neck. There are a number of fairly similar species in the same genus.

The partridge-like **Quail-doves** *(Geotrygon spp.)* of South America, and the **Australian Bronzewings** *(Henicophaps spp.)* are medium-sized terrestrial Pigeons showing a high degree of adaptive radiation—although they are superficially diverse they are still similar in behaviour and pattern. Most species are basically grey or brown but the **Bleeding-hearts** *(Gallicolumba spp.)* of the Philippines have bright red wound-like spots on their pale breasts. The **Pheasant Pigeon** *(Otidiphaps nobilis),* from New Guinea is a large aberrant species looking very much like a Pheasant with its long legs and laterally compressed tail.

Fruit Pigeons

These are entirely arboreal species in most of which the gut is specialised for digesting large fruits whole while the seeds pass through intact. The **Green Pigeons** *(Treron spp.)* however, have a well-developed gizzard for crushing the wild fig seeds on which they feed. The 23 species in Africa and South East Asia are basically green enhanced with yellow, black or brighter colours. The **Fruit Doves** *(Ptilinopus spp.)* of South East Asia and the Pacific region are highly variegated and amongst the most beautiful of all birds. Most are basically green with pinkish caps and bright breast patches and vary in size from that of a House Sparrow to that of a **Wood Pigeon** *(Columba palumbus).* The **Orange Dove** *(P. victor)* has a predominantly vivid orange male and a plainer green female. The **Imperial Pigeons** *(Ducula spp.)* are similarly distributed to the Fruit Doves but in general are larger and mainly grey, though some are brightly coloured or black and white.

Crowned Pigeons

The three species of the genus *Goura* are confined to New Guinea. They are the largest members of the family—about the size of a large Chicken. With their blue-grey or purplish-red plumage and lace-like laterally flattened crest, they are also some of the most attractive Pigeons. They are mainly terrestrial birds of thick forests.

Tooth-billed Pigeon

The **Tooth-billed Pigeon** *(Didunculus strigirostris)* is an aberrant species from Samoa. Its most distinctive feature is the massive, hooked yellowish bill. It is a ground-feeder nibbling berries into pieces with its bill and also holding food down with the feet whilst tearing at it with the bill. Both of these habits are virtually unknown in other Pigeons.

Sand Grouse

(Pteroclidae)

There are 16 Afro-Asian terrestrial species which are structurally similar to Pigeons but with tougher skin and very thick plumage. The birds are protectively coloured to merge with their desert surroundings. Their need for water daily causes most species to visit water-holes at dawn or dusk, and these may be many miles from the nest. The clutch of usually three eggs is incubated by both sexes—the female during the day and the male at night. The down-covered young leave the nest soon after birth. The food of adults consists of hard seeds of various plants and also buds and small insects.

The **Pin-tailed** *(Pterocles alchata)* and the **Black-bellied** *(Pterocles orientalis)* are the only species that occur in Europe (in the south-west), but their ranges are chiefly in North Africa and South-West Asia. The Pin-tailed Sandgrouse is about $12\frac{1}{2}$ in (320 mm) long, including the greatly elongated central tail-feathers. The upperparts are yellowish mottled with black, and underneath it has a white belly with a chestnut band across the breast. It has recently been proved that the male carries water to the young in the feathers of the belly which are specially adapted for this purpose. Another genus, *Syrrhaptes*, is represented by **Pallas's Sandgrouse** *(S. paradoxus)* of the Asian steppes which is subject to eruptive migrations which several times in the last century brought them west to the British Isles where they even bred. Due probably to climatic changes they very rarely appear so far west now.

Dodos and Solitaires

(Raphidae)

The three species were confined to the Mascarene Islands in the Indian Ocean and have become extinct in historic times. They are regarded as Pigeons which, after long isolation, lost the power of flight and grew to the size of a Turkey. They were massively built with strong feet and bills and rudimentary wings. The **Dodo** *(Raphus cucullatus),* from Mauritius is the best known because of its ridiculous shape. The **Solitaires** *(Raphus solitarius)* and *(Pezophaps solitairia)* of Reunion and Rodriguez respectively were not so bulky, with longer neck and legs.

Parrots and Parakeets

The Parrots are among the most brilliantly coloured birds in the world and some species can live 80 years.

The order Psittaciformes consists of a single family of 315 species which are very diverse but so similar in diagnostic features that all are instantly recognisable as Parrots. The hooked bill and the strong feet with two toes in front and two behind are characteristic. They occur widely in the tropics and southern temperate regions. Many are very brightly coloured and most are strictly arboreal but some have become terrestrial to a varying extent. The diet is mainly vegetable matter—various fruits, including nuts, and others feed on pollen and nectar. Virtually all Parrots nest in holes, mainly in trees, but sometimes in rocks, sand-banks or termite-mounds. An exception is the Monk Parakeet *(Myiopsitta monachus)* of South America, which builds huge communal nests of twigs in trees, each pair with a separate nest chamber. The round, white eggs, usually 1–10 or even 12, produce naked, helpless young which both parents feed by regurgitation.

Lories

(Loriinae)

There are 62 species widely distributed in Australasia and Polynesia and which are mainly brilliantly coloured in greens, blues, reds and yellows. Their tongues are edged with a brush-like fringe which is primarily an organ for harvesting pollen and pressing it into a form suitable for swallowing. Nectar is also collected but the pollen is the main source of nitrogen. Most species are very gregarious and flocks of one of the largest species, the **Rainbow Lorikeet** *(Trichoglossus haematodus)*, are a familiar sight in Eastern Australia where they feed on Eucalyptus blossom and also commonly visit bird-tables.

Cockatoos

(Kakatoenae)

This is a group of mainly large Parrots from Australasia with conspicuous erectile crests which are raised after alighting or when excited. Most species are white, some washed with pink or yellow but four species are mainly blackish. The massive bill is used for crushing hard nuts and extracting the contents. The largest species is the **Palm Cockatoo** *(Probosciger aterrimus)* of New Guinea, with a curved bill ending in a long sharp point and bare cheek patches that 'blush' deeper red when the bird is excited. The **Cockatiel** *(Nymphicus hollandicus)* is a small species familiar as a cage bird with a small bill and long graduated tail.

The Smaller Parrots

The **Pygmy Parrots** *(Micropsitta spp.)* of New Guinea are about 4 in (10 cm) in length and climb Woodpecker-like on trees in dense forests, eating seeds, termites and also slime-like fungi (the last probably unique among the food of birds). The 5 species of **Fig Parrots** *(Opopsitta* and *Psittaculirostris spp.)* are amongst the several small species of the region. They keep to the tops of tall trees feeding on figs and other fruits, and their habits are little known. The **Racket-tailed** species of the genus *Prioniturus*, the **Heavy-billed** species of *Tanygnathus* and the vulturine **Pesquet's Parrot** *(Psittrichas fulgidus)* are the largest remaining species. The widely distributed **Eclectus Parrot** *(Lorius roratus)*, has remarkable sexual dimorphism, both sexes are brightly coloured but the male is green and the female red and blue. Australia has a large variety of seed-eating, broad-tailed species, many of which are brightly coloured. The smallest, and a familiar cage-bird, the **Budgerigar** *(Melopsittacus undulatus)*, is usually green in the wild and is abundant in arid areas. Two largely nocturnal species are the **Ground Parrot** *(Pezoporus wallicus)*, and the **Night Parrot** *(Geopsittacus occidentalis)*. Both are weak fliers and are protectively coloured in yellow, brown and green to merge with the tussocks among which they feed. The Night Parrot is very rare and may even be extinct.

Afro-Asian Parrots

Of Afro-Asian Parrots there are about 46 species in six genera, of which the best known is the **African Grey Parrot** *(Psittacus erithacus)*. This is a medium-sized, short-tailed grey bird with a rich scarlet tail, found throughout tropical Africa. It feeds on a variety of seeds and fruits but particularly on the nuts of the Oil Palm. It is one of the best mimics of the family and is, therefore, one of the most familiar pet birds. The 8 species of the genus *Poicephalus* are smallish stocky African birds that are mainly green or brown, some with yellow head or shoulder patches, or bright coloured bellies. The two blackish, fruit-eating **Vasa Parrots** *(Coracopsis spp.)* are confined to the islands of the West Indian Ocean. The **Lovebirds** *(Agapornis spp.)*, of which there are nine species, are a group of small seed-eating birds in Africa and Madagascar, which are named from the apparent affection between birds of a pair. They travel in large flocks and can be a serious pest to crops. The fruit-eating **Hanging Parrots** *(Loriculus spp.)* are also small parrots which range from India east to New Britain, and roost, bat-like, hanging by their feet.

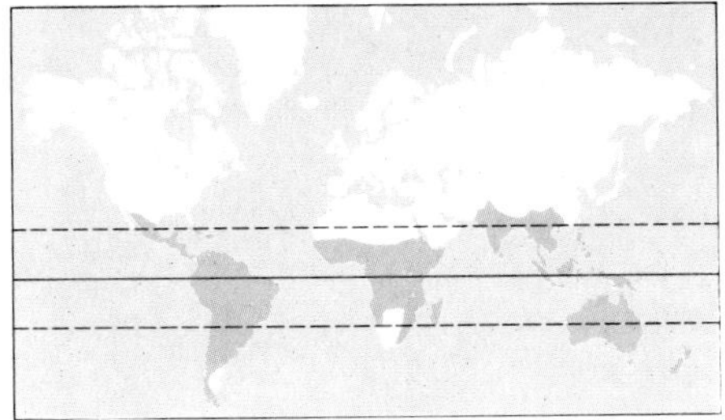

△△ The Australian Galah, found on the plains, feeds on bulbs and roots which it digs up.
△ Map showing the distribution of Parrots. Highly specialised for arboreal life, Parrots use both their beaks and feet to climb.
▷ Many birds, including Cockatoos fly in flocks as a protective measure, confusing predators with sheer numbers.

The other Oriental genus, *Psittacula,* consists of a number of medium-sized species with long narrow tails and mainly green plumage with a variety of head patterns. The **Rose-ringed Parakeet** *(P. krameri),* has an extensive range including much of southern Asia and also extending across tropical Africa.

American Parrots

The best known are the spectacular **Macaws.** These long-tailed brightly coloured species contain the largest birds in the family, over 40 in (100 cm) in length. Three of the largest species, the **Red-and-Green** *(Ara chloroptera),* the **Red-and-Yellow** *(Ara macao),* and the **Blue-and-Yellow** *(Ara ararauna)* are commonly seen in zoos. These birds with their bare faces and massive hooked bills frequent forests in Central and tropical South America, and travel in pairs or family parties, feeding on seeds and fruits in the tree-tops. There are several small species, 12–20 in (30–50 cm) long in the same genus, which are less brightly coloured but have similar habits. There are two large species in the genus *Anodorhynchus* which differ from the other Macaws in having a feathered face. Both are a beautiful deep shade of blue and the largest, the **Hyacinthine Macaw** *(Ara hyacinthinus),* is confined to a limited area from the mouth of the Amazon to central Brazil, frequenting the denser types of forests.

Amazon Parrots

The **Amazon Parrots** *(Amazona spp.)* are often kept as cage birds. They are stout bodied, mainly green birds with short, rounded tails, most of them marked with brilliant red, yellow or blue areas. About half of the species are confined to various islands of the West Indies, where a great diversity of patterns are found. Unfortunately many are threatened with extinction because of their restricted range, and two forms have already disappeared. Among the other groups there are a large number of **Conures,** mainly in two genera. These are medium-sized, slender-bodied birds with long graduated tails. The plumage varies from the large, striking, almost totally yellow **Golden Conure** *(Aratinga guarouba),* and the smaller many-coloured **Painted Conure** *(Pyrrhura picta),* to the many basically green species which usually have patches of brighter colour. Most of these species live in warm, forested areas and feed arboreally on fruits but the related small green **Andean Parakeets** *(Bolborhynchus spp.)* inhabit open temperate and sub-tropical areas even up to 12,000 feet, and feed on the seeds of low-growing plants. The **Parrotlets** *(Forpus and Touit spp.)* are a group of very small species that feed mainly on fruit high in the trees and they are, therefore, little known. Of the remaining medium-sized species the most interesting is the **Hawk-headed Parrot** *(Deroptyus accipitrinus)* with its red and blue erectile ruff.

Kea, Kaka and Kakapo Parrots

Two strange New Zealand Parrots, the **Kea** *(Nestor notabilis),* and the **Kaka** *(Nestor meridionalis),* are both fairly large brownish-green birds with heavy decurved, hooked bills. The Kea is a highland form living above the tree-line in South Island. In the summer it lives mainly on fruits and other vegetable matter and insects but it is also a notorious scavenger and has acquired the unfortunate reputation of killing sheep. Many were killed under a bounty system but research has revealed that it is unlikely that they attack any normal healthy sheep. They are now partly protected.

The **Kakapo** *(Strigops habroptilus)* is an even stranger nocturnal dweller with soft plumage, unique in Parrots. It is incapable of true flight but can glide downwards for some distance. It is a large Parrot with yellow-green plumage cryptically barred with black and an owl-like facial disc. Up to the end of the last century this bird was fairly abundant in South Island but it has declined rapidly and today it is the rarest and most threatened of their flightless birds. The introduced Stoat and Deer are probably the main factors responsible—the Stoat as a predator and the Deer as competitors in the use of the Kakapo's intricate system of forest pathways. They feed mainly on soft grass tissue and have a weird booming note likened to a Donkey in distress.

Cuckoos and their Kin

The main characteristic of the Cuculiformes is their zygodactyl feet, two toes directed forwards and two behind, a feature that they share with the Parrots, to which they are closely related, and the Woodpeckers. They also have an immovable upper mandible and the bill is slightly decurved. They are slim-bodied birds with long tails; their flight varies from being fairly strong in the migratory species to that of very weak in the terrestrial Cuckoos.

Their call is a monotonous repetition of loud notes and it is from the familiar call of the European species, *Cuculus canorus*, that the common name is derived. The order is composed of two very distinctive families; the brightly coloured, fruit and insect eating Touracos found in Africa and the duller, mainly insectivorous Cuculidae which have a world-wide distribution. Many members of the Cuculidae are brood parasitic; that is they lay their eggs in the nest of other species, the young being reared by the foster parents.

△ The special form of the Touracos' feet enables it to grip quite slender twigs. and branches. Forest-living birds, their colouring camouflages them against foliage and the crimson-lining to their wings is visible only in flight.
▽ ▷ Roadrunners can fly but appear to prefer to run about.
▽ 16 day-old Cuckoo in a Reed Warbler's nest. Cuckoo chicks rapidly outgrow the host birds and dominate their lives with demands for food.

Bird Parasitism

There are two main forms of bird parasitism, piracy and nest parasitism. Piracy is where one bird robs another of the food which the latter has obtained; it occurs occasionally in most birds, but **Frigatebirds** and **Skuas** will rob other seabirds in preference to catching their own fish.

Nest parasitism is where the eggs of one species are laid in the nest of another species. Five or six families of birds have parasitic members; the **Ducks** (Anatidae), the **Weavers** (Ploceidae), the **Weaver Finches** (Estrildidae), the **Cowbirds** (Icteridae), the **Cuckoos** (Cuculidae) and the **Honeyguides** (Indicatoridae). Parasitic birds show certain specialised characteristics in their life-history, and the Cuckoo is the most successful.

Cuckoos can lay their eggs very quickly while the host has left its nest, and the eggs are similar in colour to those of the host; a single hen Cuckoo will always remain attached throughout her life to one specific host. The eggs have a shorter incubation period than those of the host and therefore are the first to hatch. The nestling **European Cuckoo** *(Cuculus canorus)* has an extremely sensitive region in its back which it used to evict the host nestlings and eggs. In other parasitic Cuckoos, the nestlings resemble those of the host and are raised with the rest of the brood. Parasitic fledglings usually have the same pattern of colour in their gape as the host chicks, so they are instinctively fed by the parent bird.

Touracos

In the past, the **Touracos** were associated with the Galliformes, but anatomically they resemble certain non-parasitic Cuckoos, particularly the **Couas** from Madagascar. There are 20 different species of Touracos; commonly called the **Plantain-eaters.** They are all fairly large birds with long tails; the largest is the **Great Blue Touraco** *(Corythaeola cristata)* which has a body the size of a Wood Pigeon. Their bills are short and stubby; wings short and rounded and only capable of short dipping flight. Their feet are semi-zygodactyl in that the outer toes can move round at right angles to the rest of the foot; this makes them extremely agile in moving about in the trees.

A remarkable feature about this family is the unique green and red colouration; both colours being produced by specific pigments. The rich red colour is a copper complex and for centuries it was thought that the pigment was washed out of the living bird in the rain; this is untrue although the pigment does dissolve immediately in alcohol. Some species also have blue, brown, violet and yellow feathers.

Touracos live in pairs or small family parties. Their nests are flat, flimsy structures high in the trees, in which two or three white, or slightly tinted, eggs are laid. The young, when hatched, are covered with thick down and often clamber out of the nest on to the branches well before they can fly. Both the male and female share in the upbringing of the young, feeding them on regurgitated fruit pulp.

Cuckoos

(Cuculidae)

This is a large diverse family of about 127 species with a world-wide distribution and is split into six sub-families.

The typical **Cuckoos** of the Old World are all parasitic in their breeding habits. Their body shape is elongated and the tail is long and often graduated. The bill is short and slightly decurved, and the legs are also short. Cuckoos are often rather dull in colour and some show remarkable mimicry to other birds of unrelated groups—either to birds of prey, such as the **Hawk Cuckoo** *(Cuculus varius)* which closely resembles a species of Sparrowhawk, the **Indian Shikra** *(Accipter badius),* and this serves to frighten the potential host, giving away the position of the nest; or to the host species itself, such as in the **Koel** *(Eudynamys scolopacea),* which is similar to the **House Crow.** Cuckoos are arboreal birds but some inhabit the bushy savannah. Their flight is strong, particularly in those that migrate north to breed. They feed mainly on insects.

The sub-family Phaenicophaeinae is mostly non-parasitic and is found mainly in the tropics, although some are known to extend farther north; the **Black-billed** *(Coccyzus erthropthalmus),* one of the species which is occasionally parasitic and the **Yellow-billed Cuckoo** *(Coccyzus americanus)* are, in fact, migratory and breed as far north as southern Canada. The Asian species are the brightly coloured **Malcohas** *(Rhobodytes spp.)* that live in the tropical forests.

The sub-family Crotophaginae contains the **Aberrant Guira** *(Guira guira)* found in the Argentine pampas, and the three species of **Anis.** Anis have black plumage and long squared-off tails. They are gregarious in their habits and even go so far as building communal nests; they are confined to the tropical portions of America.

The **Couas** are restricted to the Malgasy region and very little is known about them, except that they are gregarious and non-parasitic. The **Crested Coua** *(Coua cristata)* is a handsome bird of the forest whereas the other species frequent the more open country.

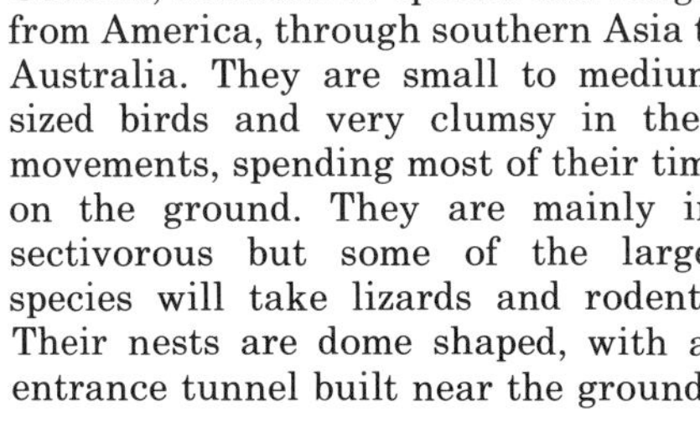

The Centropodinae sub-family, or Coucals, contains 27 species and ranges from America, through southern Asia to Australia. They are small to medium-sized birds and very clumsy in their movements, spending most of their time on the ground. They are mainly insectivorous but some of the larger species will take lizards and rodents. Their nests are dome shaped, with an entrance tunnel built near the ground.

Roadrunners

(Neomorphinae)

Roadrunners are mainly confined to southern and central America, although two species are found in Borneo, Sumatra, Thailand and Indochina. The **Greater Roadrunner** or **Chaparral Cock** *(Geococcyx californianus)* is a striking black and amber bird, with a shaggy crest and long tail. It has stout legs and can reach speeds of up to 23 miles per hour. Roadrunners feed largely on snakes and lizards, and are even able to cope with rattlesnakes, pounding them with their bills and then swallowing them head first. The remarkable feature of this sub-family is that three of the American species are parasitic; these are the **Striped Cuckoo** *(Tepera naevia),* **Pheasant Cuckoo** *(Dromococcyx phasianellus)* and **Pavonine Cuckoo** *(Dromococcyx pavoninus).* The other species of Roadrunners build nests and rear their own young.

Owls - The quiet night predators

While any Owl is immediately recognisable as such, the life-history of many of the 133 or so Owl species found in the world is practically unknown; indeed some are only known at all from a few specimens.

Owls are found throughout the world, except on some islands and in the main Antarctic land-mass. Their habitat varies but most are arboreal in some degree. They vary in size from the tiny 5 inch Pygmy and Elf Owls to the giant 27 inch Eagle Owls. Most have relatively drab, camouflaged plumage and the sexes are mainly alike in appearance—although females are generally larger than males.

Binocular Vision

All Owls are compact, large-headed birds with forward-facing eyes set in facial discs, exceptionally soft plumage, sharp hooked claws and short hooked bills. Their food is apparently exclusively animal, covering a wide range of mammals, birds, insects and other forms, and while most catch these on the ground, a few take flying prey and others are expert fishermen. About two thirds of Owl species hunt between dusk and dawn, and these show the special adaptations of the order particularly well. The soft plumage helps to make flight virtually silent, while the huge eyes with their binocular vision function exceptionally well in the poorest light, (although Owls cannot see in total darkness). Equally important is the remarkable development of the ears which are very large and lie behind the facial discs—which appear to act as reflectors, aiding hearing rather than vision. Some species have asymmetrical ears, believed to be used to help precise location of the tiniest sounds. The 'ear-tufts' which are prominent in some Owls are not in any way associated with hearing.

Planned Parenthood

Most Owls make little or no attempt at nest building. Some use the old nests of other birds, while many of the smaller species nest in holes in trees, rocks, etc. A few, like the Snowy Owl *(Nyctea scandiaca)*, nest on the ground in the open, while desert species often breed undergound in old rodent burrows. The eggs are white, the clutch size and number of young reared being closely related to the amount of prey available; more eggs are laid and more Owlets survive in years when food is plentiful. This form of natural control is best seen in species like the Short-eared Owl *(Asio flammeus)*, which are dependent on such a prey as rodents, which have a pattern of population fluctuation. These owls may suddenly appear (or 'irrupt') into certain areas in considerable numbers in good rodent years.

△ The Short-eared Owl is similar to the Long-eared Owl on shape and general colouration but its plumage is rather more marbled than barred. It has longer wings and has very short ear tufts. The Short-eared Owl lives in open country but it is a wandering migratory bird, moving across Europe in flocks.

▽ The further round the side of a bird's head its eyes are set the greater its field of vision. Woodcocks can see through a complete 360 degrees, Pigeons can see rather less. The Owls have forward-facing eyes and though they have a narrower field of vision, more of it is binocular, the images overlapping.

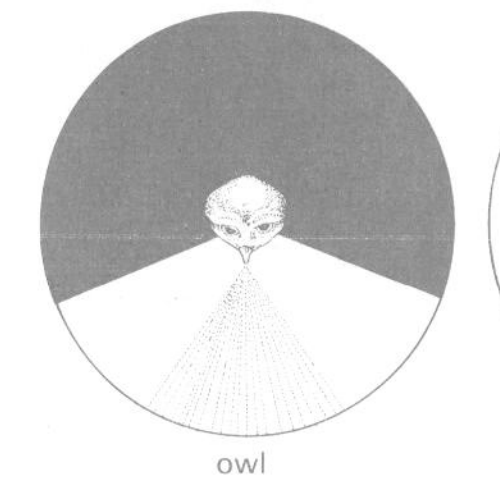

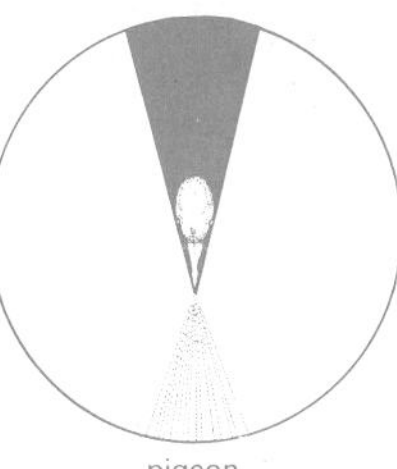

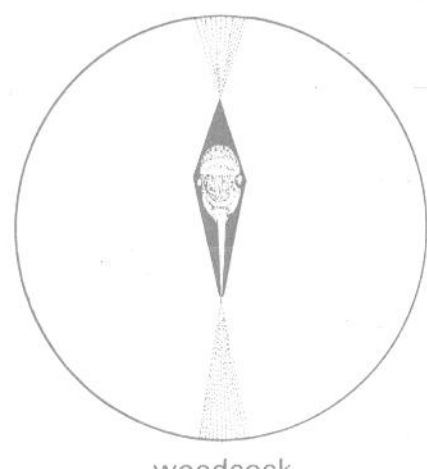

△The Eastern Owl is a species of Barn Owl and it is distinguished by the heart shaped facial disc. Like most species of owls, the Eastern Owl is adapted for silent flight with very soft plumage and flight feathers which enable the birds to fly almost without sound as it hunts for its prey at night. Barn Owls, like the Eastern Owl, feed on small rodents and sometimes on frogs, fish or small insects. Barn Owls are a partly nocturnal species and they have a distribution almost all over the world, in moderate climates but the numbers are said to be decreasing in Britain. It has a vocal repertoire of screams and snores.

Typical Owls

There is considerable variety among the 120 or so 'typical owls' of the family *Strigidae*, and three European species in particular are representative of most. All three can be replaced by allied species elsewhere in the world.

The **Tawny Owl** *(Strix aluco)*, is the most familiar British Owl, a bird of woodland which is also common in parkland, larger gardens and even cities. It is resident over much of Europe, in North Africa and in parts of Asia. The Tawny is a medium-sized Owl (1 ft 3 in long). Thoroughly nocturnal, it has typically large eyes and ears, and, being principally a woodland hunter, it has relatively short wings. Small mammals are the main prey, the bird usually sitting in wait, watching and listening, before pouncing. Birds are also taken, especially in built-up areas. Like many other woodland Owls, it is very vocal and probably best known by its song, the familiar, wavering hooting.

The **Scops Owl** *(Otus scops)*, is a small (7½ in long), 'eared' species and one of about 40 members of the widespread nocturnal genus, *Otus*. It is found in southern Europe, eastwards into Asia, in north-west Africa and widely in the southern half of that continent. In Europe, it is a bird of open woodland, parks and gardens, where it feeds on insects, Lizards, small mammals and birds. The European Scops Owls, which winter in the Savannahs south of the Sahara, are one of the few truly migratory Owls. Its song, a monotonously repeated single note, is a characteristic sound of summer in southern Europe.

With a length of up to 2 ft 4 in and a wingspan of over 4 ft 6 in, the **Eurasian Eagle Owl** *(Bubo bubo)*, is the largest and most powerful of all Owls. It is found in parts of Europe, Asia and North Africa. Although various forms occur in various habitats, the typical bird in Europe inhabits wooded, rocky country with gorges and cliffs. The prey is chiefly small to medium-sized mammals and birds, but with its great size and strength, and the opportunist hunting methods common to many Owls, it can tackle very large prey. These can include larger birds and mammals such as Fox, Wild Cat and even Roe Deer.

A huge, predominantly white Owl, the **Snowy Owl** *(Nyctea scandiaca)*, is almost as big as the **Eagle Owl.** An Arctic bird with a circumpolar distribution, living in the tundra zone, it does occur much further south in winters in irruption years. A pair has bred in Shetland since 1967. Arctic Hares and Lemmings are the main prey, (though a wide range of other bird and mammal food is eaten), and distribution and breeding success vary considerably according to the cyclic population fluctuations of these species. Geographical circumstances force it to do a considerable amount of daylight hunting. Unlike most Owls, this species shows a sex difference in plumage, the female being much more heavily marked than the often pure white male.

△Zoo specimen of a Snowy Owl. These birds live in the Arctic in the tundra zone, feeding on small mammals. They are also quite skilled at catching fish, waiting on river banks for one to come near and then seizing it in their claws.

The **Barn Owl** *(Tyto alba)*, is one of about 10 species of the family *Tytonidae*, which are separated from the 'true Owls' *(Strigidae)*, by minor structural differences, of which the Barn Owl's long, slender legs and heart-shaped facial disc are typical examples. It is a partly nocturnal species, with a worldwide distribution, and although decreasing in numbers, it is still widespread in Britain. Perhaps it is best known for its ghostly white appearance, its habit of breeding in ruins, churches, barns etc., and for its weird vocal repertoire of screams, snores and hisses. It is a hunter of open country and one which preys mainly on small rodents.

An eye of a bird

A bird's eye is undoubtedly one of its most important organs; with a very few exceptions, such as the Kiwis and some probing and filter feeders, birds depend on their sight to find food and to avoid obstacles when flying at speed. Therefore, sight is a sense which is very highly developed in most birds, and indeed their eyes are large in relation to their head size. The complexities of eye structure in birds are too many and too varied to be described: it is enough to show that these vary sufficiently to cover the amazing night-vision of nocturnal birds, the incredible long-range vision of birds of prey and the ability of many different species to use their eyes efficiently under water.

Most birds gain something like 'all round visibility' by having their eyes placed at the sides of their heads. Others, such as Owls and birds of prey, depend for accuracy and timing on the stereoscopic vision gained by having their eyes positioned frontally. Another interesting feature of birds' eyes is the nictitating membrane, which is drawn horizontally across the eye. This is transparent, and besides having moistening, cleaning and protective functions, may be modified to compensate for loss of corneal refraction.

The Most Beautiful Birds in the World

Throughout the world, in a variety of habitats, from deserts to tropical forests and in snowy places in high mountains, there are beautiful birds, distinctive in their brilliantly coloured plumage.

Birds with brilliant plumage occur in several different orders. The Apodiformes, the Swifts and the Hummingbirds are fast-flying, weak-legged birds. The Trogons belong to Trogoniformes and their toe structure is distinctly different from that of other birds. The first and second toes are pointed backwards while the third and fourth point forward. Coraciiformes are mostly Old World birds with long bills, short legs and short wings, many of them nesting in holes. Kingfishers, Hornbills, Hoopoes and the Bee-eaters are in this order.

Swifts

(Apodidae)

Swifts may resemble **Swallows** but are not related to them. They are among the most aerial of birds, feeding, and sometimes mating and spending the night on the wing. The **Common Swift** *(Apus apus)*, is a late summer visitor to Britain and leaves early on its journey to Africa. It nests in holes in cliffs or buildings, but the **Crested Swifts** *(Hemiprocne spp)*, make a nest on the branch of a tree, which is so small that it will hold only the single egg; the parent brooding the egg is actually forced to sit on the branch round the nest. Another celebrated Swift's nest is that of the edible **Nest Swiftlet** *(Collocalia fuciphaga)*, which builds entirely with salivary material. These are the nests used in the manufacture of birds' nest soup, a Chinese delicacy.

Kingfishers

(Alcedinidae)

The **Kingfisher** *(Alcedo atthis)*, is the brightest coloured British bird, yet pales before other members of the same family. They vary greatly in size, from the tiny **Little Kingfisher** *(Ceryx pusillus)*, of 4 in (102 mm) to the Australian **Kookaburra** *(Dacelo gigas)*, of 17 in (457 mm). The characteristic of Kingfishers is their short, stumpy bodies, short necks and long bills. They are not all fishers, for Kookaburras will take insects, crabs, lizards, frogs and small birds. The wild laughing cry of the Kookaburra is famous.

◁ The brilliant plumage of the Kingfisher, metallic green, azure blue and copper red, is achieved not so much by actual pigmentation in the feathers but by the structure of them, which breaks up the light to produce the effect of colours. The bird will perch on a branch, motionless, until it spots a fish. Then plunging, it takes the fish and returning to the branch, swallows the fish head first.

Hummingbirds

(Trochilidae)

There are over 300 species of **Hummingbirds** and some of them must be among the most beautiful birds in the world. They also include the smallest species, the **Bee Hummingbird** *(Mellisuga helenae),* which measures only $2\frac{1}{4}$ in (63 mm), including beak and tail.

Perhaps the most noted characteristic of the Hummingbird is its hovering flight, as it moves from flower to flower feeding on nectar; it is capable of moving backwards as well as forwards in flight. The bill is usually long and may be straight, or curved up or down. Some of these gorgeous little creatures have fantastic tails, like the **White-booted Racket-tail,** in which the outer feathers of the tail are extended like long wires with a 'racket' of black feathers on the end of each. The feathers of Hummingbirds only occur in two colours—black and rufous. The exquisite irridescent colouring is caused entirely by the diffraction of light through the feathers.

Trogons

(Trogonidae)

Trogons are birds of the trees and exhibit their beauty statically, often perching in the same position for long periods. They are gorgeously patterned, the males having breasts of red, pink, orange or yellow with metallic green upperparts and long, magnificent tails. The most famous of the Trogons is the resplendent **Quetzal,** the sacred bird of the Mayas and Aztecs. In the male Quetzal, the upper tail coverts may extend beyond the tail feathers and measure over 2 ft (60 cm) long.

Hoopoes

Upupa epops, the **Hoopoe** of the Old World, is a classic representative of its family. Its history is long, for its portrait occurs in mural paintings of ancient Egypt and Crete. It is widely distributed in Europe, Africa and Asia and is a handsome bird up to 12 in in length. The plumage is principally pinkish cinnamon in many shades and the wings, crest and tail are strongly barred black and white. The bill is long and pick-like, useful for removing insects from crevices. Like the Kingfisher, the Hoopoe is notorious for the foulness of its nest.

Bee-eaters

(Meropidae)

The **Australian Bee-eater** *(Merops ornatus),* is called **'Rainbow Bird'** with good reason and this description could apply to most of the species. The birds are long winged, long tailed and the beak is also long and decurved. They are insectivorous, their prey being mainly bees and wasps. Observations on a captive Rainbow-bird disclosed that it regularly rapped its prey on the perch to kill it and then wiped it on the perch to remove the sting. Bee-eaters nest in holes in banks, cliffs or in the ground which they excavate themselves. They are gregarious and some perform aerial evolutions, accompanied by a characteristic trilling call.

Hornbills

(Bucerotidae)

The **Hornbill** family is named for the huge bill, in some species topped by a horny casque. They are birds of the tropical Old World and there are 45 species, ranging in size from 15 in to $5\frac{1}{2}$ ft. Large species are notable for the loud noise their wings make in flight, like a railway train thundering through the forest. Equally strange are their nesting habits, for the female is walled up in her nest by the male plastering the exit with mud so that only a narrow aperture is left through which he passes food. 'Hornbill ivory' is obtained from the **Helmeted Hornbill**. This used to be exported to China as it resembled yellow jade.

▷ Malachite Sunbird with its distinctive long, curved and pointed bill. The bird feeds on insects and nectar, the nectar is rich in sugars to replace the bird's energy. The insect food is essential to supplement the diet and provide protein. Most Sunbirds are found in Africa but they have also been noted in other warm parts of the Old World.

Rollers

(Coraciidae)

The **Rollers** are stout birds, between 9½ in and 17 in in length, with short legs and long tails. There is a **European Roller** *(Coracias garrulus)*, a handsome blue bird with cinnamon wings, but it is seldom seen in Britain and has never bred there. Perhaps the most beautiful species is the **Lilac-breasted Roller** *(Coracias cordata)*, of Africa. These birds are so-called because of their aerobatics.

Riddle of Migration

The word 'migration' means to travel from one place to another and seasonal migration of some kind is the practice of many birds. The whole process is still wrapped in mystery, so far as the reason behind it is concerned, because although it has always been felt that the travelling was intended to better the creature's chances of survival, some birds (the Swift, for instance) leave their brooding areas at a time when the weather is ideal and there is food available.

To suggest that birds which nest in the north, travel south for the winter, is an over-simplification. There are also local movements where only short journeys are involved, perhaps from an inland breeding area to the coastal winter feeding place, or vertical displacement where birds which breed high up move down to the lowlands to winter. There are also transequatorial migrations, where birds fly from a temperate latitude in one hemisphere to a similar latitude in the other.

Information about migration has, in the course of history, been obtained by many methods. It used to be observed that birds present in the summer, disappeared with the onset of autumn, and some people supposed that this meant that the birds hibernated. However, it is now known that birds being observed flying out over the sea, or returning exhausted and hungry are taking part in migratory flights. They may also be seen during the spring and autumn migrations, passing lighthouses and lightships at night in great numbers; many perish by flying into the lights.

Some migrations are very dramatic, as is the transequatorial journey of the **Short-tailed Shearwater** *(Puffinus tenuirostris)*, which leaves its breeding islands off the south-east of Australia and flies to the North Pacific, as far as the Bering Straits off Alaska. Their chicks, which they leave behind in the nest, have to undertake this journey completely unaided. How do they find their way?

The riddle of migration and bird navigation is still unanswered. Many observers are working on the problem and perhaps one day the complete answer will be known.

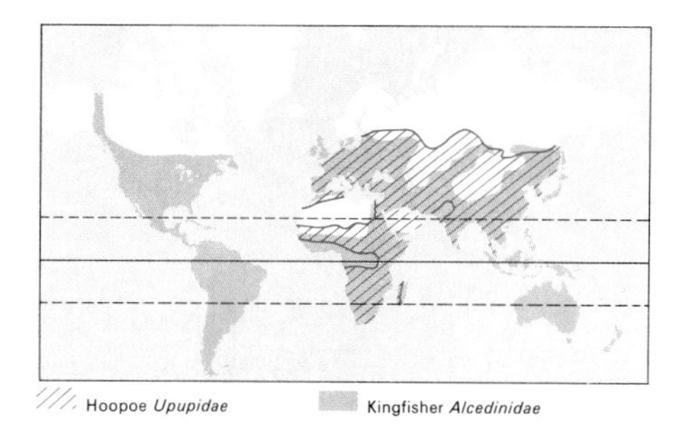

△ Distribution map of Hoopoes and Kingfishers.
▽ Red-billed Hornbill. The bills of these birds seem large for their body size. In some species the bill is hollow and seems to serve to amplify the cry of the bird, making it even more raucous.
▷ Diagrams show the construction of a bird's feather, with details of the barbules and barbs. Down feathers consist entirely of free barbs.

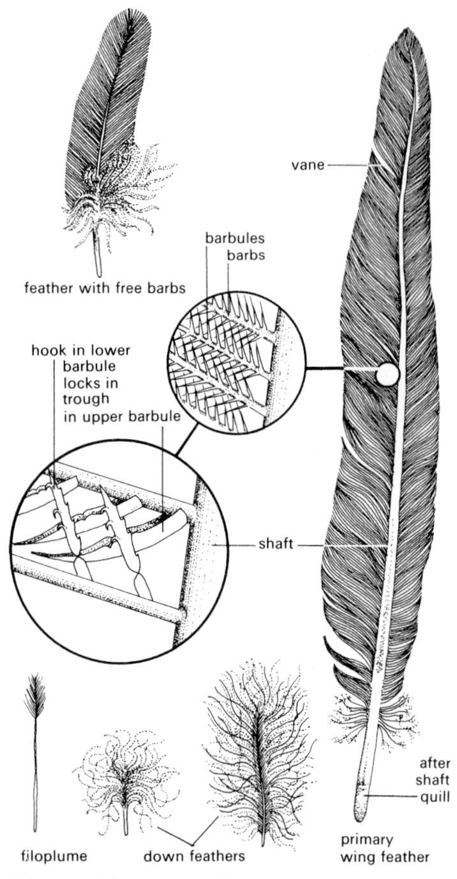

Feathers for Flying

Feathers are the main factor in the success of the flying bird. They are an adaptation of the outer skin and are formed entirely of a substance called 'keratin', a dead product of cells of the epidermis. Combined with the adaptation of the forelimb, feathers make a light and efficient structure and convert the bird into a very competent flying machine. The flight feather has a base, a long central shaft (called the rachis), and parallel barbs on either side. The barbs have tiny branches called barbules on either side.

Minute hooks on their undersides hold the barbs together and if they become disturbed, can be restored by the bird drawing the feather through its bill. This is one of the functions of preening, its importance only being realised recently. It is the contour feathers in the wing and tail which are the most important element in the power of flight. The main flight feathers are the primaries, which vary in number between nine and twelve, and the secondaries which can number anything between six and thirty-seven. The tail feathers vary in number between eight and twenty-four.

The casting of worn feathers is referred to as the moult, and most birds cope with the harmful consequences of losing their covering by shedding their wing and tail feathers in sequence so that there are not usually more than a few feathers missing from each wing at the same time. Exceptions are Ducks, Geese and Swans which go through a flightless period when they are very vulnerable to predators.

Twilight Birds

Like the Owls, the Oilbirds, Frogmouths, Potoos and the nightjars are birds which rest by day, flying and feeding at night or dusk.

The order Caprimulgiformes includes five families. These are the Oilbirds; the Frogmouths; the Potoos or Wood Nightjars; the Owlet Frogmouths; and the Nightjars.

Oilbird

The **Oilbird** *(Steatornis caripensis)*, is the only member of its family, although it bears some resemblance to the **Nightjar;** it has a hooked beak like a **Hawk** and has certain Owl-like characteristics. It is about 17 in (45 cm) long. Like the rest of the Order, this is a nocturnal bird, living in dark caves in which it nests, and from which it only flies out at night, to feast on the fruit of the forest. The young birds become extremely fat and are collected for their oil by local people. Oilbirds have been proved to find their way in the dark, like bats, by echo location, that is, by the echoes of their clicking calls bouncing off the cave walls. Their nests are built on ledges from regurgitated semi-liquid fruit material and droppings and are re-used each year. Oilbirds occur in Trinidad and in tropical South America.

Frogmouths

(Podargidae)

There are 12 species in this family, between 9 in and 21 in in length. They are tree-living, nest-building birds which feed either on the ground or by taking short flights after insects. The **Tawny Frogmouth** *(Podargus strioides)*, of Australia is typical. It is a creature of mottled plumage, and closely resembles the dead wood of the trees where it perches. When disturbed in daylight, it draws itself up so that its beak tip, surrounded by bristles, looks like the jagged end of a broken branch and its eyes close to slits.

Potoos

(Nyctibiidae)

Potoos or **Tree Nighthawks** are native to tropical Central and South America. The plumage colouration is similar to Frogmouths but they have long wings and tails. The Trinidadian name for them is **'Poor-me-one'** from their sad, weird call. Again like the Frogmouth, Potoos rely by day on camouflage. Little is known of their life history.

Nightjars

(Caprimulgidae)

Nightjar is a general name for a family of crepuscular birds, although in America the names **Goatsucker** or **Nighthawk** are used. The name **Goatsucker** is derived from the belief that Nightjars would suck the milk of goats. The **European Nightjar** *(Caprimulgus europaeus)*, is a summer migrant to Europe, spending the winter in Africa. It is well known for its 'churring' song, delivered in the dusk from a song post in its territory. As in the two previous families, the Nightjars depend on their mottled plumage for daytime concealment and are almost invisible among the leaf litter and dead wood among which they lay their eggs. The Australian **Long-tailed Nightjar** has its own peculiar call, a series of knocks like a hammer tapping wood. The **American Nightjar,** known as the **'Poor-will'**, is one of only two species of birds (both Nightjars) that are known to hibernate.

△△ White-throated Nightjar roosting during daylight, almost invisible against the background of dead leaves. At dusk, the bird hunts in the clearings of the forest.
△ Potoos have gigantic mouths, like the true Nightjars, and take insects on the wing. During the day, they perch motionless on a branch with their eyes closed.
▽ This picture shows how well the mottled colouring of the Nightjar chicks serves as camouflage against predators. The hen Nightjar lays her eggs in a hollow in the ground, usually under a protecting bush, where she sits on them until they are hatched.

Woodpeckers, Barbets and Toucans

The Woodpeckers, Toucans, Barbets and their relatives are noted for their noisiness, their raucous cries resounding from the trees where they live and feed.

The main indicative feature of the Piciformes is their feet, which have two toes pointing forward and two pointing backward, a characteristic that they share with the Parrots and the Cuckoos; they also have a distinctive arrangement of tendons in the feet. All Piciformes have a specialised form of bill which characterises each family. The colour of their plumage varies considerably from brilliant hues to more sombre tones. Woodpeckers tend to be solitary birds and occupy forests or sparsely wooded country. They are all hole-nesters, either using naturally occurring cavities or excavating their own in trees, termite nests, or in the soil. The eggs are unspotted and white; the young are hatched blind and usually naked.

Woodpeckers are virtually cosmopolitan in their distribution, but are absent in the Malagasy and Australasian regions, as well as in the very high latitudes and the remote islands. In the temperate zones some are migratory.

There are six families in the Piciformes; the Jacamars and Puffbirds which are found in the New World tropics, and the Barbets, Honeyguides, Toucans and Woodpeckers.

Barbets and Honeyguides

(Capitonidae; Indicatoridae)

The **Barbets** contain 72 species; the name Barbet is derived from the tuft of feathers around the nostrils and the bristles which develop round the beak. They are stocky, heavy billed birds, with short, rounded wings used for occasional flight. Their plumage is often very brightly coloured and the sexes are alike in most species. Barbets are found in the tropical forests, living in the tops of trees where they feed on berries and insects. They are noisy, quarrelsome birds, uttering harsh repetitive notes; one Indian species has acquired the name of **Coppersmith,** due to its metallic call. Unlike the **Woodpeckers,** the Barbets are unable to drill holes in hard wood, and therefore usually nest in decaying trees, sandbanks or termites' nests; in several species, pairs will use the same hole every season.

The **Honeyguides,** of which there are 11 species are similar in shape to the Barbets, but inconspicuous in colour. Unlike the Barbets, they have relatively pointed wings and their flight is fast. They have a special toughened skin for protection against insect stings.

Honeyguides inhabit the tropical forests and deciduous woodlands of Africa and southern Asia. Some feed on insects, but others eat bee larvae and honey, including beeswax. This eating of wax, or cerophagy, is unique among birds and they have special symbiotic bacteria in their intestine to aid digestion. These birds have obtained their name from their habit of guiding man and other mamals to the source of the beeswax. The **Black-throated Honeyguide** *(Indicator indicator)*, first draws attention to itself by calling and fanning out its tail; it then guides the man or **Ratel** *(Honey Badger)*, by calling and flying, in a series of short distances, to the Bees nest. The bird then feeds on the remains of the wax, once the nest has been broken into for the honey or grubs.

Honeyguides are parasitic in their breeding habits; often using the nest of Barbets. The nestling Honeyguides have a sharp hook at the end of each mandible, with which they fatally injure the host's nestlings; this hook drops off once the Honeyguide is about a week old and has no further use for it.

Toucans

The *Ramphastidae,* or **Toucan,** family, contains 37 species and they are confined solely to the tropical part of America; their most remarkable feature is an enormous, brightly coloured bill. Although the Toucan sometimes appears top heavy, the bill is extremely light and strong, being composed of a network of fibres ramifying through the space within the horny outer shell. The

▷▽ The long pointed tongue of the Woodpecker is sticky and has a hooked tip. It is an organ of both taste and touch. The long tongue passes to the back of the head, round inside the skull and ends near the base of the bill or near to the eye.
◁ Red and Yellow Barbet: males and females have similar colouration in most species of Barbets.
△ In spite of its size, the Toucan's bill is lightweight and is useless for defence against the bird's predators, weasels and hawks. It is likely that the colour and size of the bill are for social display purposes but it is suggested by observers that it also helps the bird to pick fruit.
▷△ Great Spotted Woodpecker carrying food for its young. Woodpecker nests are hollowed out of a tree trunk and the nest is lined with wood chips.

tongue is long, narrow and horizontally flattened with a bristly tip. The tail is long; the wings are short and rounded, flight being an alternation of flapping and gliding motions.

In addition to insects, nestlings and lizards, Toucans feed on fruit, which they pluck from the ends of the branches with the tips of their bills then, by tossing their heads backwards, they throw the food into the backs of their throats. The large, bright bill is useful when intimidating other birds, and it may also be used in protection of the nest and in courtship display.

Toucans are sociable birds, moving round in small flocks of a dozen or more. They nest in the disused hollows of trees, where the only carving to be done is the widening of the doorway. They do not line the nest but will regurgitate large seeds while they brood the eggs, and these seeds eventually form a pebbley base. The lower mandible of a newly hatched Toucan projects beyond the upper one, and they have heel pads which help support their weight on the rough floor. They are extremely slow in their development and it may be three or four weeks before their eyes are opened, and several months before the bill grows to its full dimensions.

Toucans can be divided into three groups: The largest Toucan, such as the **Rainbow-billed Toucan** *(Ramphastos sulfuratos),* which ranges through the forests from southern Mexico to northern South America; the medium-sized Toucans, or **Araçaris,** which are gregarious and roost throughout the year in the old holes of Woodpeckers high in the trees; and the **Toucanets** which are the smallest and are mainly green in colour and live in the higher altitudes; these last are the only Toucans that migrate.

Woodpeckers

(Picidae)

Woodpeckers are a large family containing 208 species divided into three subfamilies; they are small to medium-sized birds, of one or two predominant colours; they are often barred or spotted. The stiff tail of the true Woodpecker acts as a support when climbing; the other subfamilies, the **Wrynecks** and the **Piculets** have soft tails.

The **Wrynecks** *(Jynginae),* represent the ancestorial form; they rarely cling to the branches but perch, and are named for the snake-like way in which they twist their heads when frightened. They are migratory species, found in the Palearctic region and tropical Africa.

The tiny **Piculets** *(Picumninae),* are about the size of a **Wren** and are mottled green or grey in colour. They are found in the tropical areas, where they shyly climb about the trees. Their short bills are not strong enough to drill into trees, but are used on rotten wood.

In the true Woodpeckers *(Picinae),* the bill is strong and chisel-like; their tongue is extraordinarily long and mobile. Woodpeckers are able to flick their tongues out so far because the supporting bones, the hyoids, are greatly extended round the skull. The tip of the tongue is bordered with bristles or barbs, and when coated with mucous from its base, it forms an excellent implement for catching insects. The head of most Woodpeckers is large and they have exceedingly powerful neck muscles for hammering. The legs are short and strong; when climbing, the second and third toes are directed forwards and the fourth sideways. The first toe is vestigal and in some species is lost. If perching, they revert to the normal zygodactyl pattern of two forward and two back.

Most Woodpeckers are arboreal, boring into the bark for hidden larvae and carving out holes for their nests. The **Ground Woodpeckers** prefer stumps or rotten branches where Ants are found, and usually excavate their nests in banks of earth or in termites' nests.

The most familiar genus is the **Pied Woodpeckers,** which are from the northern hemisphere; very similar to these are the **Three-toed Woodpeckers,** which have a characteristic yellow crown. A widespread New World genus is the **Redheaded Woodpeckers; Sapsuckers** and **Flickers**

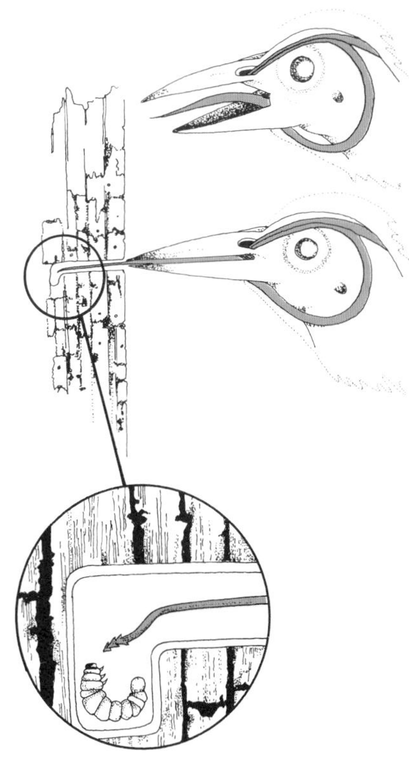

are two other purely American groups, deriving their names from their methods of feeding. The Old World **Green Woodpeckers** feed largely on Ants and fill the same ecological niche as the Flickers in the New World. Another interesting genus is the **Crested Ivory-bill.** This is the largest and most imposing of the Woodpeckers but is rapidly nearing extinction.

Perching Birds

All the popular birds, the Tits, Swallows, Larks, Starlings, Honeyeaters and Wagtails, etc., are members of the Passeriformes, the Perching Birds. Most are small-sized but they are a highly successful group.

With more than 5,000 species, over half the total number of species of birds, the order containing the Passeriformes, is by far the largest. The main characteristic of passerines is a foot with three toes pointing forward and a well-developed hind-toe. The structure of joints and the leg muscles are adapted for perching and the hind toe is not reversible. Each wing has either nine or ten flight feathers and the tail usually has 12 feathers. The passerines have a distinctive form of spermatozoa, unlike that found in other birds. They are small or medium sized birds; the largest are Ravens, (2 ft long), and Lyre birds (3 ft 3 in, including the tail).

Passerines are the most developed birds in the evolutionary scale; the finest songbirds and mimics are found among them. The songbirds form a sub-order called the Oscines; they have well-developed vocal organs, and it is into this group that the European passerines fall.

All perching birds are landbirds, although certain species, notably the Dippers (Cinclidae), have adapted to a partly aquatic life. To a passerine, the sea is a hazard to be crossed on migration, although it may provide food for certain species that frequent the tideline.

The only landmass which has no passerines is the Antarctic. Some species are found in habitats such as deserts and high montane areas, where there are few obvious perches in the way of trees and bushes.

Perching birds display a wide range of breeding and nesting habits. All young are hatched blind, featherless and helpless, which calls for a high degree of parental care.

Similarly, feeding habits depend on the particular habitat of a species. The more specialised the feeding habits, the more specialised the habitat and less widespread the species. The most successful of the smaller perching birds are the House Sparrow *(Passer domesticus)*, and the Common Starling *(Sturnus vulgaris)*, both of which will eat almost anything and have learned that Man in his wastefulness is a regular source of food; in the temperate zones wherever Man lives there are usually House Sparrows and Starlings.

▽ The male Lyrebird has a long tail of plumes, the outer pair curved like a letter 'S'. In display, the feathers fan forwards over its body. In display, the Lyrebird builds a mound of twigs upon which he stands and sings. While the female is brooding her single egg, the male sings to her. It is believed that Lyrebirds mate for life and a family may consist of three or more young birds of different ages.

Ovenbirds

There are 215 species of **Ovenbird,** all found in Central and South America. They are small and range in length from 120 mm to 280 mm. Their nests, built of mud, resemble ovens and depending on the species and its habitat, are built on the ground or in trees. Most species are predominantly brown and have the slender, often slightly curved, bills typical of insect-eaters. One group, the **Tree-runners** have stiff tails, which they use to brace themselves against tree-trunks when foraging. Other groups use burrows, some 1.5 metres in length.

Cotingas

A large group of 90 species, the **Cotingas** inhabit the forests of tropical and neo-tropical America, particularly the rain forests of Central America and the Amazon basin. Their colouring tends to be extravagant, with dramatic arrangements of feathers. The largest Cotinga, the **Umbrellabird** *(Cephalopterus ornatus)*, 508 mm in length, is a dark metallic blue, with a large forward-pointing crest and a feathered lappet, often 325 mm long, hanging from its throat. The crest of the male **Cock-of-the-rock** *(Rupicola rupicola)*, grows so far forward that it runs along the top of the bill.

Manakins

The **Manakins** are small, ranging from 83 to 159 mm. Because they are brightly coloured, alert and active, they are very noticeable in their habitat, New World tropical rain forests. The species, of which there are 59, are chubby-looking birds with short tails and wings, and thin legs and feet. Males are brightly coloured, while females tend to be olive green or brown. Manakins' bills are comparatively stubby, with a broad base and pointed, slightly hooked tip; they feed on fruit and insects. Their song is simple and short, but not unpleasant.

Tyrant Flycatchers

Amongst all the primitive passerines of the New World, **Tyrant Flycatchers** are, biologically, exceptionally successful. There are 365 species, widely distributed throughout almost every ecological niche in the New World, except in the extreme north. All but the tropical species are migratory. Although they are comparatively primitive, probably sharing a common ancestor with Cotingas and Manakins, outwardly they are similar to the more highly developed Old World Flycatchers; the two groups have evolved from different roots to fill identical niches on opposite sides of the world. In length, Tyrant Flycatchers vary between 3 in and 16 in.

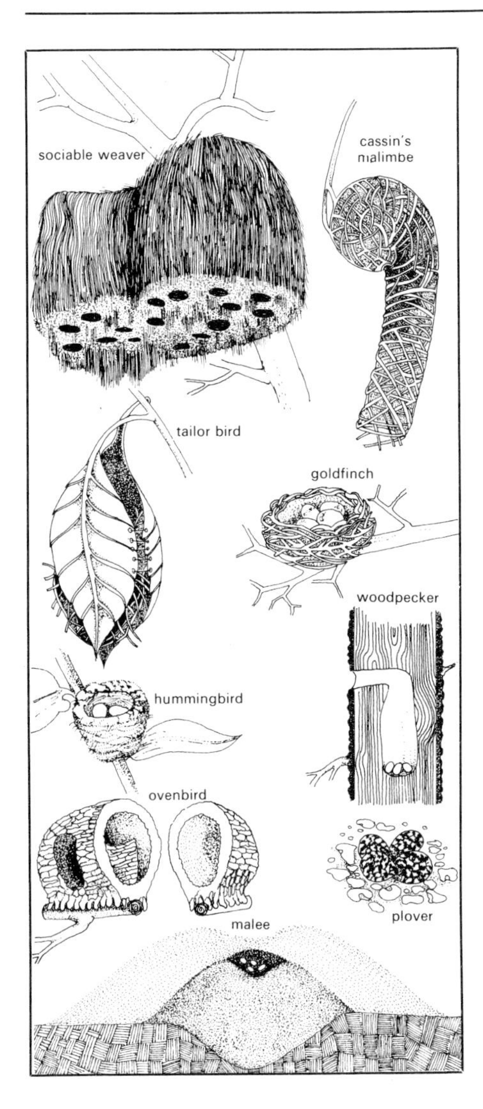

△Nests: Weaver Birds nests are woven from grasses and hang from twigs. Cassin's Malimbe builds a conical structure of vegetable fibres. The Tailor Bird 'oversews' two large leaves with strands of grass. Goldfinches build a circular, flat nest. Hummingbird nests are cuplike shapes made of fine vegetable fibres and spider's webs. The Oven Bird builds an 'oven' of clay while the Plover lays its eggs in a shallow hollow on the ground. The Malee builds a mound and covers the eggs; the Woodpecker makes a hole in a tree and lines it with wood chips.

△ A male Cock-of-the-Rock—the female of the species is a dull brown colour. These birds indulge in a communal display performance with the females looking on. It is apparently a form of selection display.

▽ Swallows build their nests from small pieces of mud and invariably choose a sheltered place near to human habitation, such as barns, outhouses, chimneys and eaves. It can take them up to eight days to collect enough mud for a nest.

Lyrebirds

The forests and scrub of south-east Australia have some strange inhabitants, including two species of **Lyrebirds.** Their ancestry and classification is obscure; at first they were thought to be Pheasants, then they were found to have vocal organs not dissimilar from the more developed passerines, the Oscines, and breastbones similar to Waterbirds. Their nearest relatives are the **Scrub-birds** of Australia, one of which, the **Noisy Scrub-bird** *(Atrichornis calamosus)*, was rediscovered in 1961, after apparently 'disappearing for 40 years. The male Lyrebirds have magnificent lyre-shaped tails which are erected dramatically during display. The **Superb Lyrebird** *(Menura superba)* has a total length of 38 in. The male builds a mound, from which it displays to attract a mate, erecting its tail, and singing. It is one of the bird world's best mimics.

Larks

Larks are numerous, with 75 species widely distributed in Europe, Asia and Africa. They are little, dull-looking brown birds at first glance, but close examination of their plumage shows an intricacy of pattern. They sing well and indulge in beautiful song-flights. The **Skylark** *(Alauda arvensis)* has been an inspiration to several romantic poets. Larks favour open country, plains, deserts and beaches nesting on the ground. Many species are migratory. Their main food is insects and seeds.

Crows

Generally unspecialised, the 100 species of *Corvid* birds, are world-wide, absent only from New Zealand and some oceanic islands. **Crows** are medium-sized birds with stout pointed bills and they will eat almost anything edible, frequenting refuse tips, eating carrion and feeding on eggs and the young of other species. Because they are general feeders, Crows are widespread, and certain species have distinct geographical races. The **Azure-winged Magpie** *(Cyanopica cyanus)*, is a specialised feeder and is limited to south-west Spain and Portugal and parts of Japan and Eastern China. On the other hand, the **Magpie** *(Pica pica)* ranges across Eurasia to Western North America. Crows have a high learning ability which accounts for their survival in large numbers despite Man's constant persecution. Generally, Man's development of the countryside has been beneficial to Crows and they have exploited developments in agriculture.

Old World Orioles

This family of rather brightly coloured birds has 26 members spread through Africa, Eurasia, the East Indies and the Philippines to Northern Australia. Males, more brightly coloured than females, are predominantly yellow, red, olive-green or brown with black heads and wings. Pairs tend to be solitary and found in woodland and forests. Most **Orioles** build cup-shaped nests of leaves, twigs and grasses, slung hammock-like in horizontal forks between branches. The birds are often wary and their presence is usually detected by their clear, flute-like calls. They feed on fruit and insects. Some species are migratory.

Swallows

Distributed throughout the world, except for the extreme north and south and some oceanic islands, **Swallows** are migratory and highly specialised for flight. Their wings are long, curved and pointed; they are very strong and agile fliers. Barely able to walk because of their short, weak legs, they are rarely seen on the ground; some species, such as the **House Martin** *(Delichon urbica)*, build nests of small pieces of mud and they will land at puddles and the edges of rivers or ponds to collect this mud. Other species, such as the **Sand Martin** *(Riparia riparia)*, will burrow into cliffs and banks to make nests. Swallows have short, pointed bills with a wide gape, developed to feed on flying insects while on the wing.

Wattlebirds

Two remnant species of **Wattle-bird** are found only in New Zealand. Both of them feature in the list of endangered species published by the International Union for the Conservation of Nature. These rather crow-like birds were once widespread in the primeval forests on the main islands, but they are now restricted to certain offshore islands. Neither species flies very well. The **Kokako** *(Callaeas cinerea)*, can do little more than glide and it moves through trees or across the ground in a series of hops, aided by its wings. The **Saddle-back** *(Creadion carunculatus)* has been introduced to certain islands where it has thrived and been successfully bred in captivity, so that its future does not seem too bleak. Already in this century, New Zealand's third and most striking Wattlebird has almost certainly become extinct. The last authenticated sighting of the **Huia** *(Neomorpha acutirostris)*, was in 1907, but there is a slim chance that like the **Takahe** *(Notornis mantelli)*, it might reappear at some future date.

Bellmagpies

These Australian and Papuan birds resemble Crows, with whom they may share a common ancestor. There are 10 species, ranging in length from 10 in to 23 in. They have large heads, compact bodies and black, or black, white and grey plumage. Like the **Shrikes**, the **Grey Butcherbird** *(Cracticus torquatus)*, uses thorns to impale and hold its prey, usually large insects, lizards or small rodents. Butcherbirds are among the best songbirds in Australia. The Western **Bellmagpie** has a territorial relationship unique among birds; groups of six to twenty birds, of all ages and both sexes, form together in a clan and defend their territories fiercely against all intruders.

Bowerbirds

Bowerbirds are the master architects of the bird world. There are 18 species, distributed through New Guinea and its nearby islands, and Northern Australia. Medium-sized (9 to 14 in), many species are ornate, but none are so ornate as their closely related **Birds of Paradise.** To attract females, the males build elaborate bowers of grasses or twigs, which they decorate with flowers, berries, pebbles and even pieces of glass. One group of four species in New Guinea are known as the 'Maypole Builders' because the bowers are built around the base of small trees. They are all closely related and look similar, being differentiated from each other by the amount of yellow or orange in their plumage. The species are the **Golden-fronted Bowerbird** *(Amblyornis flavifrons)*, **Crested Bowerbird** *(Amblyornis macgregoriae)*, **Orange-crested Bowerbird** *(Amblyornis subalaris)*, and the **Brown Bowerbird** *(Amblyornis inornatus)*. The last-named is a drab brown bird, with no bright colours, but it is his bower that is the most elaborate of the four species. It seems that the drabber the species the more ornate its bower.

lautebach's bower bird's pavilion (seen from above)
great grey bowers pavilion (seen from above)
striped gardener bower
vogelkop bower
macgregors bower
golden bower

◁ The Great Tit is distinctive among European Tits for its yellow chest with a central, black stripe. These birds nest in holes and crevices lined with feathers and moss and they are frequently to be found making use of nesting boxes.

△ Bower birds are closely related to the Birds of Paradise and are so-named because the male builds special places for courtship. The bowers are made of interwoven twigs and decorated with shells, stones, bones and flowers. After mating the female goes off to build her nest.

Birds of Paradise

Not a great deal is known about this strikingly beautiful family, because its 43 members inhabit the forests of New Guinea and Northern Australia and have not been studied in great depth. The males display a wide variety of plumages in many colours. These are put to dramatic effect in all manner of contortions during mating displays, including in some species, hanging upside down from a branch. In most species, the females are dull-looking and they will visit the display-ground, mate and return to nest and rear their young unaided. In species where the males do not have extravagant plumes and are similar in plumage to the females, there is pair-formation and the male helps with rearing the young. The feathers have had great attraction for Man for a long time; Bird of Paradise feathers form part of the traditional costume of native headmen and for some years, at the turn of the century, American and European ladies of fashion decorated their hats with feathers from Birds of Paradise, as well as from other species.

Tits, Nuthatches and Creepers

The **Tit** family has 65 species, distributed throughout the Old World, except Madagascar, New Guinea, Australasia and Polynesia, and, except for the extreme north, throughout North America, south to Guatemala. They are small birds (3 in to 8 in), rather stout and have short, pointed bills. They have strong legs and are agile, often hanging upside down to find food. Some

△ Long-tailed Tits build their nests of feathers—more than 2,000 feathers have been counted in one nest. They can take up to 3 weeks to build. Like all Tits, the Long-tailed Tits have large families sometimes laying as many as 20 eggs but the survival rate of the hatched young is low.

▽ Tailor Bird feeding its young. The Tailor Bird nest is made from leaves which the male bird literally 'sews' together—hence the name. It makes holes in the leaves with its bill and then threads through strands of vegetable matter and grasses, using a kind of oversewing technique.

species visit garden bird-tables regularly. Many are hole-nesters and will use nest-boxes; others, such as the **Long-tailed Tit** *(Aegithalos caudatus)*, make complex domed nests of feathers, mosses and spiders' webs.

Closely related are the **Nuthatches,** of which there are 17 species found throughout North America and most of Eurasia. With the exception of two **Rock Nuthatches,** all species are dependent on trees for their food of insects, nuts and seeds, and nest-holes. They are amusing to watch and can often be seen spiralling head-first down a tree trunk. They nest in holes in trees around the entrance of which they plaster mud to decrease the size of the entrance.

Tree Creepers are also small birds, 120 mm to 177 mm, and generally associated with trees. They have down-curved bills and strengthened tails which they use to steady themselves as they move up trees, searching for insects.

Dippers

Dippers are the only habitually aquatic passerines. They feed in fast-flowing streams on insects, molluscs and crustaceans. They swim well above and below the surface, and will walk on the stream-bed. Their plumage is dense with a thick layer of down next to the skin to repel water. They also have an enormous preen gland that produces oil for waterproofing. There are only five species, all of which are small (5 in to 7 in), but they are spread throughout Europe, across Asia to China and Japan, and through western America from the Yukon to Argentina.

Babblers and Bulbuls

The **Babblers** are a huge family of 282 species, found in the forests of Eurasia, Africa and Asia. Some scientists are doubtful that all the species should be assigned to one family. There is certainly great diversity in membership. Lengths range from 3 in to 16 in and the shapes of bills vary greatly. All seem to have strong legs and short rounded wings.

Bulbuls, which are found in Asia and Africa range between 5 in and 11 in in length. They have shown an ability to make the fullest use of agricultural development. This, with their noisiness and apparent energy, make them familiar birds. There are 109 species, some of which are crested and many of which are boldly marked on the head. All have hairlike feathers on the back of the neck.

Wrens

Wrens are small, most between $3\frac{1}{2}$ in and $8\frac{3}{4}$ in long, the largest being the **Cactus Wren** *(Campylorynchus brunneicapillus)*. They tend to be brown and have stubby bodies, short tails and thin, slightly decurved bills. The two sexes almost always look alike. They have short, rounded wings and are weak fliers. They are very active and often forage on the ground or in undergrowth. Their song is well developed and the **Wren** *(Troglodytes troglodytes)*, sings so fast that the human ear is incapable of picking out all the notes. There are 63 species, with a number of island races, found in Europe, Asia, north-west Africa and America.

Catbirds, Thrashers and Mockingbirds

Mockingbirds are confined to America, ranging from southern Canada south to northern Argentina and Chile. They are medium sized (8 in to 12 in), with long tails and strong slender bills. They eat insects, seeds and fruit. There are 30 species, and most sing well and are excellent mimics. They appear very active and will stoutly defend their territories against intruders and are renowned for harrying domestic dogs and cats. Some species, notably the **Catbird** *(Dumetella carolinensis)*, nest in gardens and will become quite tame.

Thrushes

This world-wide family has 305 species, many of which are famous songsters. Both the **Song Thrush** *(Turdus philomelos)*, and the **Blackbird** *(Turdus merula)*, were introduced by homesick Europeans to New Zealand, the only landmass, apart from the Polynesian Islands, from which **Thrushes** are naturally absent. A few Thrushes have highly glossy plumage, sometimes with contrasting colours. For instance, the **Eurasian Rock Thrush** *(Montiola saxatilis)* has a metallic blue head, throat and back with an orange belly and white rump. The Thrush family includes smaller species such as the **European Robin** *(Erithacus rubecula)*, $5\frac{1}{2}$ in in length, which because of its peculiar tameness in Britain has become the national bird. The **American Robin** *(Turdus migratorius)*, is larger at over 9 in long and more closely related to the **Blackbird.**

Old World Warblers and Flycatchers

The family of Old World Warblers has 398 species and a number of sub-families. Fundamentally, **Warblers** are small and brown or green, with little patterning, except streaking or barring. Many species are migratory although their power of flight is generally rather weak. They are primarily insect-eaters, frequenting woodland, open ground with cover and reedbeds.

The Old World **Flycatchers** form another large family, with 328 species. They live almost exclusively on flying insects, which they catch in a hovering flight, as do the New World **Flycatchers.** When perched, they have an unmistakable stance. Colours vary from the dramatic purples and reds of the **Paradise Flycatchers** from the Far East to the dull brown of the Eurasian **Spotted Flycatcher** *(Muscicapa striata).* The **Paradise Flycatcher** *(Terpsiphone paradisi)* has long tail streamers, almost twice its body-length.

Wagtails and Waxwings

Wagtails, which also include **Pipits,** are ground birds, found throughout the world. The 48 species are divided into three genera. They are trim little birds, between 5 in and 9 in in length, with thin insect-eating bills. Wagtails have long tails, which they bob characteristically, and all species have long toes and an elongated hind toe. Many of the species are migratory and they have a strong undulating flight. The Pipits of which there are 34 species, tend to be brown. Wagtails often have a number of geographical races with different plumages within one species.

There are three true species of **Waxwing.** They have very soft, brown plumage and crests. They are found in the sub-Arctic and temperate zones of the Northern Hemisphere, but their distribution is patchy. They eat berries, seeds and insects; when food is plentiful, breeding success is high and large numbers may be seen moving south in winter.

Starlings

The **Common Starling** *(Sturnus vulgaris),* has spread throughout most of the world, except South America and it is probably one of the most well-known species in the world. There are 104 species in Europe, Asia, Africa and Australia. They have shiny plumages in metallic purples, greens and blues. Most are highly gregarious and many have associated themselves with human habitation. Starling calls tend to be harsh, but the birds are good mimics; the **Hill Mynahs** *(Gracula religiosa),* of India and south-east Asia are popular house pets because of their mimicry. On the ground, Starlings' long legs and short tails give them a rather upright appearance, as they strut about searching for food.

Shrikes

Shrikes are thick-set birds found in Europe, Asia, Africa and North America. They have long tails and are usually strikingly patterned, especially in Africa, where they are often brightly coloured. Their slightly hooked bills are used to catch and tear their prey—insects, small reptiles and mammals. Although their feet are strong, they are not as well adapted to holding prey as those of true birds of prey and will use thorns, barbed wire and forks in trees to secure their prey when eating it. They tend to be solitary and may often be seen on exposed perches looking for prey.

△ Weaver birds nests in an African tree. The nests which are truly woven from grass are pendulous and suspended from twigs. They are in separate units and sometimes in blocks, like apartments. Weaver birds are colourful and rather like Finches in appearance.

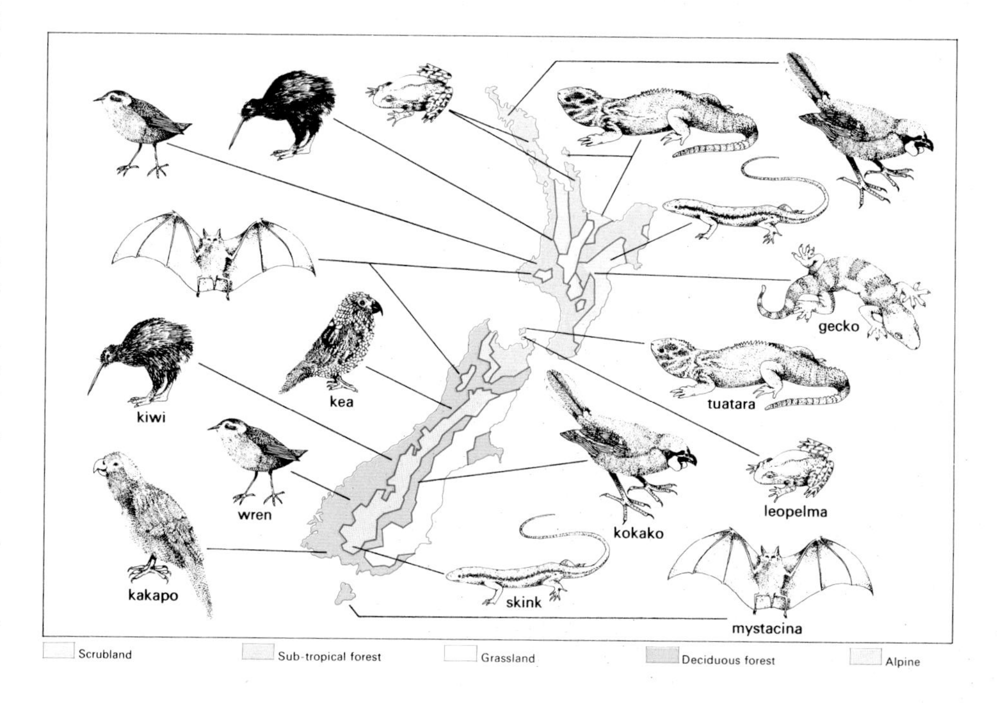

△ The diagram shows the distribution and types of mammals, birds and reptiles which were natural to New Zealand. The only native mammals were the Bats and there are few species of reptiles. Many of the birds are flightless. 35 species of birds were introduced by Europeans to compete with the native birds. Stoats, Weasels and Cats were also introduced and these preyed on native wildlife.

The animals of New Zealand are peculiar. There are no native mammals and few reptiles. A number of birds became flightless because they were not threatened by any mammalian predators and lived in vast forests of Kauri trees and tree fern that once covered both main islands.

The first humans to settle in New Zealand were the Polynesians, or Maoris, who landed in about 950 A.D. From fossil remains, it seems that there were at one time about 150 bird species.

Europeans settled in New Zealand in the early 19th century by which time the Maoris had probably accounted for the extinction of over 40 native species. For at least 50 years after the arrival of the Europeans, depredations continued. Settlers were too busy clearing habitat and shooting for food to bother about the future of the birdlife. It is sad to reflect that so many animals have been lost because humans have been concerned with short-term problems and do not think ahead to future generations.

The Europeans, or Pakehas, felled the natural forest to make way for farmland and they introduced grazing animals like deer, which prevent vegetation regenerating. They also introduced 35 species of birds to compete with native species and, perhaps worse still, carnivorous mammals like stoats, weasels and cats.

Present-day conservationists in New Zealand are skilful and energetic, but they are faced with problems. Natural forest habitat has diminished and 12 species are in danger of extinction. Fortunately successive governments have given support to conservation; for instance, when the **Takahe** *(Notornis mantelli),* was rediscovered in South Island in 1948, the government declared a closed, prohibited zone of 434,000 acres. At the Mount Bruce Native Bird Reserve experiments are being undertaken in breeding threatened species, for repopulation in the wild.

Weaverbirds

The **Weaverbirds,** closely related to the Finches, number 315 species. The most well-known member must be the **House Sparrow** *(Passer domesticus),* a European species that has followed the spread of Europeans throughout the world. Weaverbirds include other well-known species—the **Quelea** *(Quelea quelea),* notorious for its ravages on African farmland, the **Waxbills** from Africa and Asia which are beloved of cage-bird fanciers, and the South African **Social Weaver** *(Philetairus socius),* which builds vast communal nests that take up to 300 pairs. Weaverbirds are distributed throughout the world except for some Oceanic islands. They vary from 3 in to 25 in in length. They are not renowned for their ability to sing, but they are noisy, chattering and chirruping, especially when in groups.

▽ The Silvereye or White-eye has a ring of tiny white feathers round the eyes. It is a tree songbird, feeding on insects, berries and probes flowers for nectar with a brush-like tongue. Silvereyes are found in tropical areas in Africa, Asia and Australia.

Honeyeaters

Over half the 160 species of **Honeyeater** are found in Australasia, but they are also distributed through the Papuan region, South Africa and some Pacific islands. They range in length from 4 in to 14 in. Characteristically, they have fairly long decurved bills to probe flowers for nectar. The faces of some species are bare of feathers, perhaps to enable them to probe deep into flowers without becoming wet and matted. Honey eaters are usually found in forests, where their well-developed calls and song must aid identification between individuals of the same species.

Finches

No one is quite sure of the number of species in the **Finch** family, *(Fringillidae),* but it is thought to be somewhere in the region of 375, spread throughout the world, except for Madagascar, Papua, Australia and some Oceanic islands. They are not large, between $3\frac{3}{4}$ in and $10\frac{3}{4}$ in in length, and their plumage may have a combination of colours, with sexes dissimilar in some species. The do not usually nest in colonies, but often form flocks outside the breeding season, when food may be hard to find singly. All have thick bills, suitable for cracking seed husks and although they are basically seed-eaters many will feed their young with insects. Perhaps the most familiar Finch in North America is the **Cardinal** *(Richmondena cardinalis);* this almost completely red bird has a long tail and crest and is a frequent visitor to gardens in the eastern and southern States.

Sunbirds and Flowerpeckers

The **Sunbirds** are small, between $3\frac{3}{4}$ in and 10 in, and there are 104 species found in Africa, south of the Sahara, Madagascar, parts of the Middle East, India, south-east Asia, Papua and northern Australia. They are brightly coloured and have long decurved bills, with which they probe for nectar and insects. Although they look similar to humming birds, they are unrelated to this New World group.

Flowerpeckers are very small, and like Sunbirds, feed on nectar and insects. They have short, pointed bills. Typically, the plumage is dark and glossy above, but light below, often with solid areas of bright colours.

Diagram shows the way in which the bills of Darwin's Finches have adapted to fit each of the species for its particular feeding niche in the environment. Vegetarian Finches tend to have parrot-like bills while the warblers have thin pointed bills—examples of adaptive radiation.

Darwin's Finches

The Galapagos Islands, on the Equator, 600 miles west of the South American coast, have a strange and very fascinating fauna. Animals have evolved quickly and because of the lack of development by humans, the evolution is much more obvious than on the major continents where relationships are much more complex. In 1835, Charles Darwin visited the Galapagos and his study of the islands' fauna provided evidence for his theory of the evolution of life.

Among the species that he studied and discovered were the **Darwin's Finches.** There are 13 species, varying in length from 4 in to 8 in, all confined to the Galapagos. They are typically grey-brown, with black males in some species. The feature in which each species differs, is the shape of the bills, which have adapted to fit each species for a particular feeding niche in its environment. Darwin's Finches share a common ancestor but now the six separate, and very different, bill structures can be recognised—Finch-like and varying from thin to very thick; long, pointed and slightly down curved; parrot-like; smaller and parrot-like; long, pointed and quite thick like a Nuthatch; and thin and pointed like a Warbler. In addition, the **Woodpecker Finch** *(Camarhynchos pallidus),* uses cactus spines to probe the bark of trees for insects.

From their internal structure, it is apparent that Darwin's Finches share a common ancestor and yet each species has evolved to fill a particular ecological niche. They have become the world's most outstanding example of adaptive radiation, whereby species of a common root become different and spread out to fill separate niches.

MAMMALS

Mammals are those animals whose young are born alive then fed on milk produced by the mother's mammary glands. Before birth the young mammal (called a *foetus*), must be supported inside the mother during its early development. This is done by the *placenta*, a dense mass of closely interwoven blood vessels attached to the wall of the uterus. This permits an exchange of oxygen, food molecules and other essentials, between the blood systems of mother and foetus without their blood becoming intermixed.

A Cheetah in open grasslands, its natural East African habitat. One of the fastest animals alive, the Cheetah catches its prey through its sheer speed.

After birth, the placenta is discarded and the young are nourished by milk. This secretion of the mammary glands is the one feature characteristic of all mammals. There are other diagnostic features of mammals, though many of them have their exceptions. The rule that mammals give birth to live young is broken by one group, the Monotremes, which lay eggs.

Mammals also have important internal characteristics. They have a four chambered heart, which increases the efficiency of the blood system; and in most species the red blood cells are without a nucleus. Unlike reptiles, the teeth of mammals are differentiated into different kinds (incisors, canines, and grinding molars) each specialised for a particular task. Major groups, even individual species, can often be recognised by the structure of their teeth.

The mammal skeleton has its special features too: a single bone in the lower jaw, two rounded articulations between the skull and vertebral column, a simplified pectoral girdle and many other distinctive features, lacking in other groups of vertebrates.

Perhaps the most highly developed structure is the brain, which in mammals (especially monkeys, apes and whales) reaches a considerable size and complexity. Size alone is not necessarily indicative of a high level of intelligence, but the complex development of those regions of the brain associated with perception and analytical thought certainly reflects considerable mental ability among mammals in general. The brain is only a processing centre for information fed to it by sense organs and in conjunction with an elaborate brain, mammals have very keen senses. Smell—particularly in carnivores—taste and touch are well developed, though often no better than in certain other animals.

Vision is exceptionally acute, with some species being specialised for nocturnal life (by having extremely sensitive eyes), and some (especially grazing animals) have a very wide angle of vision to watch for predators whilst feeding. A few mammals (monkeys and apes) have colour vision, something we humans take for granted which is lacking in most animal groups. Mammalian hearing is also well developed; the bats and whales are even able to use sound patterns to find their way about instead of using their eyes.

The Evolution of Mammals

Many suggestions have been put forward as to why the Dinosaurs became extinct including great catastrophes such as earthquakes, volcanic eruptions and floods. One or more such events is not likely to have been the cause since it would have had to have been universal, as fossil Dinosaurs occur on most continents. Furthermore, some of the marine Dinosaurs such as the Swan-lizards, Plesiosaurs, and the Sea Monitors (Mososaurs), would have a better chance of survival in their marine environment than the land Dinosaurs.

It is much more probable that a number of factors were involved to bring about the total extinction of the Dinosaurs towards the end of the Cretaceous period, some 70 million years ago. Considerable mountain-building occurred during the Cretaceous period, (for example, the Rocky Mountain range in North America), and also, much of the land was flooded at this time. Along with these events were those taking place in the plant kingdom. Flowering plants appeared about 120 million years ago and began to displace the ferns which had provided much of the food of the plant-eating Dinosaurs. This necessitated some change in diet and in the creature's encounters with new substances from the flowering plants, they may have found some to be toxic. The climate changed and there was a general lowering of the temperature and this would have affected the reptiles since their body temperature remains close to that of the external environment. Some evidence exists to lend support to this suggestion as the microstructure of some Dinosaur egg shells has shown that the formation of the shells was interrupted, to begin again after some time had elapsed. This was perhaps the result of a period of low temperature on the reproductive activities of the animal.

Unlike the reptiles, modern birds and mammals are able to maintain and regulate their body temperature and presumably their ancestors were also able to do so, although possibly less efficiently. This ability gave them a better chance for survival under changing climatic conditions. The first birds and mammals appeared in the Jurassic period which began some 195 million years ago. Thus, members of both groups had more than 100 million years of development and evolution through the reign of the Dinosaurs before the latter became extinct.

Among the early reptiles was a group called the Therapsida. These were the mammal-like reptiles and were among the earliest groups of reptiles to expand, both in numbers and types, before even the Dinosaurs were present in large numbers. The

Therapsids possessed one of the chief mammalian characteristics—the differentiation of teeth into incisors, canines and molars.

From the Jurassic period there are fossils of animals that were true mammals. The jaws consist each of a single bone, and the teeth have cusps to help in holding and crushing food. This may have been important in enabling these animals to feed efficiently and in sufficient quantity to maintain their body temperature above that of the surroundings.

All of the animals of this period were small, about the size of present-day rats. There were several groups of animals during this time. The Triconodonts, animals whose molar teeth had three cusps, reached the size of present-day cats and were probably true carnivores. Another group was the Morganucodonts who had a primitive skeleton at the reptilian-mammalian boundary. A little known group was Docodon with a complex cusp pattern on their molar teeth. It has been suggested that there is some relationship with the egg-laying mammals, the Monotremes. The Multituberculates, a specialised but very successful group appear to be the first herbivorous mammals. Their skull and tooth specialisation were similar to those of present-day rodents. They had a long history of about 100 million years and indeed they survived until usurped, mainly by the rodents who came to occupy their ecological niche. The Symmetrodonts had three cusps on their molar teeth arranged in a symmetrical triangle. These animals were probably predaceous and about cat-size. The most important group with respect to evolution was the Pantotheres, small, shrew-like animals that fed mainly on insects. It is generally

The major groups of living mammals are derived from tiny animals that lived during the Cretaceous era. The exact lineages and inter-relationships are obscure and still much debated by scientists and naturalists. Mammals have evolved to live in a great variety of environments—underground, in trees, in water and up in the air. Even those that live on the ground have diversified to avoid competition with each other. The Giraffe, above, for instance, feeds on the tops of bushes and trees because it has become tall enough to do so, whereas the other herbivores amongst which it lives feed on the grasses and vegetation nearer the ground. Several groups of mammals, including the Pilot Whale, below, live as part of the marine fauna but still breathe air.

believed that it includes the ancestors of the higher mammals, the marsupials and the placental mammals.

The Monotremes probably arose from a little known group, Docodon. They are perhaps the most bizarre of living mammals since they still lay eggs, although the young are suckled after hatching. They have such typical mammalian features as hair, milk glands and one lower jaw element, but also have many reptilian features in their skeleton and soft parts. The two living members, the Duckbill Platypus and the Spiny Anteater, both live in the Australian region.

The higher mammals have two main divisions, the marsupials and the placental mammals, and both arose from the Pantotheres. The marsupials probably had world-wide distribution in the late Cretaceous period but became restricted to the Australian region and South America where they are found living today. All the other living mammals are members of a single major group which has been dominant since the Tertiary, the Eutheria, or placental mammals. These animals have a well-developed placenta which permits a long period of gestation so that the young are born at a more advanced stage than the marsupials.

The two orders of mammals living during the Cretaceous period increased to twenty orders, of which there are seventeen orders now with living representatives, including four orders of Marsupials. There are some 3,700 living species in the thirteen orders of high mammals. New animal groups first evolve in a limited area and their further spread may be limited by such barriers as mountains, extremes of climate or oceans. As a result various regions of the world may have characteristic (or endemic) animals which have arisen there and still live in the same region. South America was isolated from the rest of America during the Tertiary period and this resulted in the development of several groups of mammals there—the Anteaters, Armadillos, Sloths and marsupial Opossums. Outside South America, mammals diversified into various forms, many surviving until the present time. The large size and conspicuous appearance of mammals makes them assume importance out of proportion to their numbers, but they have a large biomass. One of the most recently evolved species—Man—has made many changes to the environment as he has spread throughout the world.

Mammals that Lay Eggs

Only two small families of egg-laying mammals survive. They both live in or near Australia and provide further evidence of the fascinating independent evolution of the animals of this isolated region.

The egg-laying mammals are the most primitive kind of mammals, and closely related to the reptiles. They are found only in Australia, Tasmania and New Guinea, areas which have been cut off from the rest of the world for many millions of years as a result of Continental Drift. These egg-laying mammals are the *Monotremata* (Monotremes), a name given to them because their digestive organs and their reproductive systems terminate in a single opening, or monotreme. In this they are like the birds and the reptiles, whereas all other mammals have two openings.

Only two families

The Monotremata form the only order in the subclass Prototheria of the class Mammalia. Today there are only two families in existence, the Duckbilled Platypus *(Ornithorynchus anatinus)*, and the Echidnas—the Spiny Anteater *(Tachyglossus aculeatus)*, and the Long-beaked Echidna *(Zaglossus bruijni)*.

Skeleton fossils have been found dating from the Pleistocene epoch—only about 2 million years ago compared with the 70 million or so that mammals have been known on Earth. But scientists think the Monotremes must have parted from the main stream of mammalian development soon after the first mammals evolved. A study of the brains and skeletons of Monotremes also suggests that they were related in earlier stages to the Pouched Mammals (the Marsupials), and the female Echidnas develop a rudimentary pouch during the breeding season.

Although there are many differences between the Echidnas and the Platypus, there are even more similarities. Both creatures have rib-systems and limbs that are similar to those of reptiles in basic construction. They are also like reptiles in that the female lays eggs. The female Echidna puts her eggs in her marsupial pouch to incubate them until they hatch, while the Platypus makes a nest. When the babies hatch from the eggs they are suckled like mammals, but the animals' mammary glands do not have nipples and are more like modified sweat glands.

None of the adult Monotremes has teeth, but the young Platypus has milk teeth, which are not replaced once they are shed. Both kinds of animals have bills, the Platypus a broad one resembling that of a duck—from which comes its secondary names of Duckbill or Duckmole—and the Echidnas a long, pointed beak more suited to searching out ants. The newly-hatched babies have very short bills that enable them to suck their mothers' milk.

The Platypus is found in south-eastern Australia and Tasmania. The Spiny Anteater lives in Australia and New Guinea, and the Long-beaked Echidna lives only in New Guinea.

The Spiny Anteater

(Tachyglossus aculeatus)

The Echidnas live mostly on a diet of ants, and they do most of their hunting by night, though they are often seen in daytime. The **Spiny Anteater** is a chunky animal, with a long, cylindrical beak-like snout. With this snout it probes for ants, which it catches with its long, sticky tongue. The fully-grown animal is about 46 centimetres long, weighing between 2.5 kg and 5.9 kg. The animal's back is covered with spines about 5 centimetres long, which form its main protection. When disturbed, a Spiny Anteater rapidly digs its vulnerable head and front part into the ground and stays there. It can burrow completely underground in about 9 minutes. The Spiny Anteaters spend their lives in sandy areas where ants and termites are plentiful. They appear to have a life span of anything up to 50 years. There are three varieties of Spiny Anteaters in Australia and one in New Guinea, differing from each other only very slightly.

The **Long-beaked Echidna** *(Zaglossus)* of tropical New Guinea, is a larger animal, the biggest being about 76 centimetres long and weighing about 9 kg. The spines are not so noticeable because the fur is thicker. The animals have longer noses than the Australian species, and three toes as against the five of the Spiny Anteater.

△ The narrow, toothless snout of the Spiney Anteater is clearly visible in this picture. The snout is used as a probe in the search for ants.

◁ A Duckbilled Platypus swimming under water and using its sensitive and unusual beak to turn over stones in search of food.

▽ A baby Echidna in its nest underground and before its protective cover of spines has grown.

▽▽ A Spiny Anteater swimming. Normally a nocturnal hunter the anteater is nonetheless quite active during the day.

The Platypus

(Ornithorynchus anatinus)

The **Duckbilled Platypus** is one of the most extraordinary looking animals, and for many years zoologists who had seen only skins thought they must be fakes. The Platypus is well adapted for an aquatic life, with short, close fur and no external ears. It has webbed feet with strong claws, equally suited to swimming and digging. The tail is broad and flat, to serve as a rudder when diving.

The Platypus digs a burrow in the river bank, with an entrance above or, if the water level has risen, below the waterline. A long, close-fitting tunnel serves to squeeze the water from its fur before it reaches the nesting chamber. Most of its time is spent in the water. The animals float on the surface, then dive with a loud splash of the tail. A flap of skin shields eyes and ears underwater, and the animal hunts by touch. The bill is covered with very soft skin and is extremely sensitive. A dive lasts from one to five minutes and in that time the animal catches small fish, larvae and other water creatures. A fully-grown Platypus measures about 50 centimetres long and can eat 0.9 kg of food in a day.

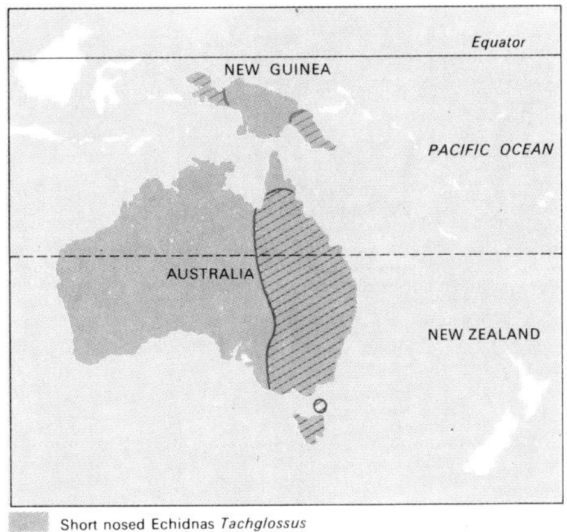

The dangerous male

Male Monotremes possess a pair of offensive weapons—a spur on each hind foot which, in the Platypus, is connected to a poison gland. Nobody knows quite what the purpose of the spurs is, but their venom can cause a painful wound. Some zoologists believe the spurs are used in courtship to inject the female with a substance that makes her more passive, but a female Platypus that was attacked by a male in a zoo nearly died of the poison.

Courtship in the Platypus world begins when the male grips the female by her tail and swims her round in circles. The female, after mating, retires to her burrow to lay her eggs. The nesting chamber is lined with leaves, which the female carries into the burrow clamped under her tail. She stops up the entrance to the burrow with earth before laying and produces up to three eggs—usually two—incubating them for about 10 days. The eggs are soft-skinned, and about 16 millimetres long. The babies when hatched are about 25 millimetres in length, blind and helpless. They remain in the burrow for about four months, by which time they have grown to more than half their adult size.

The future of the egg layers

The Platypus and the Spiny Anteater are both protected animals under Australian law. The Platypus is found mostly in Tasmania, where its numbers are increasing. The Spiny Anteater is found in New Guinea as well as in Australia. Probably the biggest protection for both these animals is that they are harmless and useless to Man.

Studying the Platypus is particularly difficult, because it is largely nocturnal in its habits and extremely shy. Zoologists studying the animals must be prepared to spend many weeks of patient and often unrewarding vigil. A number of pairs have been successfully kept in captivity, where their habits can be more closely observed, but they are not easy to catch.

Echidnas are easier to observe, because they move about a good deal during the day. In consequence they are easier to capture for study in zoos, but like the Platypus, the Echidna does not breed well in captivity.

The first Platypus was found towards the end of the 18th century, but its breeding habits were not finally established until 100 years later. It was thought at first that the eggs hatched inside the body, as in some species of reptiles, but in 1884 the British zoologist, W. H. Caldwell, proved that the eggs were laid and hatched outside the body. A German zoologist made the same discovery about Echidnas.

Pouched Mammals

Like the egg-laying mammals, the Marsupials are a more primitive kind of animal than the rest of the mammals. For the past 100 million years they have been evolving in parallel with other mammals. At one time they formed the main branch of the mammal class, and their fossils have been found in most parts of the world. They now survive in just two areas, the Americas and Australasia.

Scientists believe that competition with placental mammals—that is, animals whose young develop fully in the womb with the aid of a placenta before birth—led to Marsupials dying out in Europe and Asia, and probably in North America too. No fossils have been found in Africa. The Marsupials survived in Australia, New Guinea and South America. When South and North America became linked millions of years ago, the Marsupials spread north again. Today, just one species is known in North America. As a result of this continental link-up, most of the species in South America died out, because of competition from the placental mammals moving down from the north. In Australia relative isolation has allowed the Marsupials to undergo their own private evolution to fill the ecological roles taken up elsewhere by non-Marsupial species.

There are several basic differences, primarily in the female reproductory organs between Marsupials and other types of mammals. They have smaller brains and usually only one set of teeth. Marsupial babies are born at an early stage of development—too young for separate life. They make their way to a 'marsupium', or pouch, on the mother's abdomen; inside the pouch are the mammary glands. The baby siezes a teat in its mouth and retains it there throughout its development period.

The position of the pouch varies according to the species. Those Marsupials that jump and climb trees have the pouch opening forwards. A familiar example is the Kangaroo, whose pouch opens upwards when the animal is standing upright on its back legs. Those Marsupials that spend their time running, walking or burrowing, have the pouch opening towards the rear, so that the pouch presents a streamlined appearance as the animal runs.

There are about 230 species of Marsupials. Zoologists differ on how to classify them, but a simple breakdown gives about half a dozen main types: the insect-eaters and carnivores (flesh-eaters), such as the Tasmanian Devil *(Sarcophilus harrisii)*; the Bandicoots; the American Opossums; the Phalangers; Australian Possums and Cuscus; the Kangaroos and their relatives; and the Marsupial Mole.

▽ Many Marsupials, like this South American Opossum, with its young, do not have pouches and the young, after they have been weaned, ride around on the mother's back, clinging tightly to the fur.

The Opossums

The American Opossum is very different from the Australian Opossum, although both are Marsupials. The American animals belong to the family *Didelphidae*. They are flesh and insect eaters, and have teeth that are suitable for this diet, and not unlike those of the **Tasmanian Wolf** *(Thylacinus cynocephalus)*. There are about 70 species, grouped in several genera.

The best known is the **Virginia Opossum,** or **Common Opossum** *(Didelphis virginiana)*. This little animal, about the size of a domestic Cat, is found from Canada to Argentina. It is the only Marsupial known north of Mexico. It has a grizzled or black coat, a white, pointed snout, and a long prehensile tail which is of great value as it climbs among trees. Although the Virginia Opossum prefers a tree-borne life, it also lives in plains regions, where it burrows into the ground. Litters, ranging from one a year, in cool areas, to three in warm ones, contain about nine young, each less than 12 mm in length.

▽ △ The Native Cat, about the same size as the common domestic cat though longer in the tail, likes to spend much of its time in trees.

Opossums are nocturnal animals. They have omnivorous appetites, eating insects, small animals, carrion, fruit and grain with equal zest. When attacked, the Virginia Opossum feigns death, from which comes the phrase 'playing possum'. When the babies are weaned, they ride around on their mother's back, clinging to her fur.

The other species live mostly in South and Central America, but they are of two basic types, one similar to the Virginia Opossum and the other similar in size to Rats. The most notable of these species is the **Water Opossum** *(Chironectes minimus)*, found in Brazil and Guatemala. It is the only Marsupial that leads an aquatic life, and its habits are similar to those of an **Otter** *(Lutra)*. It has webbed feet and burrows into river banks. Other kinds include the **Four-eyed Opossum** *(Philander)*, named for the white spots over its eyes; the **Mouse Opossum** *(Marmosa)*, which has no pouch; and the **Short-tailed Opossums** *(Peramys* and *Monodelphis)*.

▽ A Tasmanian Devil, once found on the mainland of Australia but now living only on the island from which it takes its name. It is a sturdy, fierce, hunter of small mammals, though it will eat carrion, emerging from its daytime hiding place amongst the underbush to hunt at night.

The Forest Shrew

Deep in the thick undergrowth of the Andean forests of South America lives a little-known pouched mammal which is the Marsupial version of a Shrew. It is known variously as the **Selva,** the **Opossum Rat,** or the **Caenolestid.** It belongs to the family *Caenolestidae*, and there are three known genera—*Caenolestes,* living in Bolivia and Ecuador, *Rhyncholestes,* in Chile, and *Lestoros,* in Peru.

They have long heads with rounded ears, and long tails. The animals are nocturnal, and feed on a mixed carnivorous diet of birds, eggs, small mammals, spiders and insects. They are particularly interesting to zoologists because their teeth are similar to those of the Phalangers.

Marsupial Wolves and Devils

Of all the Marsupials, two species are renowned for their ferocity—the **Tasmanian Wolf** *(Thylacinus cynocephalus)*, also called the **Thylacine** or **Tasmanian Tiger;** and the **Tasmanian Devil** *(Sarcophilus harrisii)*. They both belong to the family *Dasyuridae*, a group of flesh and insect eaters. The Tasmanian Wolf is the largest of this family. It resembles a large Dog about 1.5 metres long, with a reddish-brown coat barred on the back—from which comes the 'Tiger' nickname. It hunts by scent, loping tirelessly along a trail harrying its quarry to the point of exhaustion. Tasmanian Wolves have been known to bound on their hind legs like Kangaroos. A litter is usually two to four cubs. This animal is now virtually extinct; its only habitat is Tasmania where it has been driven into territory so inaccessible that recent expeditions have found only tracks.

The Tasmanian Devil has wiry whiskers, a bear-like face, and a fierce snarl. It is stocky in build, up to 1 metre long, with a black coat marked with white. It will attack anything in its search for food, and will also eat carrion. The babies, never more than four, are less than 12 mm long at birth. When they emerge from the pouch they live in a nest which their parents build. They live only in Tasmania, but fossils have been found on the mainland.

Marsupial Cats and Rats

Before Marsupials were fully understood, many species were given names suggestive of their similarity to placental mammals, such as wolves or cats. The **Marsupial Cat** is better known by its native name of **Quoll.** There are several species, living in Australia and New Guinea. The **Eastern Native Cat** *(Dasyurus quoll)*, is the size of a domestic Cat and has a brown or grey coat marked with white spots. It spends much of its time in trees. Other Quolls, also with spotted coats, include the much larger **Tiger Quoll** *(Dasyurus maculatus)*, also known as the **Spotted-tailed Native Cat.** It is almost 1.2 metres in length, two-fifths of which is tail.

There are also Marsupial 'Rats' which are similar in appearance to their rodent equivalents but are mostly carnivorous. An agile tree-dweller to be found in the bush, the **Tuan** or **Brush-tailed Phascogale** *(Phascogale tapoa-*

tafa), is a speedy little hunter with a long, black bushy tail and a bluish-grey coat. It eats insects, small mammals, and honey, and makes its home in hollow trees and logs.

Marsupial Mice

Marsupial Mice are small, attractive-looking little animals, resembling their placental namesakes, the true Mice, in their quick movements. They are related to the Marsupial carnivores of the family *Dasyuridae*. They have only rudimentary pouches; like several other Marsupials, mother Mice carry their young on their backs once they are too large for the pouch.

Some species, such as *Dasyuroides byrnei* and the **Mulgara** or **Crest-tailed Marsupial Mouse** *(Dasycercus cristicauda)*, live in deserts. The Mulgara gets all the moisture it needs from the insects on which it feeds, and does not need drinking water. The **Flat-headed Marsupial Mouse** *(Planigale)*, lives both in woods and in open country. The daintiest of the Marsupial Mice are the **Narrow-footed Mice** *(Sminthopsis)*. They are pale grey or fawn with white underparts, with erect ears and large, staring eyes. One species is the **Fat-tailed Sminthopsis** or **Dunnart** *(Sminthopsis crassicaudata)*, which has a short, plump tail that swells up as a reserve of fat when the animal has plenty of food.

The struggle to survive

The arrival of Europeans in Australia in 1788 onwards has seriously disturbed the balance of nature, especially through hunting and the clearance of large areas for raising Sheep. The Sheep compete for food with the Marsupials, and as a result many species are in danger. Some have already disappeared.

The Tasmanian Wolf is on the verge of extinction. It has been a protected species for almost 40 years, but none has been seen for some time. It is thought a few may survive in the remote mountain woodlands of western Tasmania. The Bilby, once common over a large area of Australia, survives only in remote regions. Competition from Rabbits led to their decline. Koalas, once slaughtered by the million are now protected, and survive in reasonable numbers in the eastern part of Australia.

The Kangaroo family is having varied fortune. The larger species are surviving well, but some of the smaller Wallabies are becoming rare, and at least two species are thought to be extinct. The **Bridled Nail-tail Wallaby** *(Onychogalea fraenata)* has not been seen for many years; but several Marsupials once feared extinct, such as Leadbeater's Possum, have been rediscovered in remote areas.

The Tasmanian Wolf, a protected animal for four decades but now considered to be extinct though it is possible a few may survive in remote mountain regions.

Marsupial insect-eaters

The fallen timbers of south-western Australia's eucalyptus forests form a splendid breeding-place for Termites, and these forests are consequently the home of the **Banded Anteater** or **Numbat** *(Myrmecobius fasciatus)*. The Numbat is a small, sharp-eared, sharp-nosed animal, about 432 mm long—of which nearly half is tail. The animal has a brownish-grey coat, banded with transverse white stripes on the back. It digs out the Termites with its sharp claws and licks them up with its long, sticky tongue. It hunts only by day. The Numbat has no pouch, and the young hang by their mouths from their mother's teats. There is one other, similar species, the **Rusty Numbat** *(Myrmecobius rufus)*.

A very different kind of insect-eater is the **Marsupial Mole** *(Notoryctes)*. It is similar to the true Mole in its burrowing habits, but whereas the Mole digs a long, continuous burrow, the Marsupial Mole digs its way through the soil for several feet at a time, closing its tunnel behind it, and surfacing every so often. The Marsupial Mole, usually pale in colour, has a horny shield on its nose and two huge claws on each forefoot to help it dig. The **Bandicoots** *(Peramelidae)* comprise about 20 species in seven genera. The animals range in size from around 380–760 mm long, including their tails. They hunt through the bush and in the eucalyptus forests for worms and insects. Bandicoots make a noise something between a sneeze and a squeak. The two main types are the **Long-Nosed Bandicoots** *(Perameles)*, and the **Short-Nosed Bandicoots** *(Isoodon)*. *Perameles* has soft fur and long ears. The prettiest is probably the **Tasmanian Barred Bandicoot** *(Perameles gunni)*, whose light grey fur is marked on the rump with dark bands. *Isoodon* has a wiry coat. The **Bilbies** *(Thylacomys)*, have long ears like Rabbits, from which comes their other name, **Rabbit-Eared Bandicoots.** They also have a long tail and long pointed nose. There are also some **Pig-Footed Bandicoots** *(Choeropus)*.

Marsupial Fruit-eaters

The **Phalangers** *(Phalangeridae)* are the most widely distributed of the Marsupials; various members of the family live in Australia, New Guinea, Timor, the Celebes and the Solomon Islands. Most of them live in trees, eating leaves and fruit. The big toe is opposable, like a thumb, which gives the animals a good grip for climbing. The Phalangers are generally known as **Possums** in Australia, though this is not true of all of the more than 40 species.

The **Common Phalanger** or **Brush-tailed Possum** *(Trichosurus vulpecula)*, is more than 600 mm long. It has a bushy, prehensile tail, and nests in eucalyptus. At the other end of the scale come such species as **Leadbeater's Possum** *(Gymnobelideus leadbeateri)*, with a body length of under 150 mm and which leaps rapidly from branch to branch. The **Striped Possum** *(Dactylopsila trivirgata)*, of New Guinea and northern Queensland has a striking black and white striped coat, and a long fourth toe on each front foot which it uses for winkling grubs out of crevices.

The largest Phalangers are the **Cuscuses** *(Phalanger)*, which can grow to more than 1 m long. Their ears are so small as to be invisible under the thick fur, and the prehensile tails are rough to give them a good grip.

Flying Phalangers

There are five species of Flying Phalangers. The largest is the **Greater Glider** *(Schoinobates volans)*, just over one metre long including the tail. Like all the Gliders, they have a membrane extending from elbow to ankle which gives them a greater gliding surface. They can glide for a distance of up to 90 metres, swerving through the trees as they go. The smallest is the **Pygmy Glider** *(Acrobates pygmaeus)*, which is only 150 mm long including the tail. The other species are the **Sugar Glider** *(Petaurus breviceps)*, **Squirrel Glider** *(Petaurus norfolcencis)*, and **Yellow-Bellied Glider** *(Petaurus australis)*.

Ring-tailed Possums

(Pseudocheirus)

The **Ring-Tailed Possums** *(Pseudocheirus)*, get their name from the fact that their tails are usually curled up at the tip. They build nests of twigs and leaves, in a thick shrub or in the fork of a tree. They are found mainly in high country in dense woodland, though one species, the **Common Ringtail** of Queensland *(Pseudocheirus peregrinus)*, lives in comparatively open country.

△▽ Glider Possums, the Greater, above, and the Pigmy, below, which can glide long distances from tree to tree. They spread flaps of skin out from the sides of their bodies to make gliding surfaces and the Greater uses its long bushy tail to help stabilise itself in 'flight' while the Pigmy uses its flattened tail as a rudder. They are the Marsupial equivalent of Gliding Squirrels.

The Koala

(Phascolarctos cinereus)

The **Koala** *(Phascolarctos cinereus)*, looks more like a teddy bear than any member of the true Bear family. They are not as pleasant-natured and affectionate as they look, and they can defend themselves with their long claws if attacked. They are small animals, the largest weighing about 7 kg, yet an adult can put away more than 1 kg of food a day.

Koalas spend all their lives in eucalyptus trees, and prefer only five out of the 350 species of eucalyptus—though there are about 15 others whose leaves they will accept. They even reject their favourite trees, the manna eucalyptus, at certain times of the year, and researchers found that at such times the trees were secreting prussic acid. The animals themselves smell strongly of eucalyptus oil.

Unlike other tree-living Marsupials, female Koalas have their pouches opening to the rear. They produce only one offspring per year. The baby spends six months in its mother's pouch. Koalas, greatly reduced in numbers by hunters, are now protected.

Wombats

(Vombatidae)

Wombats *(Vombatidae)* are hairy, chunky animals that have been compared to animated bulldozers. They grow up to 1 metre long, and can reach a weight of 36 kg. They are often called **Badgers,** because of their resemblance to a European Badger.

The Wombats have teeth like rodents, that continue to grow and are worn down by hard chewing. They have an exclusively vegetarian diet of grass, bark, roots and fungi. Like the Badger, the Wombat digs itself a large burrow, with a maze of tunnels up to 12 metres long. A Wombat can dig as fast as a man with a spade. Generally Wombats live solitary lives except in the mating season, but often their burrows are found to be interconnecting. They have rear-facing pouches—essential for burrowing animals.

There are two main kinds of Wombats. The **Common Wombat** *(Vombatus hirsutus)*, has a hairless patch on its muzzle, and lives in south-eastern Australia.

Its relative, the **Tasmanian Wombat** *(Vombatus ursinus)*, has a very bristly coat and lives in Tasmania. The other kind of Wombat is the **Hairy-Nosed Wombat,** of which one species, *Wombatula gillespiei*, lives in southern Queensland, and another, *Lasiorhinus latifrons,* principally in South Australia, and in Queensland. These Wombats have hairy muzzles.

△ The Koala, once persecuted for its skin is now rigorously protected. Its loveable appearance belies its savage temperament and it will use its long claws to protect itself if attacked. It is the largest of the Phalangers and the only one without a tail.
▽ A Wombat, close to the European Badger in size and sometimes in fact called a Badger. Its habits too are not dissimilar for the Wombat also is a great burrower, digging itself a maze of tunnels in which to pursue its solitary life, a life-style changed only during the mating season. Like all burrowing Marsupials the Wombat has a rear-facing pouch.
▽▽ The prehensile tailed Cuscus.

The Kangaroos

(Macropodidae)

The **Kangaroo** family *(Macropodidae)* contains the largest, and also some very small, members of the Marsupial order. There are about 50 main species in the family. The principal characteristic of the Kangaroos is that they progress by hopping on their greatly elongated back legs, which can take them bounding over the ground at very high speed. The front legs are very much shorter than the back legs. The tail is long and heavy, and acts as a third support when the animal is sitting erect on its hind legs. When fighting or defending itself a Kangaroo can balance on its tail and deliver a powerful blow with its hind legs, not unlike a drop-kick in wrestling.

The Kangaroos have forward-opening pouches, containing four teats. Usually there is only one young, called a 'Joey', at a time. The youngster spends up to six months in the pouch, though in the later stages it makes many hunting expeditions for food.

Members of the Kangaroo family are found in Australia, New Guinea and the islands of the Bismarck Archipelago. There are three main kinds: true Kangaroos and **Wallabies** (sub-family *Macropodinae)*; **Rat Kangaroos** (sub-family *Potorinae*); and **Musk Kangaroos** (sub-family *Hypsiprymnodontinae)*.

There are three main species of true Kangaroos. The largest of all is the **Red Kangaroo** *(Macropus rufus)*, which has a body length of 1.5 metres plus a tail which may be over 1 m long. It lives mainly in the wide open spaces of southern and eastern Australia, where it can travel along at speeds over 30 mph (50 kph) for short distances. As a rule, only the males are red. In the mating season, the skin on their chests exudes a reddish colouring which they smear over their heads and backs. They are often called **Red Fliers,** while the females, which are blue-grey, are called **Blue Fliers.**

The **Grey Kangaroo** *(Macropus giganteus)*, is nearly as large as the Red Kangaroo. It lives in grassy plains and open woodlands, and is often called a **Boomer.** It has a black tip to its tail. The shade of grey varies, and the Grey Kangaroos of Kangaroo Island, off South Australia, are chocolate brown.

The third species is the **Walleroo** *(Macropus robustus)*, also known as the Mountain Kangaroo or Euro. It lives in rocky, hilly areas and is grey, black, or dark red in colour. It has black feet.

Wallabies

(Wallabia)

The larger **Wallabies** belong to the genus *Wallabia*, though sometimes they are classified with the true Kangaroos *(Macropus)*. They range from about half the size of a Kangaroo downwards. The commonest is the **Red-necked Wallaby** *(Wallabia rufogrisea)*. It lives in open woodlands and thick forests. The largest of the group is the **Pretty-face Wallaby** *(Wallabia parryi)*, which lives in hilly regions with open woodland.

The **Rock Wallabies** *(Petrogale)*, live mostly in South Australia. They bound over rough terrain with effortless ease. One of the most colourful is the **Yellow-footed** or **Ring-tailed Rock Wallaby** *(Petrogale xanthopus)*. In New Guinea the prevailing Wallabies are genus *Dorcopsis*, smallish animals with short feet and tails. One of the more unusual Wallabies is the **Nail-tailed Wallaby** *(Onychogalea)*, which has a horny tip to its tail, and swings its arms round and round when bounding along.

The **Pademelons** or **Scrub Wallabies** *(Thylogale)* are similar to *Dorcopsis*, and live in forests in Australia and many of the islands. There are many other species, some very small indeed. These include the **Rat-Kangaroos** (sub-family *Potoroinae)*, often only 60 to 70 cm long, including their relatively stubby tail.

Marsupials of long ago

In the Mesozoic Era Marsupials were a dominant group and evidence suggests that they were spread throughout all of the existing continents, their size, speed and elusiveness enabling them to survive amongst the great Reptiles. They began to disappear in Europe near the end of the Cretaceous Period—about 80 million years ago—at about the same time as the Dinosaurs died out. Their survival and development in Australia are due to that country's isolation from the rest of the land-masses. They survived in South America because that continent was also comparatively isolated.

Among the many Marsupials now known only as fossils, was a sabre-toothed animal *(Thylacosmilus)*, not unlike the **Sabre-Tooth Tiger,** a placental fossil species. There was also a Marsupial **'Lion'** *(Thylacoleo)*. These animals died out comparatively recently, a little over 25,000 years ago.

Many of the fossil Marsupials were much larger than present day animals of similar types. For example, at the time when the Ice Age was gripping Europe a **Giant Wombat** *(Phascolonus)*, the size of a Black Bear, stalked the Australian forests. There were several species of giant Kangaroos, some more than 3 m tall, and *Diprotodon* was a large, heavily-built herbivore, similar in build to a Rhinoceros.

△△ A pair of Kangaroos take off in a characteristic hop, driving off their powerful back legs and using their long, heavy tail to maintain balance. Kangaroos live in open grassland and sparse bush areas and, along with the Wallabies, they are the large fast-moving Marsupial equivalents of the Horse and Antelope absent from Australia.
△ A Scrub Wallaby, one of fifty species of Wallaby.

Capricorn

Grey Kangaroo *Macropus giganteus* and *fuliginosus*
Red Kangaroo *Magaleia rufa*
Antilopine kangaroo *Macropus antilopinus*

The Tree Kangaroos

(Dendrolagus)

Up in the rain forests of New Guinea and north-eastern Queensland lives one group of Kangaroos that climbs trees. They are the **Tree Kangaroos** *(Dendrolagus)*, actually small Wallabies adapted for an arboreal life. The largest species are around 5 ft (1.5 m) long, including the tail, which is half the length. The fore and hind legs are less unequal in length than those of other Wallabies, and their paws have long claws for gripping branches. Tree Kangaroos leap fearlessly from tree to tree, and apparently think nothing of falling 15 m or more to the ground.

The **Musk Kangaroo** *(Hypsiprymnodon moschatus)* lives on the ground of the same rain forests. It is the most primitive member of the Kangaroo family, and has some characteristics of both the Kangaroos and the Phalangers *(Phalangeridae)*. It hunts by day, eating insects, worms, roots and berries.

Birth of a Joey

A Joey—a baby Kangaroo—is born at a very early stage of development. A Red Kangaroo female weighs up to 29 kg, yet at birth, the Joey weighs less than 1 g, and is 19 mm long. This tiny entity crawls through its mother's fur and into the pouch, which the mother has previously licked clean. Once in the pouch the baby selects one of the four teats and takes it into its mouth. It is too feeble to suck, but contractions of the mother's muscles squirt milk into its mouth.

At this stage the Joey has been developing in the womb for only about 35 days. It is still in an embryonic stage, with ears and eyes not yet developed, and no fur. It spends about 33 weeks in the pouch, during which its weight may increase to as much as 3.63 kg.

A Kangaroo produces only one offspring at a time. This could lead to serious depopulation during times of drought, when three-quarters of the Joeys that have recently left the pouch die. But soon after giving birth, a female Kangaroo mates and conceives again. The new embryo develops a little way and then stops. It is kept in store, as it were, for emergencies. Development resumes as soon as the Joey becomes independent or dies, and a fresh Joey is born soon after.

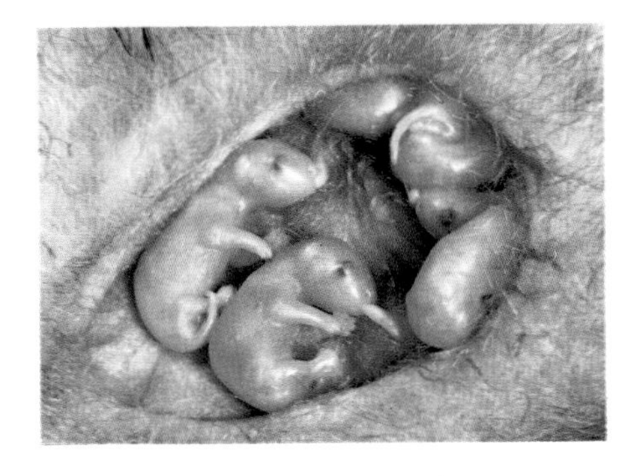

△ A Kangaroo with its Joey, born at an extremely early stage of its development and, at the time of its birth, tiny. It crawls up through its mother's fur until it reaches the pouch where it attaches itself to a teat. It remains there for two months before venturing out of the pouch in search of food. After about nine months it has developed sufficiently to go off on its own.
◁ Baby Opossums attached to their mother's teats. Even at this early stage they have had to make their own way to the pouch. In Opossums this is just a fold of skin, leaving the young partially exposed, whereas in Kangaroos the pouch forms a deep pocket completely enclosing the Joey which is always born singly.

The Insect Eaters

Few of the animals called insect-eaters live entirely on insects. Some eat other invertebrates, some eat fruit and plants, others eat fish and frogs. But although insectivores are mostly small-sized, they include some very successful mammals.

The order of Insectivora is very widespread; some of the 400 species are to be found in every part of the world except Australia, Antarctica, the Polar region and the southern part of South America. Most insectivores are small and this order has the distinction of containing the world's smallest mammal, the Etruscan Shrew.

The order Insectivora is a kind of zoological dustbin into which various odd animals have been put because they do not belong to any other order of mammals. Consequently, members of the Insectivora have few features in common except perhaps a relatively unspecialised dentition. Most have an unspecialised skeleton although Moles are very highly adapted for burrowing. Most have very poor eyesight although the eyes of Elephant Shrews are quite big. Many are spiny, (e.g. Hedgehogs and Tenrecs), and indeed, most of the spiny species of mammals are insectivores; but the majority of insectivores have normal fur and some (Desmans and Golden Moles), have beautifully soft and iridescent coats. The Shrews are one of the most widespread of all mammalian families, yet other insectivores, like Solenodons and Tenrecs, are limited to single islands. Altogether, the Insectivora constitute a very mixed group. They have some affinities with the primates (through the Tree Shrews which possess many features of both orders), and share certain characteristics with the bats. In many ways, they are more similar to some groups of very ancient mammals, long extinct, than to other living forms.

In outward appearance, many of the insectivores may resemble rodents, but unlike them, insectivores have unspecialised teeth geared to biting and tearing rather than gnawing. Some species have a musky odour and their flesh tastes unpleasant which make them unpalatable, but otherwise, apart from the spiny animals such as Hedgehogs, Insectivora have little defence against predators. Most of them are nocturnal and some, like the Mole, live in burrows for protection. Some species of insect-eaters rely on their agility and speed to escape from their enemies and for this reason, use up an enormous amount of energy. Some Shrews need to eat their own bodyweight of food each day. When frightened, their metabolic rate increases and their hearts may beat at more than 1,000 times a minute. Understandably, the creatures can easily die of fright.

Insectivores usually have poor eyesight and rely on their sense of smell and, to a lesser extent, their hearing, to track down food. Although they are primitive, and one of the earliest types of mammal to evolve, many are highly successful because they have adapted so well to their different habitats. By convergence, some species have grown to resemble quite different animals; the Great African Water Shrew, for instance, looks very like an Otter. Others are so specialised that they have a most curious look, such as the Desmans, swimming Moles which live in the banks of streams, or the extraordinary Elephant Shrews which have trunk-like noses and seem to bounce along the ground, like rubber balls.

△ The Aardvark *(Orycteropus caffer)* the only species in the mammalian order Tubulidentata and once classed with the Edentates. The name of the Aardvark's order refers to the only teeth they possess, molars made up of masses of minute closely packed, vertical tubes. They are rootless and quite different from those of any other kind of mammal. The Aardvark, the name is Afrikaans for 'earth pig', is about the same size as a pig and has pig-like bristles but donkey-like ears. Its tail is long and heavy and its fore limbs and claws are particularly powerful for tearing open termite nests. It lives on termites and ants and it has a long narrow mouth, wide nostrils and a thin tongue that can extend up to 18 inches. The Aardvark is found only in Africa, as far north as Ethiopia and down into South Africa. It is a nocturnal, burrowing animal affording a good example of the way in which the structure of animals of similar diet or method of moving about can converge.

The Fate of the Solenodons

The insectivore order includes three very rare animals which are on the World Wildlife Fund's red list of endangered species. Two of them, the Cuban Solenodon *(Solenodon cubanus)*, and the Haitian Solenodon *(Solenodon paradoxus)*, are the only surviving members of the family Solenodontidae which are now confined to remote brushland and forest areas of their native West Indian islands. In appearance Solenodons resemble large, fat Rats about 12 in (30 cm) long. Both species are usually dark brown in colour, though the Haitian has coarser fur than the Cuban Solenodon, and is distinguished by a white spot on the back of the neck. They have long, bewhiskered snouts and large curved claws used for tearing into rotten wood in search of insects. They waddle around on their toes with a curious 'drunken' motion and are so clumsy that they often trip over their own feet when pursued. As well as insects they eat Lizards, Frogs and birds and may poison their prey with toxic saliva produced by a gland at the base of a tooth. Solenodons are vulnerable creatures and have survived only because they have had few natural enemies. With more of the land coming under cultivation, they are beginning to suffer the depredations of imported predators such as Dogs, Cats and Mongooses. Only one or two young are born, twice a year, too few to replace the declining numbers, although it is hoped the establishment of reserves may help to conserve them.

The third endangered species, the Pyrenean Desman, is found only in a few mountainous areas of the French Pyrenees, Spain and Portugal. This swimming Mole is threatened not by predators but by pollution of the streams in which it lives. Fortunately, attempts are now being made to protect its habitat from further damage.

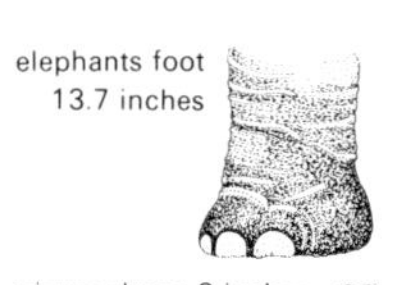

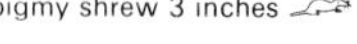

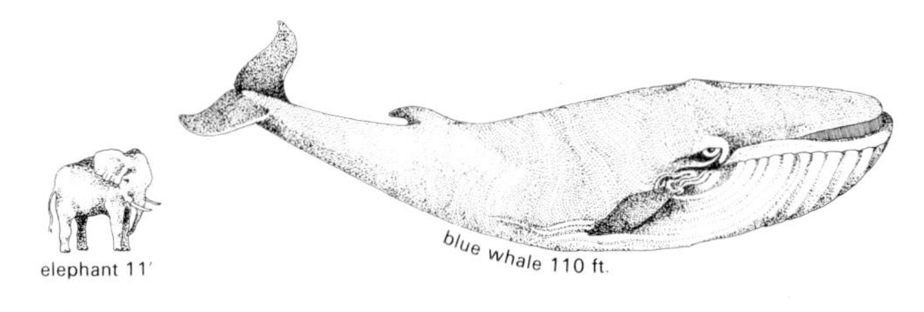

◁△△ The Spiney Tenrec, one of the larger species of this group of Insectivores found only on the island of Madagascar where they are the only insect eaters.

△ A European Hedgehog, found as far north as the limit of deciduous trees in Europe, east into Russia and successfully introduced into New Zealand at the end of the 19th century.

△△ The vast range of mammal sizes can be clearly seen from the scale drawing of the Pigmy Shrew, one of the smallest, and the foot of the Elephant, the largest land mammal, dwarfed in turn by the Blue Whale, the largest of all the Mammals.

Tenrecs

(Tenrecidae)

Madagascar, the large island off the South-east coast of Africa, is the only home of the Tenrecs. These are the only insect eaters on the island, but they have developed in many different ways so that some resemble Hedgehogs and others Moles or Water-Voles. Among the spiny members of the family, the **Common Tenrec** *(Tenrec ecaudatus)* is the largest, measuring 10 to 15 in (25 cm to 40 cm) long and covered in a mixture of hair and spines. The Common Tenrec hibernates in a burrow throughout the six-month dry season. It has also the distinction of producing the largest families among the mammals. Fourteen young in a litter is average and 36 has been recorded. The **Madagascar Hedgehog** *(Setifer setosus)*, has developed spines like a European Hedgehog and can roll itself up in a ball in the same way. The **Rice Tenrec** *(Oryzorictes hova)* looks rather like a Mole with its pointed nose, velvety fur and large digging forefeet which it uses to burrow in the damp banks of rice paddies. The rare **Web-Footed Tenrec** *(Limnogale mergulus)* is adapted for an aquatic life and swims in search of fish and water-plants, using its powerful tail and webbed hind feet.

The only relations of the Tenrecs in Africa are the giant **African Water Shrews** or **Otter Shrews** *(Potamogalidae)* of the west and central areas. The **Giant Water Shrew** *(Potamogale velox)* measures more than 24 inches (60 cm) including the tail, and looks very like an Otter. Like its cousin, the **Small Water Shrew** *(Micropotamogale)*, it lives in burrows in river banks with entrance holes below the water level and comes out in the evening to feed on fish, Crabs and Frogs.

Also distantly related to the Tenrecs are the rare **Solenodons** of the West Indies.

Hedgehogs

(Erinaceidae)

The Hedgehogs, and their relations, the **Moon Rats** or **Gymnures,** are found in Africa, Europe and Asia. There are none in the New World. Characteristic of Hedgehogs is their thick covering of spines. Although other mammals (Porcupines) are spiny, in the Hedgehogs these modified hairs reach a greater degree of complexity than in any other groups. There are also modifications in the skin musculature and skeleton which permit these animals to roll up into a ball, entirely enclosed in their spiny coat. In cooler regions, Hedgehogs retreat to burrows or well-built nests and hibernate over winter, consuming the enormous stores of body fat laid down the previous summer. In hot dry regions, (e.g. Middle East), the desert Hedgehogs go torpid during the dry season.

The three genera of Hedgehogs are *Hemiechinus*, (the long-eared Hedgehog of desert regions of the Middle East and India); *Paraechinus*, (widely distributed in Africa) and *Erinaceus* (the European Hedgehog which was also introduced into New Zealand in the 1890's).

Erinaceus europaeus is common in Britain except on moorland, mountains and marshland. The young are born in litters of about four, with soft, white spines that harden after two or three weeks. They live in leaf nests, burrows or crevices.

Moon Rats or **Gymnures** *(Echinosorex gymnurus)* are members of the Hedgehog family but are hairy rather than spiny and they have long, tapering tails, not found in other members of the family. They rely for protection on their highly unpleasant smell, which has been described as being like rotten meat. They are found in the damp forests of South East Asia, are semi-aquatic and eat fish and Frogs as well as insects. Equally smelly is the lesser **Gymnure** *(Hylomys suillus)*, which resembles a large Shrew and is found in China and South East Asia, and has a short, stubby tail.

Golden Moles

(Chrysochloridae)

Another very intriguing family of African insectivores are the **Golden Moles** *(Chrysochloridae)*, several species of which are found from east and central Africa to the Cape. Like real Moles, they have very reduced eyes, live in burrows and eat Beetles, grubs and worms. But their unique feature is the peculiar metallic sheen of their fur—a gold, green or even purple iridescence, to which they owe their name.

They have enlarged claws on the forefeet to aid digging but also a horny pad on the nose, used to bulldoze soil aside. True Moles lack this pad and do not use the nose in digging. Golden Moles have small families, usually only two young at a time, born in a nest built in the complex burrow system. They are small animals, 4–8 inches (10–20 cm) long and very squat and chubby in appearance, with no ears or tail visible externally, the whole being covered by the long dense fur.

Elephant Shrews

(Macroscelididae)

The **Elephant Shrews** are named not only for their comparatively large size—some species are over 12 inches (30 cm) long—but because of their trunk-like snouts, flexible and sensitive, which help them sniff out their food. They move not by running but by jumping along with a bouncing movement and sometimes they will hop along on their hind legs rather in the same way as Kangaroos. There are about 18 species of Elephant Shrew living in southern and central Africa and, unlike most insectivores, they are mainly diurnal and love basking in the sun. They live in burrows they dig themselves, or take over from rodents. Large species eat mainly beetles, smaller ones, ants and termites which they root out with their very mobile snouts. Only one or two young are born at a time, with fur and open eyes and they can move about almost from birth. They are mature in about four weeks and have a lifespan of under two years.

Shrews

(Soricidae)

The Shrews are the most numerous of the insectivores. There are over 200 different species distributed over most parts of the world except for Australasia, the Polar region, the West Indies and the southern part of South America. The **Common Shrew** *(Sorex araneus)*, the **Pygmy Shrew** *(Sorex minutus)* and the **Water Shrew** *(Neomys fodiens)* are found in the British Isles. The two main sub-families of Shrew are distinguished by the colour of their teeth; the red-toothed *Soricinae* and the white-toothed *Crocidurinae*. The third sub-family are the armoured Shrews *(Scutisoricinacae)* of Africa which are characterised by their very strong backbones—reported to be able to stand the weight of a man without breaking.

In appearance, Shrews somewhat resemble Mice, with long pointed muzzles which are constantly on the move, sniffing and searching for food. They eat mainly insects and other small invertebrates and some species have a mildly poisonous saliva with which they immobilise their prey. In temperate climates, Shrews breed from spring to autumn but in tropical zones they breed all the year round, producing litters of between 2 and 10. The young are born hairless and blind, in a nest of leaves or grass. They are usually independent by the age of four weeks. Most Shrews have a short life-span of 18 months or less. Their prodigious appetite makes them invaluable in controlling insect pests and their highly developed hearing and smell compensate for their poor eyesight in finding food.

Shrews lead a solitary life except in the breeding season and are nervous and aggressive animals. An angry Shrew will stand on its hind legs and chatter with rage.

Some Shrews live above the ground, but most live in burrow systems in loose soil, leaf litter and thick vegetation. Some Water Shrews are specially adapted for an aquatic life, with feet either webbed or fringed with stiff hairs, to aid swimming. One **American Shrew** *(Sorex palustris)*, can appear to run across the water, using the air bubbles trapped by these foot hairs, as floats.

The tiny Mexican **Desert Shrew** *(Notiosorex crawfordii)*, is small enough to get into a beehive using the entrance made by the bees. Smallest of all, and the smallest mammal in the world, is the **Etruscan Shrew** *(Suncus etruscus)*

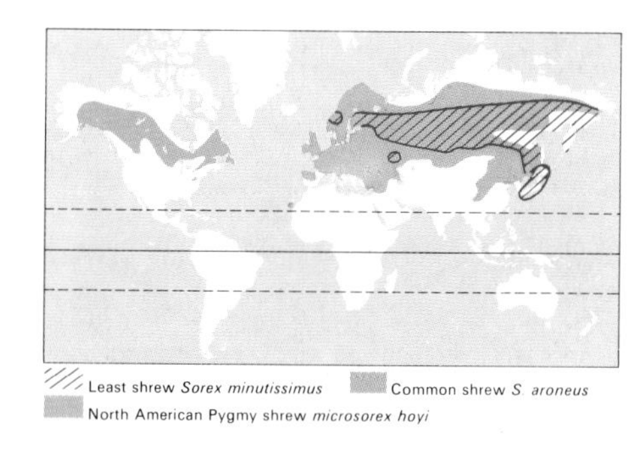

▽ The Elephant Shrew found in Africa and different from ordinary Shrews, especially in having very large eyes and ears. Internal features suggest that Elephant Shrews may be more closely related to Tree Shrews than to other insectivores.

from Mediterranean countries and North Africa. This minute creature weighs only 2 g (about $\frac{1}{15}$ of an ounce) and measures about $1\frac{1}{2}$ in plus 1 in for the tail.

Shrew-moles and Desmans

(Talpidae)

Shrew-Moles are members of the Mole family which are not so strongly modified for a subterranean life as the true moles. The **Asiatic Shrew-Mole** *(Uropsilus soricipes)* has a long tail and small claws, living above ground in damp forests in remote areas of the Far East and South East Asia. It roots among fallen leaves for insects and small invertebrates. The **Japanese Shrew-Moles** *(Urotrichus)* look like very furry Shrews and live partly above ground, though they also make burrows. The little **American Shrew-Mole** *(Neurotrichus gibbsi)*, has soft fur growing backwards, combined with the big head and large teeth of a Mole, and lives both above and below ground.

Another curious member of the Mole family is the **Russian Desman** and its cousin the rare **Pyrenean Desman** *(Galemys pyrenaicus)*. Desmans are water dwellers, feeding on fish, crustaceans and frogs in ponds and streams. They have long, flat channelled flexible snouts to sniff out their food, and dense, plushy coats like an Otter. Russian Desmans are trapped for their beautiful russet fur. They are good swimmers, using their hair-fringed, partly webbed feet and strong flattened tails to propel them, and live in burrows in the banks of streams.

△△ The large ears and the different shape of the head are two of the more obvious ways in which the White Toothed Shrew differs from other Shrews, but, as its name suggests, the principal difference is in the colour of the teeth, which for the major group of Shrews are red. White Toothed Shrews are widely distributed in the old world.
△ The powerful shovel-like fore-limbs of the European Mole and the family's unique velvety fur, which can be stroked in all directions and allows the Mole to move backwards through its tight-fitting burrows, can be clearly seen in this picture. The Mole's naked snout is covered with very delicate sensory organs to detect worms and other prey and to compensate for poor eyesight and hearing.

Moles, the excavators

(Talpidae)

The molehills which are such a familiar sight in the British countryside are made by the European Mole *(Talpa europea)*. This Mole is typical of the true Mole family *(Talpidae)* with its dense velvety fur that can be stroked in all directions. The fur does not slope backwards as in other mammals and the mole can thus run backwards in its tight-fitting burrows without the fur jamming against the walls. It has tiny, almost sightless eyes, no external ear and a short neck linking the large head and the strongly muscular body. The forefeet have developed into powerful excavators and project sideways from the body—ideal for tunnelling. The sensitive naked snout can smell out worms and insects from a considerable distance away. Moles have voracious appetites and they can die of starvation in 12 hours. Thus much of their time is spent digging and patrolling their tunnel systems searching for food. They may nip the heads off worms to store them for future use.

Three species of American Moles are very similar in appearance to Old World Moles. The **Hairy-tailed Mole** *(Parascalops breweri)* is found in North America and, as its name suggests, has a hairy tail about 1 inch long. Its fur is not as smooth as the **Western American Moles** *(Scapanus)* which are found right down the Pacific coast from British Columbia to California. Yet another area is covered by the **Eastern American Moles** *(Scalopus)*, also called **Topos** where they occur in Mexico. These are brown rather than the usual velvety black.

Unique Star Nose

Most extraordinary of the New World Moles is the Star-nosed Mole *(Condylura cristata)* of Canada and the United States of America. In this species the muzzle is ringed by 22 pink fleshy appendages which surround the two nostrils and look rather like a star. Most likely, this star assists the Mole by improving the sensitivity of its nose. The Star-nosed Mole is more sociable than other members of his family and pairs often live together throughout the winter. They are semi-aquatic, live in marshy areas and some of their tunnels lead to ponds and streams where they swim and dive for food, catching small fish and aquatic insects as well as land worms and insects. In winter and early spring, their short tails grow fat, storing food for the breeding season. Litters of four to six young are born between April and June and have their strange nose stars fully developed at birth.

Bats- the Ultra-sonic Night-fliers

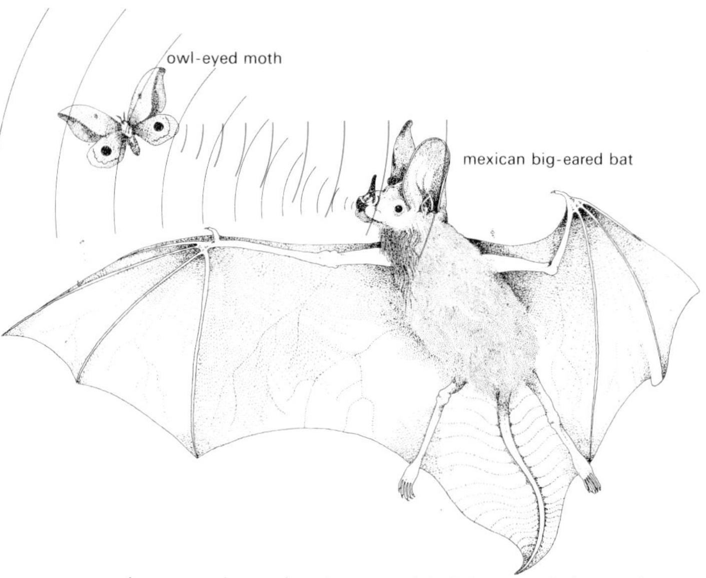

The bats form a very distinctive and rather uniform group of mammals. There are about 800 species, making the Chiroptera the largest order of mammals after the rodents and twice as large as any other order. Despite their abundance they are poorly known and many extraordinary aspects of their biology have only recently been discovered.

The word chiroptera means hand-wing and refers to the bats most distinctive feature. The wing is literally an extended hand, with a thin web of skin stretched between enormously elongated fingers and stretching back to the body and hind legs. The wings make bats the only mammals capable of sustained flight, enabling them to travel great distances on migrations and to colonise remote islands (e.g. New Zealand and the Philippines) inaccessible to less mobile mammals without human assistance. They have been known to fly to Shetland and Bermuda and even occasionally turn up on oil drilling rigs a hundred miles out to sea.

All bats are relatively small in size, the largest being only 15 inches (40 cm) long and weighing just over 1 kg. They also have poorly developed hind legs and pelvis, so most of them are rather clumsy walkers and some rely so much on their wings for locomotion that they cannot walk on land at all.

Only two or three families of bats have colonised the temperate zones of the world where cold winters cause a seasonal shortage of insect food. To escape starvation, these bats usually hibernate, going torpid for long periods, with their metabolism slowed right down, and sustained by their reserves of fat. Hibernation is a very complex physiological event and is made easier if the bat chooses a place where the temperature and humidity conditions are just right. Such places often include caves, old mines and cellars and protection in these hibernation sites is vital if the bats are to survive.

Many bats seem to be extremely rare, being known from only a handful of specimens collected years ago. Others, particularly in Europe and North America, are becoming distinctly scarcer as a result of persecution, destruction of their insect food (by pollution and insecticides) and loss of suitable places in which to breed and hibernate. However, some bat colonies are enormous; the free tailed bats in the southern United States for example, form vast aggregations in the summer. One cave in Texas is estimated to have once contained 20 million, the largest single gathering of mammals anywhere in the world. Many of these huge colonies produced so much bat dung that it has been extensively mined in the past for use as fertiliser, and in the American Civil War to produce gunpowder!

The order Chiroptera includes 17 families of bats, usually grouped into two sub-orders, the Megachiroptera and the Microchiroptera. This is a very unequal subdivision because the Megachiroptera comprise only a single family Pteropidae, the True Fruit Bats. These are found in the Old World Tropics and stand clearly apart from all the other bats in having large eyes and well developed visual capabilities instead of the ability to echolocate. They feed mainly on fruit and also have a number of structural distinctions. All the remaining families are lumped together in the sub-order Microchiroptera which, despite its name, does not comprise only small bats—some have wingspans of 35 inches (90 cm) or more—any more than the Megachiroptera contains only large ones.

Another major division among the bats is seen in their distribution. Only three of the 17 families are found in both the New World and the Old World; Vespertilionidae, Emballonuridae and Molossidae. All the others are found in Europe, Africa, Asia or Australia or in the Americas. In some cases there are interesting parallel developments in different families where individual species have evolved to perform the same ecological role and have come to resemble each other, although they are not at all closely related. In South America, the major family of bats is the Phyllostomidae which provides one of the best examples of adaptive radiation in the entire Animal Kingdom. They have undergone a kind of private evolution, which has taken place in the South American continent, where members of this one family have evolved to fill all the major feeding roles performed by separate families elsewhere. In this family alone insectivorous and carnivorous bats, nectar, pollen and fruit feeders are found, all with similar but quite independently evolved adaptations to their way of life as found in their distant relatives in the Old World.

Echolocation

Experiments made over 150 years ago demonstrated that blindfolded bats could still fly without bumping into obstacles, yet they became helpless if their ears were plugged. It is only in the last 40 years that these phenomena have become properly understood.

In flight, bats emit powerful bursts of sound and listen for echoes to detect the presence of obstacles in their path. The sounds are too high pitched for human ears to detect, but the bat's hearing is specially adapted to respond to high frequency noises. In practice the system works rather like radar and can detect objects in total darkness (enabling bats to catch food at night and also helping them navigate the blackness of caves).

The sound pulses are emitted very rapidly, 20 or more per second and the rate increases as objects are approached. Under experimental conditions, small bats can detect, intercept and eat three tiny fruit flies per minute, using echolocation instead of vision. They can also detect and avoid wires no thicker than a human hair, whilst flying at 20–30 k.p.h.

Different families of bats use their ears and voice in different ways; for example, the Horseshoe Bats emit their sound through the nose, while Vesper Bats 'shout' through the mouth. Only the Microchiropteran families are capable of echolocation, the Megachiroptera (Fruit Bats; Pteropidae) use their sensitive eyes. The exception to this rule is the Fruit Bat *(Rousettus)*, which *can* echolocate, but makes only low pitched noises by clicking the tongue.

△ ◁ A Bat's echolocation system works rather like radar. They emit high frequency noises which are bounced back to them off objects in their path.

The Fruit Bats

(Pteropidae)

The True Fruit Bats are confined to the Old World, although some New World bats have independently evolved the necessary adaptations to feed on fruit. The Pteropidae includes the **Flying Fox** *(Pteropus giganteus)*, largest of all the bats, with a wingspan of nearly 6 feet (2 metres). Flying Foxes live in large colonies, hanging like furled umbrellas among the branches of large trees. Within these colonies there is often a distinct social structure, with complex behaviour patterns not found in most bats. *Pteropus* is a strong flyer and may travel 90 km or more in a night to a particular feeding place.

Not all the pteropids are large: some, like the **Epaulletted Fruit Bat** *(Epomophorus)* of Africa, have only a 20 in (50 cm) wingspan; some are even smaller. This particular bat gets its name from the prominent white tufts of hair on the shoulders in the male. In

△ The Long-nosed Bat of Central and South America, one of those that has reduced teeth and a long brush-tipped tonge for feeding on pollen.

▷ The Flying Fox which, with a wingspan of six feet, is one of the largest Bats. Flying Foxes are powerful flyers and may fly many miles in a night in search of food. True Fruit Bats navigate and find their food using their eyes rather than the echolocation system used by other types of bats.

appearance, these are like the big golden shoulder pads seen on a soldier's dress uniform, but in the bats they conceal special scent glands. Another African **Fruit Bat** is *Eidolon helvum*, a large colony of which adorns the trees in the main street of Uganda's capital, Kampala.

Not all Fruit Bats feed on fruit; some, like the long-tongued Fruit Bats *(Macroglossus)* which live in South-East Asia, actually feed on nectar and pollen. These bats have a very long tongue which they can thrust deep into flowers to lap up nectar and collect pollen on the special brush of hairs at the tip of the tongue. Their close relatives also feed on fruit juice, licked out from squashy, over-ripe fruits. Such food requires little chewing so they have very tiny teeth, weak jaws but a long snout to house the tongue.

Most Fruit Bats do indeed feed on fruit, usually choosing soft things like mangos and guavas. They have big, rather flat-topped teeth, but usually just squash the fruit against special sharp ridges on the palate, swallow the juice and spit out the remainder.

Fruit Bats have a keen sense of smell to help seek out their favourite flowers and ripe fruits. They also have large eyes and guide themselves visually in flight. The other families of bats *(Microchiroptera)* navigate by emitting high-pitched sounds and listen for echoes to betray the presence of obstacles. Their eyesight is much less important.

Vesper Bats

(Vespertilionidae)

The ordinary bats familiar to people in the northern parts of Europe and America belong to the family Vespertilionidae. This is the largest family of bats and one of its constituent genera *(Myotis)* is one of the most widespread of all mammalian genera, being absent only from the polar regions and the more remote oceanic islands.

Vesper Bats are a varied family and have few features common to all, except that all lack the special distinctions of other families. A fairly typical species is the **Mouse-eared** *(Myotis myotis)*, the largest European bat with a wingspan of about 18 inches (45 cm). It has a white underside and like most bats has grey-brown fur on its back. The species forms large colonies in Central Europe, but though occurring in Britain also, it is one of the rarest mammals. It is sometimes thought that the small bats seen are all pipistrelles, but although the genus *Pipistrellus* occurs throughout the Old World and North America, there are many other small species present with which it is often confused. 15 types of bat exist in Britain, of which the pipistrelle is the smallest, but at least 6 other species are little bigger, having wingspans of 10 inches (25 cm) or less.

Most bats produce only one young at a time, but among the vespertilionids, several give birth to twins and the **American Red Bat** *(Lasiurus borealis)*

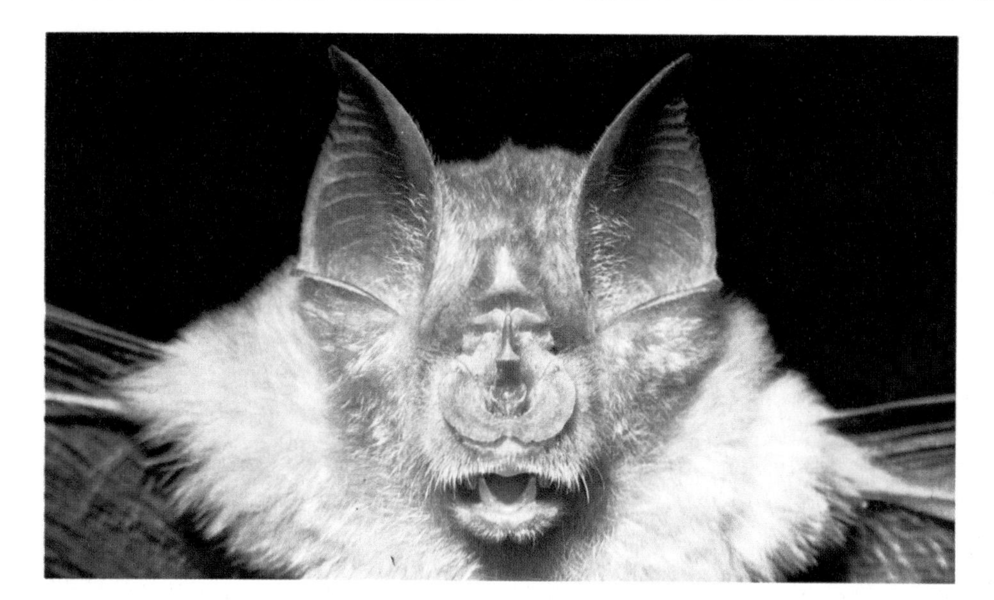

is the only bat in which family size may occasionally exceed three.

The **Long-eared Bats** *(Plecotus)* of Europe and North America are perhaps the most distinctive vespertilionids. They have enormous ears which are over half the length of the head and body; the largest ears of any mammal. The **Pallid Bat** *(Antrozous pallidus)* of the western United States of America also has big ears, but in addition to the normal diet of flying insects, this species often pounces on Beetles and Grasshoppers sitting on the ground. It even catches Scorpions this way and somehow escapes fatal retaliation.

Free-tailed Bats

(Molossidae)

Members of this family have a short thick tail that is not enclosed in the usual tail membrane stretched between the hind legs. They also have a broad, rather square-ended muzzle, which is the inspiration for their other popular name of **Mastiff Bats.** The family occurs in all the warmer parts of the world (except Australasia) and sometimes in great numbers. All have long, very narrow wings which are adapted to rapid flight. Some, like the **Guano Bat** *(Tadarida brasiliensis)*, migrate long distances; in this species a thousand kilometres or more from the southern United States to Mexico each autumn.

In Africa, *Tadarila pumila* often choose to live in the roof, a habit shared by *Molossus major* in Central America. In such places the roof is usually made of corrugated iron which is roasted in the tropical sun all day, until the loft temperature may exceed 50°C, an almost unbearable heat. The **Flat-headed Bats** *(Platymops)* of East Africa prefer to live under stones and in narrow crevices in rock and tree bark. As an adaptation to squeezing into tight spaces, the skull is only about one third as deep as it is wide; so flat that it looks as if it has been accidentally squashed.

Mouse Tailed and Sheath Tailed Bats

(Rhinopomatidae and Emballonuridae)

The four species of **Mouse-tailed Bats** *(Rhinopoma)* live in the dry areas of north-east Africa, Arabia and India. They are the most primitive of living bats and are all about 3 in (70 mm) long in the body. The tail is of a similar length; long, thread-like and not enclosed in the tail membrane.

Sheath-tailed Bats such as *Peropteryx* have prominent scent glands forming pockets in the front edge of the wing membrane. These gave rise to the vernacular name of **Sac-winged Bats** for this family. The **White-lined Bats** *(Saccopteryx)* are unusual among chiroptera in having distinctive patterns in their fur. Both these genera occur in South America; but the emballonurids are represented in the Old World by the **Tomb Bats** *(Taphozous)* which inhabit caves and cool underground spaces as their name suggests. All **Sheath-tailed Bats** are characterised by having a tail which pokes through the tail membrane about half-way along its length. The membrane slides freely along the tail as though forming a thin fleshy sheath.

Fisherman Bats

(Noctilionidae)

There are only two species in this family. Both live in Central and South America, but only one *(Noctilio leporinus)* feeds on fish. This it does by detecting ripples in the calm sea and quiet backwaters as fish swim near to the surface. The Bat then swiftly scoops its hind feet into the water in an effort to spike the fish on its specially elongated claws. If the attempt is successful, the fish will be carried to a favourite roost and eaten there. Sometimes **Fishermen Bats** misjudge their manoeuvres and fall into the water, but they can swim well and can take off

△ ◁ The face of a Horseshoe Bat showing the distinctive horseshoe shaped noseleaf surrounding the nostrils. This serves to concentrate the echolocation sounds which Horseshoe Bats emit through the nostrils.
△ Greater Horseshoe Bats hang up like furled umbrellas when they hibernate in caves.
▽ A baby Serotine Bat. Baby bats have large, well-developed feet to hang on to the roost walls.
▷ △ A Greater Horseshoe Bat in flight. The broad, rounded wings enable them to manoeuvre in confined spaces.

from the water quite easily. Their fur is specially short, sparse and greasy and does not become waterlogged, quickly drying afterwards.

Bats with extra-ordinary noses

Five families of bats have special fleshy structures built up around the nostrils. Sometimes this noseleaf forms a simple spike, but in certain species it is extraordinarily elaborate. In the **Horseshoe Bats** *(Rhinolophidae)* of the Old World the noseleaf forms a perfect cone with a triangular spike sticking up from it between the eyes. The noseleaf of

these bats plays a vital role in their echolocating abilities. Horseshoe Bats and Vesper Bats are the only families which have representatives living in the cooler latitudes of the world; all the other families are tropical.

Leaf-Nosed Bats *(Hipposideridae)* are like Horsehoe Bats, but the noseleaf is usually rather simple and square-topped. However, the **Trident Bats** of this family get their name from a three pronged addition to the rim of the noseleaf, and the **Flower-Faced Bats** *(Anthops)* have an extraordinary mass of 'petals' forming the noseleaf.

The **Slit-Faced Bats** *(Nycteridae)* have a simple noseleaf which can be closed to protect the nostrils from dust; a useful adaptation since these bats often live in very dry dusty places, especially in Eastern Africa. These bats have soft grey fur and large ears, characteristics also seen in the **False Vampires** *(Megadermitidae)*. In this family the noseleaf is a simple fleshy tongue projecting up between the relatively large eyes. The Australian representative is the **Ghost Bat** *(Macroderma gigas)*. It is 5 inches (130 mm) long with a wingspan of about 3 feet (1 metre) and is the largest of all the Microchiroptera. It is also carnivorous, preying on Rats, Mice, nocturnal reptiles and also other bats.

Members of these four families all tend to hang up by their hind feet and wrap their wings around the body when at rest. They often hang free from tree branches and if living in a cave, they hang from the roof like large pear-shaped fruits. Other bats tend to hide in crevices, hang against walls or tree trunks and usually fold their wings along their flanks. All these four families inhabit the Old World tropics.

The **Spear-Nosed Bats** *(Phyllostomidae)* live in Central and South America. One species, the **Big-Eared Bat** *(Macrotus waterhousii)*, penetrates into the United States of America. Most of them have a simple fleshy spike standing upright from the tip of the snout. Bats of this family vary in size from *Vampyrum spectrum*, with a wingspan of up to 35 in (90 cm) the largest American bat, to tiny pollen feeders like *Glossophaga*, a mere 2 inches (50 mm) long. Within this one family, a great variety of feeding habits has evolved and some species parallel the True Fruit Bats of the Old World *(Pteropidae)* in their special adaptations for feeding on fruit. *Artibeus* for example has broad teeth and jaws, a ridged palate and often makes itself a nuisance in fruit plantations.

Vampires

(Desmodontidae)

Vampires are the only mammals that feed exclusively on blood. They make a swift incision with their needle-sharp teeth into the skin of a sleeping victim, then lap up the blood as it flows out. A special anticoagulant in the vampire's saliva delays clotting, so that the blood continues to flow and the vampire may drink so much it can hardly fly. The Vampires inhabit the warmer regions of Central and South America and the **Common Vampire** *(Desmodus rotundus)* has undoubtedly benefited from the introduction of cattle, horses and other farm animals, which now provide them with many more potential sources of blood than would the relatively sparse wild mammal population.

Vampires are much smaller than their reputation. Only 4 inches (80 mm) long, with no tail and a wingspan of perhaps 12 in (30 cm), they are scarcely the demons that fiction describes. However, many of them carry rabies, and in their nocturnal forays may bite several animals and so transmit this fatal disease. Consequently, vigorous efforts are made to exterminate Vampires in many areas.

Minor bat families

Of the 17 families of bats, 5 are very trivial containing only 12 species between them. Nevertheless, each has some special peculiarity though they are all rather restricted in their distribution and some are exceedingly rare. The **Short-Tailed Bats** *(Mystacinidae)* for example, have peculiar wings and feet and only occur in New Zealand. In two of the families there are suckers on the wings, small round pads attached to the wrists and, in one case, capable of supporting the entire weight of the bat. Strange although these features are, they seem to have evolved quite independently, since one family *(Thyropteridae)* occurs in Central America and the other *(Myzopodidae)* is confined to Madagascar.

Smoky Bats *(Furipteridae)* live in Central and tropical South America. They are small and have such a tiny thumb that it is hidden in the skin of the wrist. The **Funnel-Eared Bats** *(Natalidae)* are found in the same general area. They are delicate, slim, long-legged animals often bright yellow or reddish brown. They are close relatives of the Smoky Bats and both families are insectivorous.

Primates Man's closest relatives

Primate, from the Latin 'primus' meaning first, is the term describing the highest order of the mammals. The primate families range from the family Tupaiidae of 20 species of Tree Shrews, to the family Pongidae of nine species of Ape, which are Man's closest relative. Man himself is the sole living species in the primate family of Hominidae.

With the single exception of Man, all primates are native to tropical and sub-tropical countries and are arboreal in habit, though some Baboons are more usually found among rocks than in trees. Again with the exception of man, primates are essentially climbing animals with the four feet adapted for grasping.

Primates have limbs terminating in five digits which are usually furnished with nails instead of claws. The eyes are near to the front of the head and set close together so that both focus in the same direction giving stereoscopic vision and providing the animal with good judgement of distance. The primates' senses of hearing, touch and vision are in general better developed than is the sense of smell.

Except in Man, the primate great toe is opposable to the others, though the thumb is often imperfectly so. The upper halves of the limbs are free from the body and not embedded in it. In female primates, the mammae or teats are normally only two in number

and situated in the breast and never in the abdomen. A more or less erect attitude has been adopted by the higher primates, while the dentition differs essentially from that of the carnivores and herbivores, indicating a mixed diet. The higher primates' intelligence, both inherent and acquired, is superior to that of other mammals, and the brain is more developed both in size and elaboration of structure.

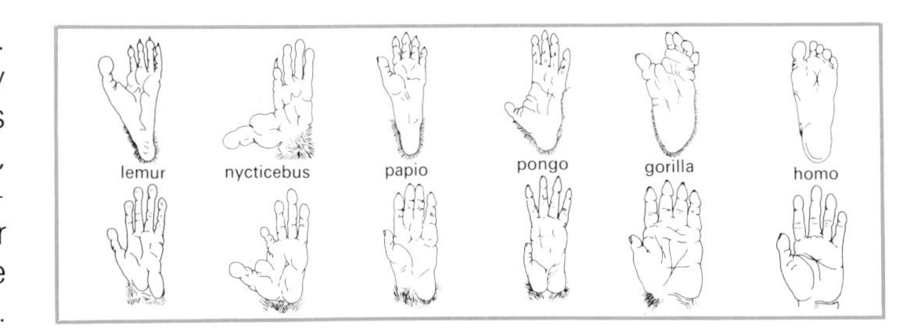

◁ ▽ Almost all Primates, like this White-handed Gibbon swinging through the trees, are arboreal in habit. Unlike Monkeys, Gibbons have no tails.
◁ The Tree Shrew is one of the most primitive of living Primates having many features in common with the Insectivores. Its long nose and many internal features are unlike those of monkeys.
△ The hands and feet of Primates are usually furnished with nails instead of claws and both the thumbs and big toes are opposable.

Tree Shrews

(Tupaiidae)

Few mammals have been the subject of so much controversy over classification as have the **Tree Shrews.** At one time zoologists insisted that they should be classed as insectivores, but it is now thought that Tree Shrews are primitive primates. This is based on both their general habits and arboreal characteristics, and certain anatomical distinctions.

There are between 15 and 25 species of Tree Shrew, the most important being the **Pentailed Tree Shrew** *(Ptilocercus lowi),* the **Smooth-Tailed Tree Shrew** *(Dendrogale murina),* the **Mindanao Tree Shrew** *(Urogale everetti)* and *Tupaia.*

Tree Shrews are rather Squirrel-like in appearance but with pointed snouts. They vary in size, according to species, from 20 to 41 centimetres in length, of which about half is tail. Most species have fairly bushy, prehensile tails except for the smooth-tailed species *(Dendrogale* and *Ptilocercus),* which have a naked tail terminating in a white, feathery tuft. The body fur is made up of long guard hairs and a soft under-fur. Colours vary from brown to grey and reddish and there are sometimes lighter or darker markings on the head.

In spite of being called Tree Shrews, the animals spend quite a large part of their time on the ground. Some species forage for food on the ground and sleep at the roots of trees. With the exception of the Pentailed species, which bounds or hops, Tree Shrews can run quickly across the ground and are expert climbers among the branches of trees. Tree Shrews are mainly solitary, though they occasionally live in family groups.

Tree Shrews range from India and south-western China, eastwards and south through Malaysia to Borneo and the Philippines. They eat seeds, leaves, pulpy fruits, worms and insects. Occasionally, they have been known to eat Lizards and Mice.

Tree Shrews appear to have no defined breeding season and, after

Primate hands and feet

Arboreal life, more than anything else, conditioned the development of the primates, and so of Man himself. The first requisite for living in trees is a modification of the forelimbs for climbing, and to climb quickly and efficiently, the hands need to be specially adapted to grasping. Thus the primates developed the opposable thumb.

Primates came into existence in the trees where they successfully developed through their long evolution. Primates are essentially tree-dwellers, and today Man is the only primate who has totally abandoned an arboreal life. By the adaptation of being able to wrap their digits round a branch instead of simply driving claws into it, as do practically all other tree-dwelling mammals, the primates made themselves undisputed masters of the trees.

All primates have hands with moveable fingers, but these differ considerably in structure and variety of use. The process of transforming clawed animals into grasping ones began with the **Tree Shrew,** which developed long, thin fingers which could be spread out in a rudimentary grasp. Even so, the Tree Shrew's 'hand' is little more than an elongated paw, able to perform the relatively elementary function of holding on to a branch, while at the same time digging in with its claws.

Somewhat more advanced in the evolution of the hand is that of the **Slow Loris**'s pincer-like grip. The **Tarsier,** which is essentially a jumper, has well-developed hands, with discs on the fingers, to help it to hold on after a leap.

The **Marmoset** has the most primitive hands of the Monkey family. Its grasp is of necessity awkward as the fingers all move in the same plane and have to be assisted by pressing against the heel of the hand. The **Chimpanzee** and the **Macaque** can be much more dextrous with their hands. Both are able to move their thumbs so that they can pick up an object between thumb and finger. Curiously enough, the Chimpanzee's thumb is less efficient than that of the Macaque because it is shorter. The Chimpanzee's superior brain enables it to make better use of its hands.

Today, there is no fundamental difference between the hand of the higher Ape and that of Man. The chief difference is in the thumb and in the degree of its opposability.

mating, the male builds the nest from leaves, an unusual occurrence among mammals. After a gestation of 45 days or more, one to three young are born, blind and naked. The female spends the first two hours after giving birth, cutting the umbilical cords, cleaning up the birth membranes and giving the babies their first feed. After this, she visits them every 48 hours to feed and does not visit them between feeds. The babies seem content without her, giving no cries of distress and, within a few days, are grooming themselves. Young Tree Shrews are sexually mature at four months.

△ The tiny Dwarf Lemur, nocturnal, arboreal and with a prehensile tail for gripping tree branches.
◁ A young Slow Loris which gets its name from being slow moving in the trees where it lives. Lorises, which have no tails, sleep by day and hunt insect prey with almost painstaking deliberation by night.
▽ A Bush Baby, sometimes called a Night Ape because of its nocturnal habits, with the finger nails, which most Primates have rather than claws, clearly visible. The Bush Baby has powerful hind limbs for jumping from tree to tree and though the long bushy tail is not prehensile it helps the animal maintain balance when leaping. All three of these Lower Primates are nocturnal and have large eyes for night vision and all live in trees.

Lemurs

(Lemuridae)

Zoologically, **Lemurs** are to some extent intermediate between insectivores and Monkeys and have been described as half-Monkeys because of their hair-covered Fox-like muzzles. They are distinguished by soft, woolly hair, long forelegs or arms and very big, but inexpressive eyes. Lemurs are true primates: the thumb and big toe are opposable to the other digits and they have binocular vision, although their sight is not as well developed as that of monkeys and they are more dependent on their sense of scent.

Lemurs have a large, sharp claw on the index toe which is used for grooming. The lower incisors and canine teeth are flattened and are used for combing the fur. Beneath the tongue is a horny filament, called the sublingua, which is used for cleaning dirt out of the 'comb' teeth.

There are fifteen species of Lemur and all are native to Madagascar. The **Ring-Tailed Lemur** *(Lemur catta)*, lives in thinly wooded rocky country, whereas most other species inhabit dense forests. The animal is about a metre long, half of which is tail. It has greyish fur and a long, ringed tail. Ring-tailed Lemurs, as with most of the bigger species, live in social groups of twelve or more. The males have a strongly enforced rank order.

The **Ruffed Lemur** *(Lemur variegatus)*, is just under a metre in length, half of which is tail. It has piebald, thick, woolly fur and has a white ruff. The **Mongoose Lemur** *(Lemur mongoz)*, is about the same size as the Ruff Lemur, but its general colour is fawn with a black ring on the forehead. The **Crowned Lemur** *(Lemur coronatus)*, is yellow, grey and white with a reddish-gold line across the forehead.

Dwarf Lemurs *(Cheirogaleus)*, of which there are about six species, rarely exceed 31 cm in length and have red or grey coats with white or yellow undersides. The **Mouse Lemur** *(Microcebus murinus)*, is about the size of a small Rat.

Woolly Lemur

(Indriidae)

The woolly Lemurs, of the genus *Avahi*, with only one species, *Avahi laniger*, live in open spaces as well as in forests, and come from eastern Madagascar. They have a thick, woolly, grey-brown coat and are about 30 cm long with a tail 45 cm long. Avahi live alone or in couples and are nocturnal.

Lemurs are sexually inactive for most of the year, coming into season about every 35 days during the cooler months. Gestation varies between two and four months according to the size of species and only a single young is born at a time.

The Aye-Aye

(Daubentonia madagascariensis)

Although related to the Lemurs, the **Aye-Aye** differs from them sufficiently for it to be the only member of the genus *Daubentonia*. Its close-set, sloping incisors are without enamel on the inner surfaces and look like those of a rodent, and caused naturalists to classify it wrongly.

Approximately the size of a domestic Cat, the Aye-Aye has thick, silvery fur and a long bushy tail. It has exceptionally long and bony fingers, the third on each forepaw being not much thicker than a wire and the animal uses this for hooking insects out of holes in tree bark. The gnawing incisors are used for cutting bamboo and sugar cane.

The Aye-Aye lives deep in the forests of Madagascar and as it is a nocturnal creature, very little is known about its habits. Only one baby is born at a time, in a spherical nest, but mating patterns and gestation periods remain mysteries.

A Sifaka mother and baby. Sifakas are found only on the island of Madagascar and they have long legs and tail and short arms. Their fur is thick and silky almost concealing their ears and the brachial membranes which to some extent function as parachutes and enable Sifakas to jump great distances. They prefer to eat unripe fruit which they always peel with their teeth, their hands, though well adapted to grasping branches, being almost useless for holding objects.

The Sacred Indri

(Indriidae)

With its long hind legs, small thumbs and inch-long stump of a tail, the **Indri** differs somewhat from the woolly Lemurs, with which it is sometimes classed. It measures approximately 60 cm from nose to rump, and its fur is in contrasting patterns of black and white. It normally walks and stands upright and is an exceptionally agile climber of the trees in eastern Madagascar. The Indri is diurnal and lives in small communities. The female bears a single young. The natives of Madagascar consider the Indri sacred and refuse to kill it.

Another member of the Indriidae Family is the **Sifaka** *(Propithecus diadema)*. The legs are joined by hair-covered membranes which the animal can stretch to form a kind of parachute; this enables it to glide from branch to branch, for distances of up to 10 metres. The fur is black on the back and orange-yellow below.

The Slow Movers

(Lorisidae)

The **Potto** *(Perodicticus potto)*, is one of the slow-moving, Lemur-like primates. It is native to the African forest belt extending from Guinea in the west to the Rift Valley in East Africa. The adult Potto is, on average, 35 cm long and has thick dark brown, woolly fur. Along the back of the neck are four or five spines which were once thought to be defensive but more probably act as sensors, like the domestic Cat's whiskers.

The Potto has well-developed fingers which enable it to clamp itself firmly to branches of the trees in which it lives. The animal is an exceptionally slow and cautious mover, letting go of one hand and one foot at a time as it creeps along branches, stalking insects, Lizards and young birds. It also eats fruit and leaves. The female comes into season every 40 days and has a single young at birth. As the Potto's large eyes suggest, it is a nocturnal creature. **Lorises,** of the same family as the Potto, inhabit the forests of Ceylon, India and south-east Asia. Like the Potto, they are slow-moving, live in trees and have broad, grasping hands and feet. The **Slender Loris** *(Loris tardigradus)*, has very thin legs while the **Slow Loris** *(Nycticebus coucang)*, has short, thick limbs. Adults of both species are up to 38 cm in length. The colour of the animal is brownish-grey with a black, forking stripe on the back. Lorises are nocturnal and feed on fruit, insects, birds and reptiles.

The **Angwantibo** *(Arctocebus calabarensis)*, is often confused with the Potto, but it is smaller, has a longer muzzle, and its eyes are less prominent. The Angwantibo is much more agile than others of the Loris family, as it moves amongst the branches which it clutches with its caliper-like feet.

Bush Babies

(Lorisdae)

Sometimes called **Galagos** or **Night Apes, Bush Babies** are native to Africa south of the Sahara. There are four species, the best known being the **Senegal** or **Moholi** *(Galago. senegalensis)*. It is 40 cm long including its bushy tail; the head is round and the muzzle short. The eyes are large and the big, flesh-coloured ears can be folded at will. The hind legs are long and the front ones shorter. The ends of the long fingers are flattened and there are pads of thick skin on the undersides which give a good grip on slippery surfaces. The tips of the fingers have flattened nails.

Bush Babies are nocturnal and move about at night in pairs, searching for the flowers, fruits and insects which they eat. Although lethargic by day, they are extremely agile at night, their long hind legs enabling them to jump, like acrobats, for distances of up to 4.5 metres from branch to branch. The female comes into season twice a year, bearing one young at a birth and gestation is about four months.

Tarsiers

(Tarsiidae)

The Tarsier is a curious evolutionary mixture. Some of its characteristics are highly advanced while others are very primitive. Its flattened face, round skull and the orbits of the eyes closely relate the Tarsier to the true Monkey. There are three species, all of which are restricted to certain islands in the East Indies.

Horsfield's Tarsier *(Tarsius bancanus)*, is 15 cm long. It has reddish brown fur and its 25 cm tail is naked except for a terminal tuft. The **Celebes Tarsier** *(Tarsius spectrum)*, is 13 cm in length with long, dark grey fur and a heavily tufted tail. The **Philippine Tarsier** *(Tarsius syrichta)*, measures 15 cm and has greyish fur and a bare, untufted tail.

Tarsiers live singly or in pairs in coastal forests and plantations. It is the only primate which is entirely carnivorous, eating Frogs, Lizards, insects and grubs. It is nocturnal and moves in leaps, using its long hind legs, which have a lengthened ankle section, or tarsus, which gives the animal its name. The paws have tactile pads on the toe tips which enable the animal to cling to smooth branches. Tarsiers breed throughout the year, the female having a monthly cycle. Gestation lasts six months and the single young is born with its eyes open, its body fully furred and able to climb on its own.

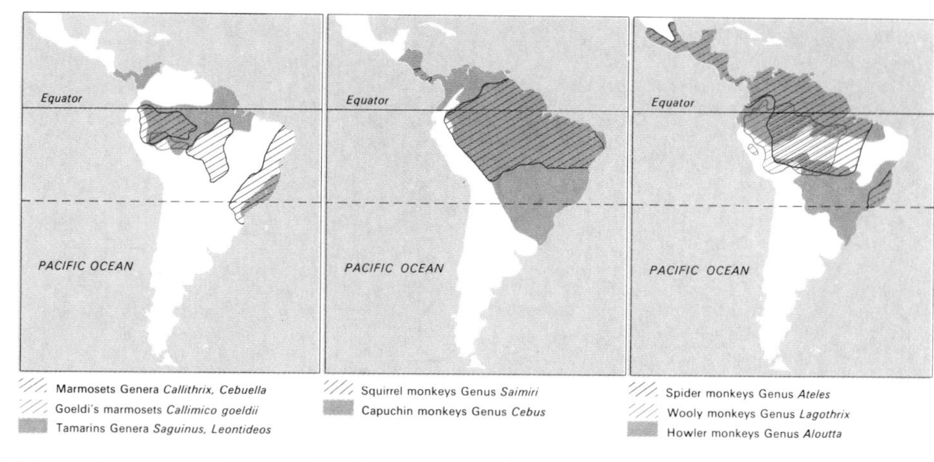

◁ The Squirrel Monkey is the most widely distributed of all South American Monkeys. They are considered the most engaging of the Monkeys and the young are looked after equally by both parents. Their brain is, relative to their size, larger than that of Man. The tail is long and only slightly prehensile and Squirrel Monkeys prefer scrub woodlands to the tall trees of the deep forest.
▷ The Woolly Monkey, so called because of its fur, is one of the few Monkeys to have a prehensile tail.

Monkeys of the New World

(Cebidae)

In the vast forests which range from the western mountain ranges of South America to the Atlantic, and from the southern edge of Mexico to the northern edge of the Argentine, there is an immense population of Monkeys, divided into 26 species. They differ markedly from the Old World Monkeys. Their faces are bare and tend to be flatter; they have a relatively large cranial capacity, their eyes are small, and their fur is thick and woolly. They have long and usually prehensile tails and most species have nails on all fingers and toes. The average South American species is less intelligent than those of the Old World and their movements are less jerky. Only one species is nocturnal.

Nigh Monkey or Douroucouli

(Aotus trivirgatus)

The only nocturnal species of the higher primates, the **Douroucouli** is 46 cm long and its furry, non-prehensile, tail with a tufted tip is 39 cm long. The body colour is russet brown or reddish grey, and a ruff of long hair surrounds the short face, which is marked by three black lines and a white band. The short ears are concealed in the soft head fur, so giving the name *Aotus* or 'earless'. Because the animal has large, saucer-like eyes, it is sometimes called the **Owl Monkey.** Throughout the night, the Douroucouli is at its noisiest and most active as it roams the trees in search of insects. By day, it hides in the thick foliage of trees, where daylight cannot damage its sensitive eyes. The Douroucouli lives in colonies which share a single tree.

The **Titi Monkey** *(Callicebus)*, is closely related to the Night Monkey, but its body is somewhat longer and it has smaller eyes. It hunts by day for fruit and nuts in the Brazilian jungle.

Blushing Monkeys and Howlers

There are three species of **Uakaris** or **Blushing Monkeys,** a charming name which they acquired because of their red-skinned faces, which glow when the animal is excited. The three species are: the **Blackheaded Uakari** *(Cacajao melanocephalus)*, the **Red Uakari** *(Cacajao rubicundus)*, and the **Bald Uakari** *(Cacajao calvus)*. Adults measure up to 46 cm and the body is covered with dull, brown hair. The tail is only 15 cm long, and this distinguishes the species as the only South American Monkeys with tails shorter than their bodies. Shaggy coats belie a slim and muscular physique and the Monkeys can leap distances of up to 6 metres.

All three species are confined to a comparatively small tract of forest in the Amazon basin, but although living in close proximity, the three species never overlap. They live in small troops, usually in the tree tops, and feed on fruit, leaves and seeds. They come down to the lower branches only when travelling through the forest. The Uakari is highly intelligent and physically resembles its Old World relative the Orang-Utang.

Saki Monkeys *(Pithecia)*, of which there are three species, have profuse hair on the head and bushy tails. They grow up to a metre or more in length

and are active tree-dwellers living on fruit. Fur colour varies from golden brown to black according to species.

Howlers *(Aloutta palliata),* are the biggest of the South American Monkeys. There are four species, the chief differences being the colour of the fur. They grow to a length of 1.25 metres including the long prehensile tail. They get their name of 'Howler' from the loud call of the male, which has greatly enlarged hyoid bones in the throat, as big as an orange, forming a resonating sound box. The howling is most often heard at sunset, sunrise and before the onset of rain.

Squirrel Monkeys and Capuchins

The **Squirrel Monkey** *(Saimiri sciureus),* is not only the most widely-distributed of the South American Monkeys, but it is one of the most charming and beautiful. About the size of a Squirrel, the animal has a yellowish-green coat. The tail is only slightly prehensile. It is remarkable for its large skull and the fact that, relative to its size, its brain is bigger than that of Man. It is gregarious in habit, and feeds on lizards, birds' eggs and insects.

The **Capuchin** or **Ring-tailed Monkey** *(Cebus),* is the most intelligent of the South American Monkeys and the commonest in captivity. Slightly smaller than a domestic Cat, the coat varies in colour from black, brown, yellowish or golden according to species or sub-species, of which there are a dozen or so. The long, hairy tail is prehensile. The distinguishing feature which gives the Capuchin its name is the hair on its head, which forms a peak, rather like a monk's cowl.

Spider Monkeys

These are among the most specialised of South American Monkeys, but they are less Man-like than the Old World Apes. There are several species. The **Common Spider Monkey** *(Ateles paniscus),* is about 50 cm in length and the coarse, wiry hair may be black, greyish or buff coloured. The prehensile tail is at least half a metre long, and has on the underside a completely hairless area about 8 cm long which is wrinkled with ridges. The tail serves as a fifth hand, with which the Monkey can suspend itself from a branch, leaving its four limbs free. The Monkey's hands have long, narrow palms with long curved fingers and no thumbs. It travels by running along the branches of trees or by jumping from branch to branch or tree to tree. In a flying leap, a Spider Monkey can cover a distance of 9 metres or drop 6 metres to the ground.

Humboldt's Woolly Monkey *(Lagothrix),* is much larger than the Common and Woolly Spider Monkeys, It has a prehensile tail and thumbs on its hands. The coat is grey.

South America's Near-Monkeys

(Callithricidae)

In South America, there is a group of primates which have been described as 'near-Monkeys'. These are the **Marmosets** and **Tamarins.** There are several species of each but all have bodies under 30 cm and their tails are not prehensile. Instead of having nails on their paws, like the Monkeys, these animals possess hooked claws, except for the great toe on which there is a true nail.

The *Callithricidae* have preserved more primitive features than the true Monkeys and the females bear a litter of several young at each birth. These are tree-dwelling animals and they are exceptionally clumsy on the ground. A few are solitary, but most travel in troops, looking for the fruit and insects on which they feed.

Marmosets and Tamarins

The **Lion Marmoset** *(Leontocebus leonicus),* with its shimmering yellow-gold coat and mane, is probably one of the most brightly-coloured of all living mammals. The body is about 20 cm in length and the fairly bushy tail is between 20 and 23 cm long. It is native to the Amazon basin. A somewhat bigger specimen is the **Silky Marmoset** *(Leontocebus rosalia),* which is orange in colour and with a brown mane. The **Common Marmoset** *(Callithrix jacchus),* is 23 cm long with a tail 30 cm long.

An unusual characteristic of the Common Marmoset is that the female merely suckles the young, the male being left to carry the youngster about—an example of shared parental responsibility.

Other species of these attractive little creatures are: the **Black-Tailed Marmoset** *(Callithrix argentata),* and the **Pigmy Marmoset** *(Cebuella pygmaea).* The combined tail and body length of the latter is only 20 cm.

Tamarins are closely related to the Marmosets. Tamarins comprise two genera and 21 species. Most are 15 to 24 cm long with tails 24 to 40 cm. They differ from Marmosets chiefly in their teeth, the canines being long and the incisors short, instead of all being of equal length.

Old World Monkeys

(Cercopithecidae)

The tailed Monkeys of the Old World, of which there are 60 species, are in many respects higher in the evolutionary tree than are their relatives of the New World. They are found in all the warmer zones of the eastern hemisphere except Madagascar and Australasia. Old World Monkeys have the same number of teeth as Man, while their close-set nostrils are more human-looking than those of the New World Monkeys. They also appear to have some degree of colour vision. Although the tails of most species are long, they are never prehensile. Old World Monkeys have a well-developed thumb and they are in general more intelligent than are their New World relations.

The Family *Cercopithecidae* includes all the Old World primates except the Lemurs, Lorises etc., and the great Apes and is usually divided into two groups. The first group feed mostly on leaves, and to enable them to digest a vegetable diet, their stomachs are complex. In general, this group are tree-dwellers.

The members of the second group have a more varied diet consisting of fruit, vegetable matter, insects, small birds, crabs and eggs. They are mostly tree dwellers, but some of the larger species are at home on the ground.

Both groups have arms and legs of nearly equal length, moving normally on all fours, with the body practically horizontal. When walking, the palm of the hand is placed flat on the ground while the heel of the foot is raised. The hands, with their broad nails, are singularly human-looking.

Macaques

Macaques *(Macaca)* are the most numerous of all Monkeys, and are the hardiest and most widespread. They

are found in the hottest parts of India and in regions where the temperature is below freezing.

Best known of the Macaques is the **Rhesus** *(Macaca mulatta),* which is widespread in northern India and south-east Asia. It is from 46 to 64 cm long with a tail about 23 cm long. The long-haired coat is greyish and the face is pink. There are red callosities on the buttocks. The adults are very powerful and agile and, generally, are gregarious.

The **Lion-Tailed Monkey** *(Macaca silenus),* is native to the dense forests of southern India. It derives its name from its lion-like tail tuft. The body fur is black.

The **Japanese Monkey** *(Macaca fuscata),* is the most northerly-living Monkey in the world and one of the few mammals peculiar to southern Japan. Communities of up to 200 members live on rocky hillsides and mountains, feeding on nuts, fruit and insects. The Japanese Monkey is closely related to the Rhesus, which it resembles in size, but its coat is more of a grey-brown. The Japanese Monkey was the original model for the little statue of three wise Monkeys.

The **Barbary Ape** *(Macaca sylvanus),* is not a real Ape because it has a rudimentary tail. It is the only Macaque native to Africa. It inhabits the wooded mountains of Morocco and Algeria. A small colony live in a semi-wild state on Gibraltar. The Gibraltar

△ Baboons, the collective name for a number of species of large Monkeys, have adapted themselves to living and hunting on the ground but they are agile climbers and frequently sleep and feed in trees.
▽ Olive Baboons drinking at a water hole. Baboons always move on all fours, hunting and living in groups under the leadership of the old males.
◁ A colourful Stump-tailed Macaque, fonder than other Macaque species of spending time on the ground though it is still a good climber and has been trained by botanists to collect specimens from tall trees.
▷ ▷ A group of Long-tailed Macaques—females and young—engaged in mutual grooming, removing parasites from each other's fur, an important social activity in many of the Higher Primates. Sometimes known as the Crab-eating Monkey, the Long-tailed Macaque is a good swimmer and diver, often crossing fairly long stretches of water to establish itself on isolated islands.
▷ Rhesus Monkeys, in which the Rh blood factor was first discovered, are the best known of the Macaques. They move in troops of 10 to 30 and live in a carefully constructed society.

Barbary Apes may be the survivors of a population whose fossil remains have been found in several parts of Europe. If so, the Barbary Ape is the only non-human primate native to Europe. The Barbary Ape is about 63 cm long

and the coat is thick, coarse and brown. The tail is vestigial. The Apes feed on fruit, leaves and insects. They generally live in rock crevices.

Black or Celebes Ape

(Cynopithecus)

Like the Barbary Ape, the **Black Ape** of the Celebes has a rudimentary tail, which means that it is not a true Ape but a Monkey. The Black Ape resembles the Baboon in size and shape.

Mangabeys

(Cercocebus)

Although named after Mangabey in Madagascar, Mangabeys are not native to that island but to the west and east coasts of Africa. Mangabeys are slender Monkeys about 60 cm long with tails 45 cm long. The fingers and toes are webbed at the base and the animal has white lids on its eyes. They inhabit

the tree-tops of dense forests. There are two distinct groups of species: the crested and the uncrested. All have long, black faces with deep hollows under the cheek bones. The **Sooty Mangabey** has dark, speckled fur with white or yellow underparts.

Baboons and Drills

Baboon *(Papio),* is the collective name for several species of big Monkeys which have adapted themselves to living and hunting on the ground. They move on all fours and have a much more highly developed sense of smell than have Monkeys in general. Baboons are distinguished by long, Dog-like muzzles, terminal nostrils, strong, large teeth and the ability to distinguish colours. The eyes are set in front of the face and directed downwards so that the animal must lift its heavy eyebrows to look upwards. They live and hunt in groups of up to 200 members and feed on a variety of plant and animal food.

The several species of Baboon are all closely related and vary in size from that of a Labrador Dog to a Mastiff. Fur colour varies from species to species and in most, the thinly-haired tail is of moderate length. In some species the males have long hair on the head and shoulders which forms a mane.

The **Sacred Baboon** *(P. hamadryas),* is native to Somaliland and the Red Sea coast. The males have a pale silver coat and heavy mane. The South African **Chacma Baboon** *(P. ursinus),* is grey-black with a black face and is without a mane. The **Yellow Baboon** has an olive-brown coat and comes from Central Africa, while the **Anubis Baboon** *(P. anubis),* belongs to West Africa.

The **Mandrill** *(Mandrillus sphinx),* is said by some observers to be the ugliest and the most brutal of the Monkey tribe and is native to Cameroun, Gabon and the Congo. The adult male's thick-set body is about 82 cm long and has a stumpy 10 cm tail. The general colour of the hair is dark brown with white cheek whiskers, yellow beard and a dark-brown crest on the crown of the head. The nostrils are broad and the long, deep muzzle of the male has long sabre-like canine teeth with which the Mandrill can inflict terrible and fatal wounds on other, larger animals. The Mandrill's chief habitat is the rain forests, but they spend most of their time on the ground, searching for fruit, roots, insects and reptiles. Mandrills are unique in their gaudily coloured faces, buttocks and genitalia.

Drills *(Mandrillus leucophaeus)*, are closely related to Mandrills but their strong thick-set bodies are slightly smaller and the tail is no more than a stub.

The Guenons

Guenon is a name given to any of the 10–20 species of the Monkey *Cerco-*

pithecus. These range in length from 40–60 cm. They have extremely efficient grasping hands and feet. Their heads are round with snub muzzles. Species of Guenon are found in most parts of Africa and the colour varies according to species. The **Grass** or **Green Guenon,** which is also variously called the **Vervet** and **Grivet,** according to habitat, is a green colour with a black face and white throat and whiskers. The **Talapoin,** which is also green, has white hairs radiating from the cheeks, and is the smallest of the Old World Monkeys.

The **Patas,** which spends much of its time on the ground, has a fast turn of speed which enables it to take refuge up trees in the event of danger. When on the ground, the Monkey sits propped up by its tail like a Kangaroo. The **Moustached** Guenon is olive-green with a bright blue face and a white mark like a moustache across the upper lip. Other varieties of Guenon are the **Owl-faced, Diadem, Mona**—from West Africa but since introduced into the West Indies—and **White-nosed.**

Colobus or Long-Haired Monkey

The **Colobus Monkey** or, as it is sometimes called, the **Guereza,** is the Beau Brummell of the Monkey world with its flowing 'robes' of silky fur. There are three species and all have vestigial thumbs. Colobus Monkeys inhabit dense forests in Africa, from Senegal across to Ethiopia and down to Angola. They live in family groups of about 20 and seldom come down from the trees. Colobus feed oncoarse leaves.

The **Black Colobus** has short black fur with long plumes of white running down his sides and on the tail. Chin, cheeks and forehead are white. The **Red Colobus** has a black body, long white hair on the arms and legs and long chestnut streamers on the back. There is also a grey-green variety.

Although slow-moving, the Colobus can on occasion make long leaps from tree to tree. Their long hair closely resembles some of the lichens that hang from the boughs of the trees and this acts as camouflage to protect them from view. Towards the close of last century, Colobus fur became popular as trimming for women's clothes and the animals were almost hunted to extinction.

Proboscis and Langur Monkeys

The **Proboscis Monkey** *(Nasalis larvatus)*, is a true oddity of the Monkey world. The male has a swollen, pendant nose, usually about 8 cm in length, hanging down over the mouth. Sometimes it curves inward and almost touches the chin. Only the male has this appendage: the females and the young have snub noses. The exact function of

Life in layers

Because different species of African Monkeys live at different levels—arboreal stratification—they can share the same forest and trees without competing with each other.

Tropical rain forests grow very tall, but some of the trees are higher than others. From the view of the Monkey population, their forest home is divided into four main layers. First comes the lowest layer consisting of a tangle of herbaceous and woody vegetation near the ground. Above this is the first layer of trees forming what is known as the lower canopy, between 7 and 12 metres above the ground. Next is the middle canopy, some 15 to 24 metres high. At the top is the upper canopy, which is between 24 metres and 30 metres, made up of the crowns of the highest trees and separated by considerable gaps.

Guenon Monkeys provide a good example of living in layers. At the lowest level live the **De Brazza** species, which are often on the forest floor searching for food. The **Red-tail** species sleep in the middle canopy but descend to the ground by day. **Blue Guenons** live in the upper canopy but often descend to the lower one, while the **Diana** spend most of their time in the upper canopy, though coming down to the middle one during the day. The **Green Guenon** generally prefers the forest floor. Most species prefer a higher canopy at night than the one where they spend the day looking for food.

◁ A Vervet Monkey, known as a Grivet in Kenya, a green Monkey in West Africa and a Tantalus Monkey in the Sudan. Vervets seldom move far from the trees in which they sleep though they will travel up to a few hundred yards away to forage during the day.
▽ The long tail of the Langur Monkey is clearly visible here. Although the Langur's long fingers and toes are ideally suited to climbing they spend a lot of time on the ground.
▷ A male Proboscis Monkey with its pendant nose which gives the species its name. The function of this strange appendage, usually about three inches long, is not known and it does not appear in the females and young which have snub noses. The Proboscis Monkey is found in Borneo where it is known by the name meaning 'white man'.

The disappearing great Apes

The Anthropoid population of the world is declining so quickly that many authorities believe that there is a likelihood that in the future the only ones in existence will be those in zoos. Gorillas particularly are becoming more rare in the wild every year.

Man is the only major predator of the Gorilla. Being a vegetarian, the Gorilla finds an easy meal where there are farm crops. Naturally, the farmers dislike this and organise Gorilla hunts. Another threat to the Gorilla is the hunter who captures young Gorillas for zoos. The mother is usually killed in the process. These threats to the Gorilla's existence are serious because of the animal's slow rate of breeding.

Laboratory demands are also responsible for a decline in the Chimpanzee population. Most of the Chimpanzees caught for laboratories are far too young, while their housing by dealers and their method of being transported take a heavy toll of captured Chimpanzees.

As more and more of the lowland forests, which form the habitat of Orang-utans, are felled to provide agricultural land, these Anthropoids are being forced into smaller and smaller areas. This has resulted in a rapid decline of the Orang-utan population. The animals also suffer reduction at the hands of hunters providing zoo specimens. Often the females are killed to capture the young. Moreover, losses to Orang-utan populations would be difficult to make good as the female bears only one young after a gestation of approximately nine months.

the male's long nose is unknown, but it does cause him to utter a nasal, honking cry. Proboscis Monkeys live in trees in the hot rain-forests of Borneo. They prefer to be near to water and are good swimmers, propelling themselves with a kind of dog-paddle. Adult Monkeys are about 60 cm and have reddish hair with some grey on the lower limbs and back.

The **Langur Monkey** *(Presbytis)*, is widespread through Asia. There are a number of species and one, the Himalayan Langur, braves the cold 12,000 feet up in the mountains. Langurs are of slender build with long fingers and toes, ideally suited for jumping about among trees. The tail is about 76 cm in length and is not prehensile. Most Langur Monkeys are brightly coloured and the colours of the young are unique. The young are dark brown and do not assume the parents' silvery coat until they are about four months old.

The young of the **Spectacled Langur** have a golden-orange coat which gradually changes to the adults' black.

One of the most colourful adult Langurs is the **Douc** from Laos. It is speckled blue-grey with white forearms, black hands and a white tail. The throat is white and the forehead has a red band.

Anthropoid Apes

(Pongidae)

Anthropoid means 'Man-like' and the Anthropoid Apes are the primates nearest to Man in respect of anatomy, physiology and psychology. They are more robust and sturdily built than are most Monkeys and many of their behavioural characteristics seem almost human. The chief anatomical features which distinguish Anthropoids from other primates, and at the same time link them with Man, are their more or less upright posture, their lack of a tail and the greater development of their brains and dental features.

Although Anthropoids often walk on all fours, they have developed to a stage beyond that of the quadrupeds. With the exception of the Gibbons, Anthropoids, when on all fours, do not support their weight on the soles of the feet and the palms of the hands, as do the Monkeys. Only the outer edge of the soles and the knuckles of the second digits rest on the ground.

The structure of anthropoid limbs show a similarity to those of Man rather than to those of the Monkey. Some Anthropoids occasionally walk erect or very nearly so.

Although the cranial capacity of Anthropoids is inferior to that of Man, it shows a high degree of development. Moreover, the basic disposition of the convolutions of the human and anthropoid brain is almost identical. Anthropoids, and again particularly the Chimpanzee, have a well-developed ability to learn. This indicates a developing intelligence comparable with that of man.

Anthropoid Apes are now represented by two groups native to Asia, the Gibbons and the Orang-utans, and two groups found in Africa, the Gorilla and the Chimpanzee. In view of the striking similarity between Man and the Anthropoids, the habitats of the Anthropoids seem significant in terms of human evolution. Most of the fossil and other relics of early man have been found in Africa and Asia.

Gibbons and Siamangs

The **Gibbon** *(Hylobates)*, is the smallest of the Apes but the most agile of the mammals. Standing upright, the Gibbon is about a metre tall. It appears very awkward on the ground and moves with difficulty as it hurries forward, holding its arms over its head to maintain balance. The arms are so long that they nearly touch the ground when the animal is in an erect posture. However, in the trees, the Gibbon moves with grace as it swings from branch to branch, and walks along some by

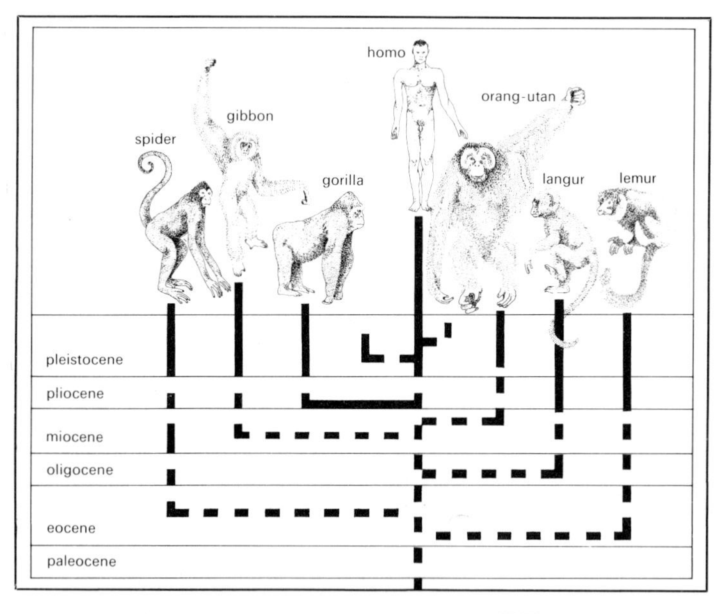

△ Hooked fingers, long and strong, and an extremely mobile thumb enable Gibbons to hang easily and to grasp onto branches as they swing through the air. Gibbons are the only Primates which regularly travel by brachiatation—swinging hand over hand with the legs drawn up clear of any obstacles.

gripping them with its great toes. The fingers are long and the thumbs appear to be so because they are deeply cleft from the palms of the hands. The thumbs are exceptionally mobile so that the animal is expert at manipulating any object it grasps. The nails are clawlike.

There are six species of Gibbon and all are native to south-east Asia. The largest species, the **Siamang** *(Symphalangus syndactylus),* stands 1.25 metres high when upright. The Siamang has a throat sac that can be inflated to give its characteristic call, which sounds like a combination of a dog bark and a grouse hoot. Gibbons live in family groups and feed on fruit, leaves, plant shoots, birds' eggs and nestlings.

Man's place in the Animal Kingdom

There is only one living species of **Man** *(Homo sapiens),* and he is intellectually the highest of the primates. He probably derived from a group of primitive Apes, *Dryopithecus* and *Palaeopithecus,* which existed some 25,000,000 years ago. It is thought that *Homo sapiens* branched off from the Apes about 1,500,000 years ago.

As Man slowly evolved from his Ape ancestors, he was living in a hostile world and the wonder is that he survived. Of all mammals, Man was a puny creature and the least protected by nature. He was without body fur to keep him warm, without massive teeth and claws with which to defend himself and without wings which would enable him to fly from the thousands of dangers that constantly threatened him.

But if Man was not developing the strength of mammals many times his size, he was more adaptable and his brain was becoming more highly specialised than that of any Ape. Man learned how to make fires to keep himself warm; he taught himself to make weapons and tools which were far more efficient than claws and teeth. He also developed speech; something which no other mammal has succeeded in doing. In spite of Man's puny physique, he was to make himself master of the earth. Other animals were there simply to provide him with food and clothing.

Although Man is a single species, ethnologists classify him into eight geographical races:

European, whose original homes were Europe, western Asia and Africa, north of the Sahara. The males tend to have considerable body and facial hair. The average height is between 1.5 and 1.8 metres.

African, who inhabit most of Africa south of the Sahara, The skin is dark and the body height generally exceeds that of Europeans. The teeth and jaws generally protrude and the teeth are on average bigger than those of other geographical races.

Asiatic, who inhabit much of Asia southward to Burma and Java, and eastwards to the Philippines and north-east to Japan and the Kurile Islands. There are small local pockets in the Aleutians, Alaska, northern Canada and Greenland. Asiatics are characterised by a fold of skin (the Mongoloid fold), overlapping the inner angle of the eye; straight black hair and wide projecting cheek bones.

Indian, who inhabit the sub-continent of India. There are many sub-races and local races, but in general, the whole race is dark-skinned with straight to wavy black hair. The height varies as much as that of the Europeans.

Australian, whose habitat is limited to the island-continent of Australia. Australian aborigines are sometimes described as primitive whites. They often have considerable hair on the face and body and there are occasional blondes. Prominent brow ridges suggest that the Australian aborigines may be Neanderthal survivors.

Malaysian, who are the natives of New Guinea and the Melanesian Island. They have certain Negro-like characteristics, such as dark skins and deeply-waved and often mop-like hair. They vary in stature from very tall to almost pygmy.

Polynesian, who inhabit a broad area from New Zealand to Easter Island. They appear to be closely related to the Asiatic race.

Amerindian, inhabiting the Americas from southern Alaska to the tip of South America. The straight black hair, Mongoloid folds and broad and projecting cheek bones are indicative of an Asiatic origin. The 90 per cent frequency of blood type M distinguishes the Amerindians from all other geographical races.

Dividing Man into geographical races, blurred by 2,000 years of intermarriage is a purely ethnic convenience.

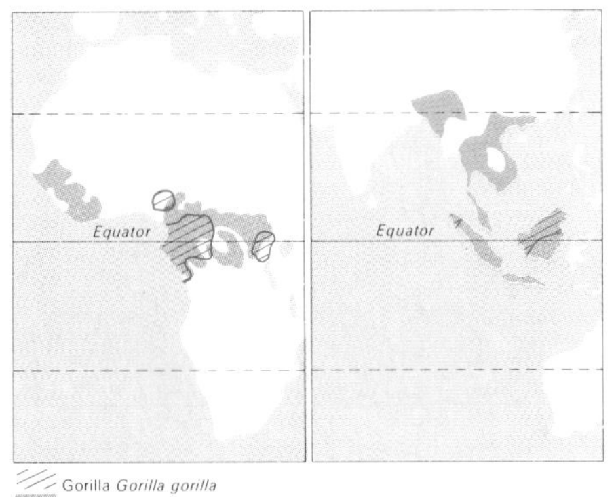

Gorilla *Gorilla gorilla*
Chimpanzee *Pan troglodytes*
Orang-utan *Pongo pygmaeus*
Gibbon Genus *Hylobates*

◁ ◁ A chart showing the evolution of Apes and Man.
◁ A White-browed Gibbon which, in common with the other Gibbons, has adapted so well to a tree-dwelling life that when running on the ground its long arms will be held above its head so that it can avoid treading on them.

The Orang-utan

(Pongo pygmaeus)

The **Orang-utan** is a heavy-weight amongst the Apes. The adult male stands 1.5 metres high and weighs up to 100 kg. Females are 20 cm shorter and seldom weigh more than 50 kg. In spite of its relatively short stature, the Orang has an arm-spread of as much as 2 metres. The arms are longer than the legs, and both hands and feet are long and narrow, to give the animal a good grasp as it swings from branch to branch and from tree to tree.

The Orang's reddish coat is so sparse that in many places the coarse grey skin shows through. The face is virtually hairless except for a moustache and beard. The male develops large cheek pouches, the function of which is unknown, and both male and female have larynx pouches which give the neck a swollen appearance. Orangs build themselves nests for sleeping—sometimes a new one each night. Their food is chiefly buds and fruit.

The Orang-utan is found in the forests of Borneo and Sumatra. The adult male, left, displays the characteristic goitre-like throat-sac and the large cheek pouches on an almost hairless face. The shaggy, chestnut coloured fur is sparse in places and thicker on the male than the female.

Orangs are sexually mature at ten years. Gestation is about $8\frac{1}{2}$ months. Only one young is born and the baby is breast fed for eighteen months. Adult males live by themselves except in the mating season. 'Orang-utan' is Malay for 'man of the woods', and certainly the animal's facial resemblance to Man is marked, the upward development of the head producing a high, arched forehead.

Orangs are native to the low-lying forests of Borneo and Sumatra. The **Sumatran** *(Pongo pygmaeus pygmaeus)*, is slimmer and lighter-coloured and does not have such big cheek pouches as that of the **Borneo Utan** *(Pongo pygmaeus abeli)*.

The Chimpanzees

There is only one species of Chimpanzee *(Pan troglodytes)*. The colour of the coat ranges from jet black to faded brown or grey-black. Some walk almost erect while others lope. It is the most widely spread of all the Anthropoids, its habitat extending from Sierra Leone and along the west coast of Africa to the Great Lakes east of the Congo.

The average male Chimpanzee is about 1.5 metres tall with big, turned-out ears and particularly expressive eyes.

Primate intelligence

Primate intelligence, particularly that of the Anthropoid Apes, is based on perpetual curiosity, the ability to perceive the importance of a fact, and the ability to initiate speculative reactions.

Chimpanzees are the most gifted intellectually of the Anthropoid Apes. The average young Chimpanzee is as intelligent as a human child before the child learns to take its first steps. But that is the limit of the Chimpanzee's natural intelligence.

Its natural intelligence renders the Chimpanzee less dependent on tooth and claw for its defence than are other animals. When a Chimpanzee is cornered in the wild, it has been seen to pick up stones and throw them at the enemy.

A further measure of Anthropoid intelligence is that Chimpanzees are the best tool-users apart from Man. For instance, stones are used as hammers to crack open nuts. The use of the 'tool' represents an extra stage in the action of food collection. The stone used for cracking nuts or the twig used for poking food out of the ground is no use as food but its acquisition implies reasoning and forethought, the ability to analyse beyond the immediate problem of picking up food. These are very special attributes, biologically far superior to simple automatic behaviour and using tools is indicative of these abilities.

Much of Anthropoid intelligence depends upon an ability to imitate. This is seen amongst baby Chimpanzees which have been observed to play with the elementary tools discarded by their elders after having watched them being used.

It is the Chimpanzee's capacity to imitate allied to its natural desire to please that makes it such a rich subject for laboratory-controlled intelligence tests. Such tests are, however, illustrations of acquired intelligence, and many other animals can acquire intelligence. Where the Anthropoids score on acquired intelligence is that their forepaws can be used like hands, enabling them to perform much more elaborate tasks than could the hooves of a cow, no matter how intelligent the animal was.

◁ A female Chimpanzee with its young clinging on tightly uses its arms to climb down out of a tree. Chimpanzees spend a lot of time on the ground and live in loose communities without fixed leaders. Clearly visible in the picture are the characteristic large ears.

Family life of Chimpanzees

Much of our knowledge of Chimpanzee family life in the wild is derived from observations made by the English naturalist Jane Goodall. Miss Goodall began her studies in 1961, when she set up camp in Gombe National Park, Tanzania. There were between 150 and 200 Chimpanzees in the mountain section of the park where she established her main observation post.

All the Chimpanzees were wild, intensely timid of humans, and initially anxious to hide from the observer. Over a year elapsed before Miss Goodall, by persistent tracking, succeeded in getting within 12 metres of them. Many more patient months of watching were spent before she was able to get close enough to watch them carrying out their normal activities.

One important result of Miss Goodall's observations was to establish that Chimpanzees have a code of manners. When two Chimpanzees meet in the forest, they frequently rise erect on their hind legs, clasp each other's hands and shake them vigorously. That there is class distinction among Chimpanzees was proved when one Ape, apparently of high status, held out his hand to some of his fellows of lower status. The underlings then solemnly pressed their lips on to the back of the superior animal's hand.

But Miss Goodall's greatest discovery was that Chimpanzees hold a kind of ritual sacrifice. One day, she observed a Red Colobus Monkey sitting in a tree. A young male Chimpanzee climbed an adjacent tree and made a slight movement that attracted the smaller Monkey's attention. The Chimpanzee then remained very still as the Colobus looked towards it.

Some minutes later, another young male Chimpanzee climbed the tree in which the Colobus was sitting. Running quickly along the branch, the Chimpanzee caught the Monkey in its hands and broke its neck.

Immediately the Colobus had been killed, the rest of the Chimpanzee group climbed up the tree, where the successful hunter tore his quarry into pieces and shared it with his fellows.

Miss Goodall later observed similar killings of young Bush Bucks and Bush Pigs. She has suggested in her writings that, in her opinion, killing and eating animals may have some ceremonial significance to Chimpanzees. Normally, the gathering of the fruit and berries which form the Chimpanzee's normal diet is done at random by all members of a group.

The coat is long and coarse and generally black or mahogany brown, except for a white patch near the rump. The face, ears, hands and feet are naked and flesh-coloured. The black face is also naked. The arms are longer than the legs and the animals normally run on all fours, although they can walk upright with ease. Females are generally one-fifth shorter than males.

Chimpanzees live in small parties, in trees where they make themselves sleeping quarters of vines and branches. Males are part of a social order, the older males having dominance over the younger. The animals' diet consists of leaves, roots, insects, eggs and small chicks, and fruit—a Chimpanzee can dispose of up to 30 bananas at one time. They occasionally kill other animals to obtain meat.

Chimpanzees are completely promiscuous; when a female comes on heat, all the males in the group will mate with her. Gestation is about nine months and a single young is born. A young Chimpanzee is dependent upon its mother for at least two years after birth, and female Chimpanzees are generally good, affectionate mothers.

The **Bonobo,** or **Dwarf Chimpanzee** *(Pan paniscus)*, whose habitat is the south Congo, is regarded by some as a member of the same species as *Pan troglodytes* and is about one-third of the size. The Bonobo fur is darker.

Gorilla: The greatest of the Apes

The **Gorilla** is the giant of the Ape world and the biggest and heaviest of all the primates. There is only one species, the **Western Gorilla** *(Gorilla gorilla gorilla)* of the Congo, Gabon and south-east Nigeria, and two sub-species: the **Mountain Gorilla** *(Gorilla gorilla beringei)* of Mount Kahuzi in the Congo; and the **Eastern Lowland Gorilla** *(Gorilla gorilla graueri)*, of the eastern Congo and the area of the central African lakes. Basically, there is little difference between the three.

The average adult male Gorilla stands 2 metres high when erect and can weigh up to 225 kg. The male has a large sagittal or bony crest on top of his skull giving a helmet-like effect to the head. The chest is broad and the neck short and muscular. The Western Gorilla's hair is a greyish-black or very dark brown, while that of the Eastern is jet black. The nostrils are broad and the ears are small. The hands and feet are strong and broad with the great toe less widely separated from the other toes than is usual in Apes. The Eastern Gorilla has an extraordinarily Man-like foot. Female Gorillas are about 28 cm shorter than the males and considerably less heavy.

Gorillas live in groups of 6 or so females and young, led by a dominant male, whose silver-grey markings on the back indicate his seniority.

Gorillas spend quite a lot of the daylight hours on the ground, where they normally walk on all fours, their arms supported by the knuckles of the middle fingers. At night, the group nests in a tree. The nests are made from twisted branches covered with leaves. Nests for the young are on the higher branches, out of danger, while the females and the young males nest lower down. The leader of the group settles down for the night at the foot of the tree with his arms folded across his chest, dozing, ever alert to protect his family.

Gorillas are strict vegetarians. They rarely drink, getting most of the moisture they need from their food. When they do have occasion to drink, they soak the fur on the back of the hand and then suck the water from it.

The leader of a group does not necessarily mate with the group's female members. Often a solitary, wandering male joins the group for a time and mates with the females. Gestation is about 8–9 months and the young are walking at ten months. A male is sexually mature at 8 to 9 years and the female a year earlier. The average life-span is about 28 years.

The gentle Gorilla

The Gorilla has become a symbol of terror to Man. Standing erect at full height, with its massive chest and huge arms spread, the enormous face is ferocious and frightening, with teeth enclosed in a gigantic jaw. But the Gorilla is not the King Kong of the horror films and science fiction. In fact, the Gorilla is probably the least offensive of the Anthropoid Apes.

When a Gorilla senses danger, it stands on its hind legs and beats its chest with its huge arms. Next, it runs sideways for a short distance, drops down on all fours and breaks into a full-speed dash. At the same time, the animal wrenches branches from trees and slaps violently at everything in reach. Finally the Gorilla sits down and thumps the ground with the palms of his hands.

When the display is all over, the Gorilla stares at the intruder and again beats its chest. If the intruder, animal or human, does not move the Gorilla turns and lumbers away. It rarely, if ever, attacks even if advanced upon. Certainly, a Gorilla would never make an unprovoked attack.

The Gorilla's display of ferocity in the wild is not so much a display of aggression as a substitute for it. They use the display even with their own kind. Gorillas bluff and threaten each other, but very seldom actually fight.

Among tribes that hunt Gorillas, it is considered a disgrace to be wounded by the Ape, because a normal Gorilla will only attack the intruder who turns and runs.

The Giants of the Ocean

Intelligent, friendly and sadly put-upon (the Right Whale got its name from being the right whale to kill) the Whales, Dolphins and Porpoises are found in waters the world over.

There are altogether about 90 different species of Whales, Dolphins and Porpoises and they are to be found in waters all over the world.

They fall into two clearly defined groups: the Whalebone Whales (Mysticetes), of which there are 12 species; and the Toothed Whales (Odontocetes), about 80 species.

Whalebone Whales have no teeth. Instead their upper jaws are equipped with whalebone or baleen. This is a series of closely packed horny outgrowths from the roof of the mouth. The main diet of these creatures is plankton, minute marine organisms. These are taken into the Whale's mouth in great gulps of water. The tongue is then raised to squeeze out the water through the baleen, which acts as a sieve, leaving the plankton behind to be swallowed down the Whale's throat. The largest Whales are members of the Whalebone group, which includes the giant Blue Whale and the other Rorquals, the Right Whales and the Gray Whale.

Toothed Whales, as the name suggests, possess teeth. They feed mainly on a variety of fish and some of the bigger invertebrates such as Squid or Cuttle-fish. Dolphins and Porpoises belong to this particular group, the largest member of which is the massive Sperm Whale.

No other group of mammals is so distinct in character as Whales. They provide a striking example of total adaptation to their environment. They swim by up-and-down strokes of a horizontally flattened tail, and while they still need to breathe air at the water surface, they are also able to remain totally submerged for periods of up to an hour.

Right Whales

(Balaenidae)

There are five different species of **Right Whales: Greenland, North Atlantic, Pacific, Southern** and **Pygmy**—all members of the Whalebone *(Mysticeti)* group. Like all Whalebone Whales they have no teeth and filter their food through baleen plates growing down from the roof of the mouth. Except for the Pygmy Right which only reaches 20 feet (3.5 metres) in length and has a dorsal fin, the Right Whales are large animals without grooves under the throat, or a dorsal fin.

The biggest Right Whales can grow up to more than 50 feet (16 metres) in length, and the head takes up one-third to one-fourth of its body.

The Right Whales were amongst the earliest of the large Whales to be hunted by men. This is because they swim slowly and float at the surface of the sea when they are dead, and do not sink as the Rorquals do. The North Atlantic Right Whale was hunted in the Bay of Biscay by Basque fishermen from the 10th to the 16th century for its oil and the long and abundant baleen plates. Then after the discovery of Spitzbergen in 1596, the European whalers turned their attention to the more numerous Greenland Right Whale in that area.

Right Whales are found in polar and temperate seas.

◁ A Minke Whale breaks surface. Like all Mammals, Whales need to breathe air.
▽ Whales fall into two distinct groups. The Sperm Whale, top, is a Toothed Whale and the Right Whale bottom, is a Whalebone Whale. The latter strain planktonic food from the sea using the filter arrangement of horny plates in the mouth. Toothed Whales feed mainly on fish and squid. In both these two the rudiments of the hind-limb skeleton and pelvic girdle are visible.

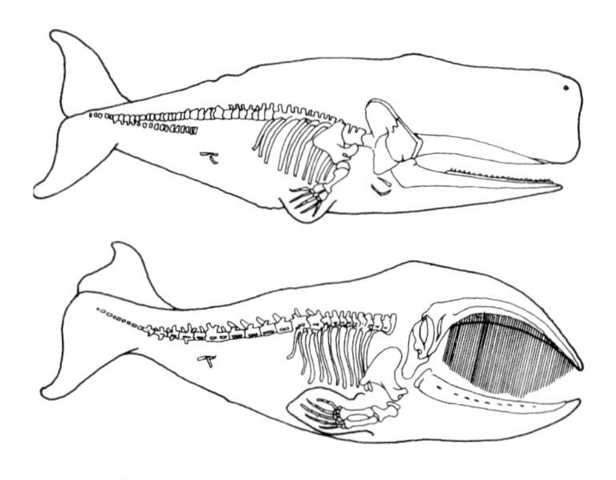

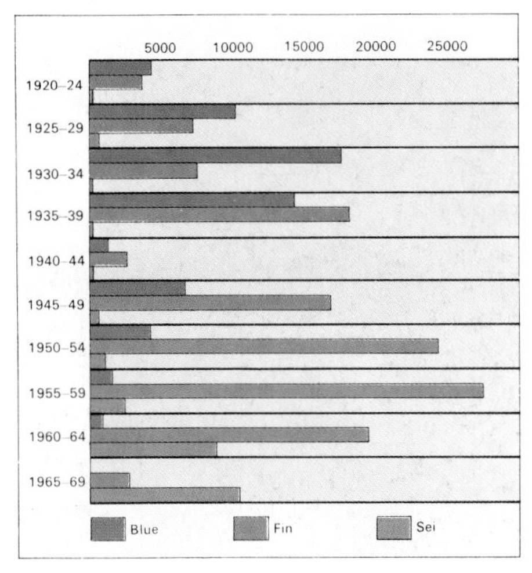

△ The chart shows the number of Whales killed in five year periods. First the Blue, then the Fin were hunted till their populations declined. Then it was the turn of the Sei.

The Rorquals

(Balaenopteridae)

Rorqual Whales are sometimes known as **Finners.** There are six species: the **Blue Whale, Fin Whale** (or **Common Rorqual**), **Sei Whale, Minke Whale** (or **Lesser Rorqual**), **Bryde's Whale** and **Humpback Whale.** They are all members of the Whalebone group but easily distinguished from the Right Whales by the deep parallel grooves on the throat and by the presence of a fin on the back set far towards the tail.

The Blue Whale is the largest of all mammals. It reaches a length of nearly 100 feet (30 metres) and weighs well over 100 tons. The heaviest Whale accurately weighed was an 89 foot (27 metres) animal of 127 tons, so the biggest Blue Whales probably approach 190 tons in weight. One of the hazards in hunting the Rorquals is that when dead, they sink to the bottom of the sea, rising only when decomposition gases form inside the body.

The streamlined, long slender-bodied Rorquals, that is, all the species except the Humpback Whale, are very fast swimmers. The Sei Whale is reputed to be the fastest swimmer, and is said to travel for short periods at a speed of 35 knots.

Most Rorquals tend to congregate and feed in the cold waters of the Arctic and Antarctic, which are rich in food, in summer. They breed in the sub-tropical and tropical oceans in the winter. Consequently they cover vast distances in their yearly migrations between these areas.

Save the Whale

Although man has been hunting and killing Whales for centuries—Eskimos were known to have been catching them as early as 1500 BC—the whaling industry as we know it today, did not really start developing properly until the latter part of the 19th century.

This resulted almost directly from the Norwegian invention, in the 1860's, of the harpoon gun mounted on a steam powered catching vessel and firing an explosive harpoon. Until then killing Whales had been a long-drawn-out and bloody process: now they could be dispatched quickly and efficiently. And they were—in their thousands. In the 1920's factory ships operated on the high seas for the first time, instead of working in anchorages. This meant that even more Whales ended up under the flensing knife.

In 1946, a group of 15 countries, alarmed at the rapid decline in the world's Whale population, set up the International Whaling Commission. Immediately the Commission laid down certain regulations. These included setting overall catch limits for some species and imposing limited hunting seasons.

The number of Whales caught annually is still high—about 30,000. There is mounting pressure from many quarters to restrict the whalers further. Otherwise, it is felt, the time might come when some of these great creatures of the sea could disappear for ever. However, the present catch limits are set on the best scientific advice, and this should ensure that the most depleted species increase in numbers. At the same time the more plentiful species are now being sensibly harvested.

The Gray Whale

(Eschrichtidae)

The **Gray Whale,** a member of the Whalebone group, has characteristics between those of the Right Whales and the Rorquals. Like the Right Whale, it has no dorsal fin and it is also a slow swimmer, although it is reasonably streamlined in shape. It grows to a maximum length of about 15 metres.

Gray Whales are now found only in the North Pacific. They breed in the shallow lagoons along the coast of Mexico in the winter months, and then move to the summer feeding grounds off Alaska. This round trip of some 11,000 miles each year is the longest made by any whale.

Remains of Gray Whales have been found in the North Atlantic. There was a danger that they would be wiped out in the Pacific, but fortunately international protection has allowed the numbers to increase to safe levels again.

Sperm Whales

(Physeteridae)

The **Sperm Whale** is the most numerous of the large Whales, with a world-wide population of about a million and a quarter animals. Almost one-third of the Sperm Whale's body is taken up by its head. This contains a mass of tissue filled with spermaceti—a waxy substance used to lubricate delicate machinery and included in the best quality candles.

As well as spermaceti, the Sperm Whale also yields another valuable substance—ambergris—found in its intestines. This is used in the manufacture of expensive perfumes.

The female Sperm Whale does not usually exceed 40 feet (13 metres) in length, but the male can grow up to 60 feet (20 metres) and reach a weight of 50 tons. It is one of the Whale family's best divers. It can remain submerged for over an hour, thanks to its slow heart beat and low oxygen consumption. The Sperm Whale is also apparently an extraordinarily deep sleeper and floats near the surface for hours. This is suggested by the number of ships which have collided with them.

It feeds mainly on Squid. These are sometimes 20 feet (6 metres) long and the scars found on Sperm Whales' heads show the tremendous battles that must have taken place.

The two Pygmy Sperm Whales are not common, but are widely distributed. They grow to a maximum length of 13 ft (4 m) and feed on Squid.

Beaked Whales

(Ziphiidae)

Beaked Whales possess the long jaws of their ancestors but not their teeth.

▷ Whales migrate seasonally between warm and cold areas of ocean. The map shows their routes.
▽ The Killer Whale, characteristically marked in black and white, is a fierce marine carnivore and frequently feeds on seals, seabirds, fish and even other whales. They hunt in packs and are sometimes called the wolves of the sea.
▽▷ Some of the different species of Whales.

Apart from one species they have, in fact, usually only one or two pairs of visible teeth in the lower jaw. They feed by swallowing their prey whole.

There are altogether 18 members of the Beaked Whale family. These include **Sowerby's Whale,** the **Strap-toothed Whale, Cuvier's Beaked Whale, Blainville's Beaked Whale** and the two **Bottlenosed Whales.**

The Bottlenosed Whales, which are found in both northern and southern hemispheres, feed on fish and Squid. They are gregarious creatures and often travel in closely-knit groups of between five and ten individuals.

When fully grown, the Bottlenose Whales reach 30 feet (10 metres). Mature animals have a very prominent forehead containing a substance similar to the Sperm Whale's spermaceti.

The Narwhal and White Whale

(Monodontidae)

There are only two members in this particular family, the **Narwhal** and the **White Whale** or—as it is sometimes known—the **Beluga.**

The Narwhal is a curious creature which grows up to 15 feet (3 metres) long. The males develop a spirally twisted tusk which caused ancient writers to name it the sea unicorn. This tusk was believed to possess magical properties and was ground into powder for medicinal purposes up until the mid-17th century.

The Beluga, like the Narwhal, has a rounded head and no dorsal fin. It is a coastal species, often moving up rivers in large schools. It is valued chiefly for its oil and skin, which is tanned into 'Porpoise-hide', a kind of leather.

Both the Beluga and the Narwhal frequent Arctic and northern waters.

The Killer Whale

(Orcinus orca)

The ferocious **Killer Whale** is sometimes named the Wolf of the sea. This is because it hunts in packs of 20 or more animals, and attacks Seals and young or weak Whales.

The normal diet consists mainly of fish and Squid, but the Killer Whale is the only member of the Cetacean group which preys upon other sea-going mammals. A Killer Whale caught in the Bering Sea was found to contain the remains of 32 adult Seals in its stomach, while another had parts of 13 Porpoises as well as 14 Seals. Like all Odontocetes, (Toothed Whales), the Killer Whale never chews; it either swallows its food whole or in large chunks. There are about 20 teeth in both the upper and the lower jaws which interlock when the mouth is closed to give a strong bite.

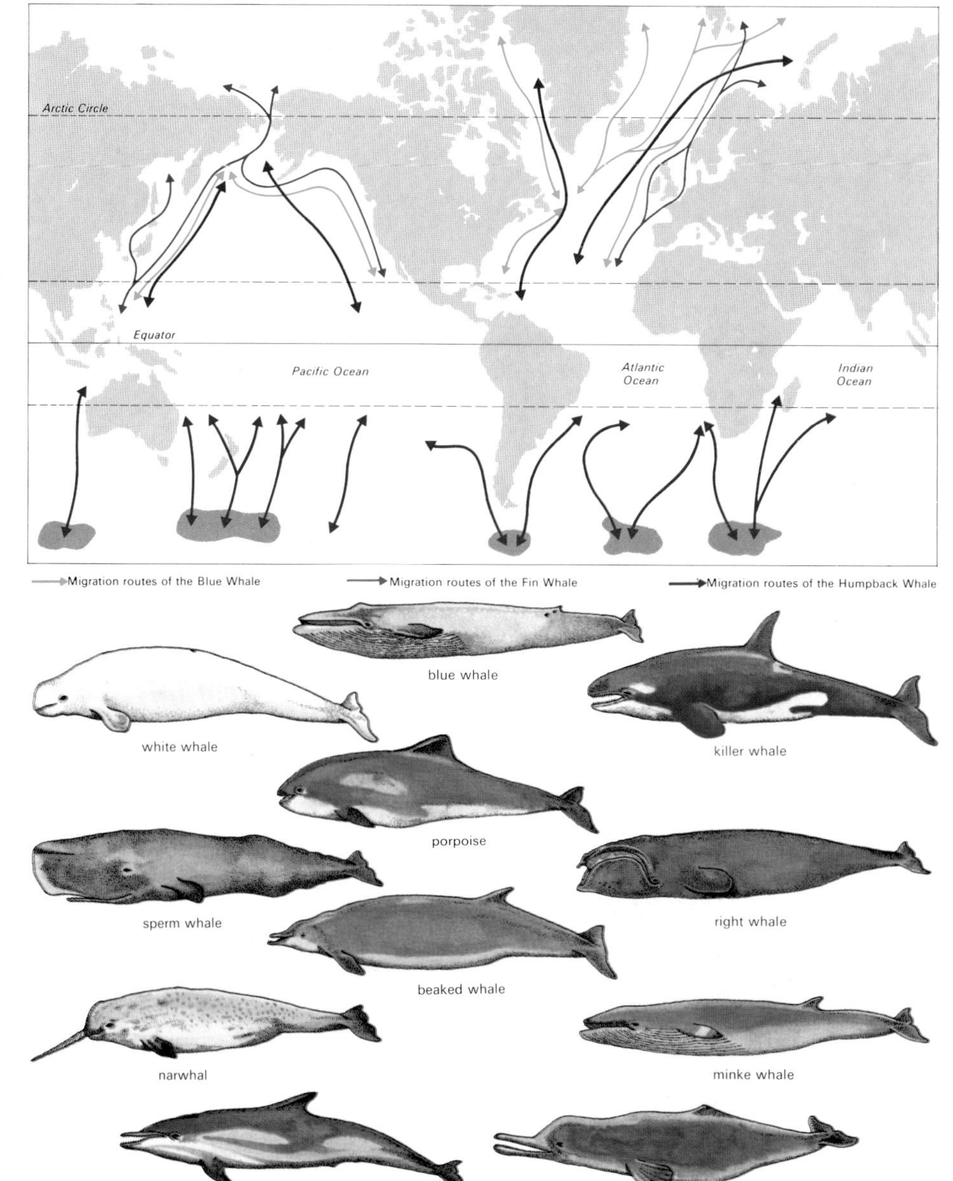

Intelligence levels

It has been suggested that Dolphins have a level of intelligence second only to man. Whether they have or not is debatable. But they do appear to possess remarkable abilities in various areas—particularly in vocal communication.

They share this capacity to 'talk to each other' through sound signals with other members of the cetacean group. Claims are made that Dolphins in captivity have been taught to count from one to ten, call for more fish, and utter the trainer's name. As well as their verbal prowess dolphins seem able to perform any number of tricks to order, including ringing bells, jumping through hoops and knocking balls into nets.

All the cetaceans appear to have a natural built-in sonar system which enables them to navigate by 'echo-sounding' in much the same way as Bats.

Cetaceans certainly seem to regard each other with almost human affection. They have been observed supporting wounded companions and even making rescue attempts when one of their number has been caught in a net.

One reason given for the apparent intelligence of Dolphins and other cetaceans is that they are carnivorous. As such they have to think about catching their food, unlike herbivores who merely eat where they find it.

The Killer Whale, which is found in waters all over the world, is easily distinguished by its black colour above and white below. There is also a characteristic white patch behind and above each eye. The males grow up to 30 feet (10 metres) long—about half as big again as the females—and have larger, more pointed and vertical dorsal fins.

In spite of their fearsome reputation it has proved relatively easy to train Killer Whales in captivity.

Dolphins and Porpoises

These two creatures are often confused with one another due to the fact that their general shape is so similar.

However, **Porpoises** are generally smaller than **Dolphins,** reaching a maximum length of about 6 feet (2 metres). Porpoises' noses are blunt—though their jaws are well furnished with teeth—while Dolphins' noses are long and beak-like. Their dorsal fins are also shaped differently. Dolphin fins curve back into a sharp point: the porpoise fins are more triangular.

The Common Dolphin

The **Common Dolphin** is less able to adjust to a life in captivity than the **Bottlenosed Dolphin.** It lives in schools of 100 or more and moves its long streamlined body through the sea at speeds approaching 25 knots. Like many members of the Dolphin family, it is a voracious fish-eater.

The Bottlenosed Dolphin

(Tursiops truncatus)

This is the creature which delights crowds all over the world with its amazing acrobatic feats in aquariums. It is larger than the Common Dolphin, and is very easily tamed once caught. Bottlenosed Dolphins have been shown to communicate with each other by whistles and other sounds, and it is also claimed that they can imitate the human voice.

▽ Pilot Whales coming to the surface to breathe. Like most Whales they usually travel in small family groups called 'schools'.

▽▽ A White-sided Dolphin, a Pacific Ocean species, seen in captivity where it can be taught to perform quite complex tricks. US Navy divers have taught Dolphins to patrol anchorages against intruding frogmen bent on sabotage.

River Dolphins

(Platanistidae)

There are four members of the Dolphin family which inhabit fresh water; the **La Plata Dolphin** from the region of the River Plate; the **Amazon Dolphin** from the River Amazon; the **Yangtze Dolphin** from the Chinese river of the same name and the **Gangetic Dolphin** which frequents the Rivers Ganges, Indus and Brahmaputra in India.

The ancestors of the River Dolphins lived in the sea at one time, and were probably isolated by geological changes. River Dolphins feed on fish and Shrimps.

Toothless Armoured Mammals

When the Age of Reptiles had passed, the great growth period of mammals began. In the Afro-Asian continents the Placental Mammals were most successful whilst the Marsupials were strongest in Australasia. In the American continent, however, a further group of mammals developed. These were the Edentates or 'toothless ones', their Afro-Asian counterpart being the Pholidota.

Most Edentates are now extinct but a few are still in existence; the Armadillos, the Anteaters, and the Sloths. The last two are known as the 'Hairy Edentates', the Armadillos being armoured in varying degrees.

The Anteaters, despite their name, prefer to eat termites, but are equipped to break open the rockhard ant-hills and do eat these insects. Ants have a tough, chitinous outer covering which is very indigestible, but the stomach of the Anteater copes by grinding its contents rather as the crop of a bird.

The Sloth, known derisively to the Spaniards as 'nimble Peter', is renowned for the slowness and deliberation of its movements. A Sloth clearly in a hurry has been timed and found to move at about 4 metres a minute.

Its very lethargy makes the animal difficult to spot amongst the branches and this probably contributes to its survival in a jungle teeming with predators. They are also extremely hardy, recovering from injuries which would kill most other animals.

Although Sloths seldom leave the trees they are surprisingly good swimmers and, due to their slow metabolism, can remain submerged for up to half an hour without harm.

Armadillos are found all over South America except the extreme south and in the heights of the Andes. They are armoured with plates of bone covered with horny scales, the plates being separated by bands of tough, flexible skin. This arrangement permits freedom of movement combined with effective protection. Unfortunately, some Armadillos are fond of sunbathing, lying on their backs and exposing their unprotected abdomens. Often, whilst doing this they fall asleep and become easy prey for a predator.

Glyptodonts were an ancient species of Giant Armadillo with a one-piece carapace, similar to that of a turtle. These animals were large and cumbersome, growing to some 3 m in length. Although these creatures have been extinct for thousands of years, many shells have been found, with bones and artifacts under some of them, indicating that they were used by early Man as ready-made shelters.

◁ The Tamandua is an arboreal animal using its long claws and prehensile tail to grip tree trunks and branches. The long snout is a toothless tube housing a long sticky tongue which is used for collecting ants and termites. The Tamandua uses its powerful front claws to break open the termite nest before extending its tongue through the nest's tortuous passages, taking a great number of insects at one time. It is the second largest of the new world anteaters.

The Giant Anteater

(Myrmecophaga)

The **Giant Anteater** *(Myrmecophaga)* is a remarkable looking animal with its tapering, tube-like snout, long bushy tail and short but powerful legs.

The front feet are equipped with four long claws which are used to rip open termite nests and anthills. These claws are non-retractable and their length causes the animal to walk on the sides of its front paws, giving it a most awkward gait.

When an insect nest has been opened the long snout comes into action, probing the galleries. A sticky saliva coats the tongue, trapping the ants and termites and holding them whilst they are transferred to the mouth.

The Giant Anteater is a ground dwelling animal but is capable of climbing. It prefers grassland or open forest areas and yet is a powerful swimmer and can cross wide stretches of water. It is not an aggressive animal but, if it is attacked, it will defend itself by rearing up on its hind legs, using its long, sharp claws as effective weapons.

Tamandua

The **Tamandua** is a close relative of the Giant Anteater, but is smaller, about 90 centimetres in length. Its snout, whilst remaining a toothless tube, is less elongated.

Together with *Cyclopes*, the squirrel-sized **Pygmy Anteater,** the Tamandua is an animal as much at home in the trees as it is on the ground. Both these creatures have prehensile tails which they use for both grip and balance.

They are nocturnal animals, Tamandua spending the days in bushes, Cyclopes remaining in the tree-tops. At night they emerge to feed on insects. Tamandua in particular has an astonishing facility with its long tongue, following termites along the galleries of the nest and eating thousands of insects in a single raid.

Anteaters appear to be unworried by the retaliatory attacks of ants or termites, and despite their size, both the Tamandua and the Pygmy Anteater can defend themselves strongly.

△The Two-toed Sloth spends all its time in trees where it hangs upside down. Its limbs and feet are adapted for this upside down existence and its fur slopes from the belly towards the back, the reverse direction to that of other Mammals. This is to prevent the fur from becoming waterlogged.

Two-Toed Sloths

(Choloepus didactylus; Choloepus hoffmanni)

There are two kinds of **Sloths,** the **Three-fingered,** with three digits on all its paws, and the **Two-toed,** with two on its forefeet and three on its hind limbs.

Within the Two-toed genus *Choloepus didactylus,* comes from Brazil, and *Choloepus hoffmanni,* is found northwards from Ecuador. Two-toed Sloths are peculiar in that they have only six neck vertebrae, whilst most other mammals have seven.

The toes of the Sloth are sheathed in a sleeve of skin and muscle, the fore digits terminating in large claws curved like stevedore's baling hooks. These claws anchor the animal to tree branches where it spends the greater part of its life hanging upside down. Despite being very slow, the Sloth climbs easily, but on the ground it is almost helpless, moving only in an ungainly crawl.

The diet of the Two-toed Sloth is made up of fruit and leafy branches from whatever tree the animal is occupying. Although it is an Edentate, it is equipped with cheek teeth capable of chewing the toughest leaves.

South American dilemma

Early man, during his migrations through South America, was responsible for the extermination of entire species of animals. Modern man, in his rush to exploit the continent is endangering many others.

The development of natural resources for the benefit of a growing population is, of course, of first importance to the Government of any country. South America is potentially the richest territory on Earth, but the standard of living of the average citizen is very low.

Good housing, sufficient food, and work depend on clearing jungle, creating farms and settlements out of wild country, and the exploitation of mineral resources. All this, unless the greatest care is taken, is incompatible with nature conservation. The wealthy countries can support conservation programmes; the poorer cannot, and find it difficult to justify the expenditure of large sums of money on wildlife conservation when the people are hungry and homeless.

However, most South American countries have programmes, with National Parks, nature reserves and forest areas. Brazil and Argentina have the most, many of them up to International Union for Conservation of Nature standards. Venezuela is probably the most advanced and is, in many ways, the headquarters of the South American conservation movement.

Given this sort of Governmental goodwill, South American fauna has a brighter future than at any time since the arrival of mankind.

Three-Fingered Sloths

The **Three-fingered Sloths** *(Bradypus)* have three digits on both fore and hind paws, each equipped with long curved claws typical of the Edentate. They are found in most of the rain forest areas of South America, also inhabiting parts of Central America. Three-fingered Sloths have nine neck vertebrae.

According to many authorities, the Three-fingered Sloth is most exacting in its diet, eating only the leaves of the Cecropia tree and little else. However, recent research indicates that its diet may in fact vary quite widely.

Because of the upside-down existance of the Sloth, its hair grows backwards, from the belly towards the back, giving a considerable water shedding capacity. The coat has a greenish hue, rendering the creature difficult to see amongst the trees. This colouring is not natural pigment but is due to algae which grow in grooves in the strands of hair. Small moths, which feed on the algae also inhabit the coat of the Sloth, which is never cleaned or groomed.

Mammalian protection

Most mature mammals are capable of some defensive action. In some cases this takes an active form such as modified attack with teeth and claws; in others the behaviour is passive, the avoidance of damage to themselves.

Edentates employ various methods of defence, from the wielding of the large front claws in the face of attack to the motionlessness of the Sloth, to avoid the notice of the would-be attacker.

Throughout the mammalian species, various means of protection have evolved. Natural equipment, such as the armour of the Armadillo, allows the animal to withstand quite determined assaults, whilst the spines of the **Porcupine** or **Hedgehog** make the creature painful to molest.

Other mammals, although not having armour or spines are well protected by a tough skin.

Some animals, not having sufficient physical equipment for protection have, over the ages, developed behavioural

The Pangolin

The **Pangolin,** an anteater, is the scaly Afro-Asian counterpart of the South American Edentate. Once classified with the Edentates, toothless, ant-eating and with similar large claws, and tapering head the Pangolins are of a completely separate Order, Pholidota. It is thought that the resemblance is due to the convergence of evolutionary lines, perhaps as the result of similar feeding habits.

methods. One typical example is that of the **Common Opossum** which feigns death, 'playing possum', allowing itself to be mauled until the attacker loses interest.

Skunks and **Pangolins** employ anal glands which secrete and eject a foul-smelling fluid. This is extremely effective, persuading all but the most desperate predators not to attack.

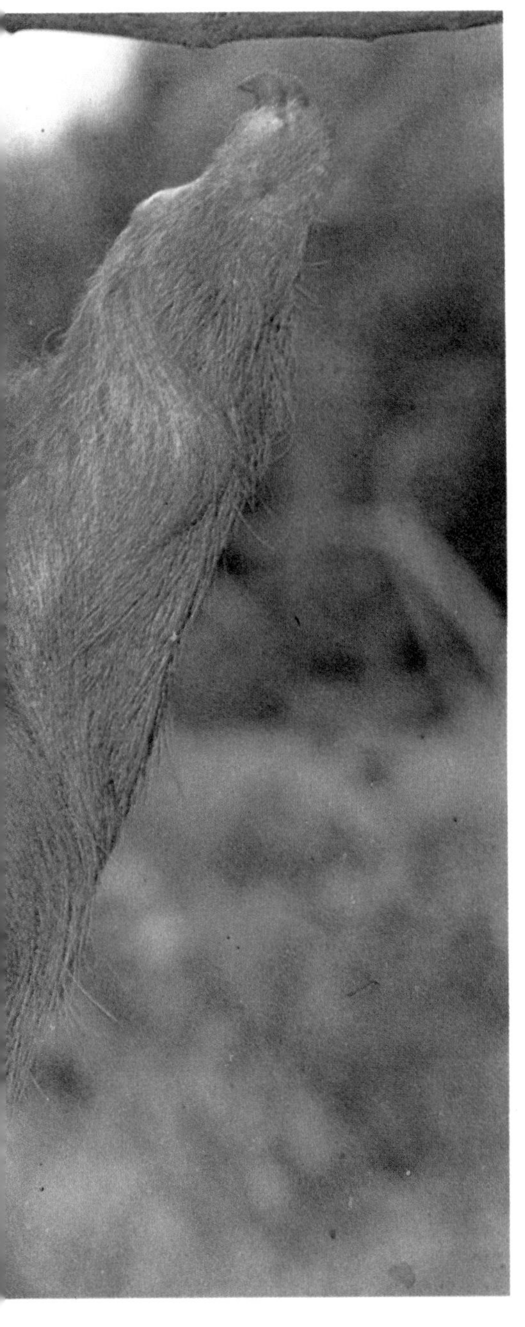

The hair of this animal has, over the ages, evolved into large flattened overlapping scales. The underside of the body is unarmoured and the Pangolin rolls into a tight ball when attacked. As a further defence it ejects a foul-smelling fluid from anal glands.

There are several species of Pangolins, some with prehensile tails live in trees, others on the ground. The biggest, the **Giant Pangolin** *(Manis gigantea)* weighs up to 30 kg.

Pangolins are widespread in Africa, India and South East Asia.

The Giant Armadillo and Relatives

Armadillos have prospered in Central and Southern America and the northern part of this region is inhabited by the **Giant Armadillo** *(Priodontes giganteus)*, with twelve moveable bands between its scapular and pelvic shields, the **Six-banded Armadillo** *(Euphractus sexcinctus)* and the **Tatouay** *(Cabassous unicinctus)* which is distinguished by the scales covering its tail. In the region south from Argentina is the **Hairy Armadillo** *(Euphractus villosus)*, so called because its scales have masses of fine hair growing between them.

The Giant Armadillo is the largest of the species, weighing about 50 kg and heavily armoured from head to tail. Each foot has five digits, the front ones carrying curved claws. The front third claw is the largest claw in the animal kingdom and is used to break open the ant-hills and termite colonies from which the Armadillo obtains its food.

During his South American travels Charles Darwin discovered, on a Patagonian beach, fossil remains of the ancient and huge **Glyptodont,** which he concluded 'seemed to be related to the extant species'. One of these fossils can be seen in the Natural History Museum in London.

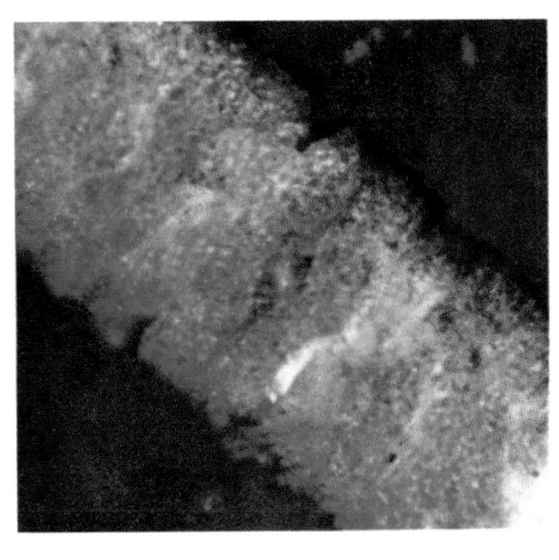

◁ The hooked claws of the Two-toed Sloth, which has two claws on its forefeet but three on its hind limbs, make it next to impossible to dislodge it from its upside-down perch.
△ The hair of the Three-toed Sloth—a single hair is seen here in close-up—is often thickly encrusted with algae giving it a greenish colour and providing beneficial camouflage.
▽ ◁ Some Pangolins are tree-dwelling while others like this one live on the ground, but all have the distinctive covering of horny scales, made up of compressed hair, which provide flexible protection.

Three bands or nine

The bands of armour carried by Armadillos vary from one species to another, although it is not known how this evolved from the original solid carapace.

Both the **Three-banded** *(Tolypeutes)* and the **Nine-banded** *(Dasypus)* are nocturnal animals, foraging at night for grubs, worms and berries.

The Three-banded have armour consisting of a head plate, a carapace covering the shoulders, and an overhanging shell protecting the hind quarters. Between these sections of body armour are the three flexible bands which allow the animal to curl into a tight ball for protection.

The Nine-banded does not roll up when threatened, but burrows and wedges itself into a hole. The armour covers the entire body of the animal, except the underbelly, making the creature quite heavy for its size. Should it be forced to swim, it floats by swallowing air and inflating its intestines, obtaining a buoyancy it would not normally possess.

The young of the Nine-banded Armadillo are always identical quadruplets. They are well-developed at birth and are capable of independence at an age of about three months.

The Armoured Ones

Armadillos, the 'armoured ones', are inhabitants of South and Central America. They are widespread, with one species, the nine-banded, also being found in the southern United States.

The **Giant Armadillo** is the largest, over a metre in length whilst the smallest, the **Fairy Armadillo,** is only a few centimetres long.

Altogether there are just twenty species of Armadillos, the only mammals to have a shell. They are of many sizes but share the typical Edentate characteristics of rudimentary teeth and large front claws. Most Armadillos are grassland dwellers but some prefer forest areas. Others live subterranean lives seldom venturing into the open. All Armadillos are expert burrowers and dig themselves into hiding with remarkable speed, even in hard, gravelly soil. It is probable that they are sensitive to ground vibrations which enables them to be fore-warned of the approach of enemies. In the burrow the animal arches its back, wedging itself in securely and with its armoured rear protecting the rest of the body.

The Fairy Armadillos

(Chlamyphorinae)

Fairy Armadillos *(Clamyphorus truncatus)* or **Pichiciegos** are rare creatures, seldom seen even in captivity, and not too much is known about them.

The Fairy Armadillo is a tiny pink-shelled animal 12 centimetres long with white, silky hair on its abdomen and tail. Like all Armadillos the Pichiciego is an expert burrower. It lives a subterranean existence and has ears and eyes similar to those of the Mole. Its diet is believed to consist of roots and insects.

The Fairy Armadillo has a separate plate covering its rump.

The Carnivorous Cats

The Cats and their relatives range from those, like the Lion and Tiger, admired and feared by Man, to the Hyaena, despised as a scavenger, and include that expert in ambush, the Leopard. But amongst them are also the numerous, little-known, but equally interesting, species such as the little marauding Geoffroy's Cat; the nocturnal, insect-eating Aardwolf; and that arch snake-killer, the Mongoose.

This Lion, only the males have the distinctive main, sitting in the shade of a tree is showing his true colours for the 'King of Beasts' is more lazy than regal. Most of the hard work is performed for him by the lionesses, including most of the hunting.

The Hyaenids, comprising Hyaenas and the Aardwolf, superficially resemble the Dog family but are in fact more closely related to Cats. The Hyaenas are not true hunters, relying mainly on the kills of other animals, whilst the Aardwolf, despite its name, lives on a diet of insects.

The Viverrids—the Genets, Civets and Mongooses, are higher on the scale of hunting skills. Genets are agile creatures, preying on birds, Squirrels and Hares, while Civets are more stealthy hunters, prowling the undergrowth for reptiles, Rodents and insects. The Mongoose is an able hunter of smaller mammals, and the Indian species is justly famous for speed and ability in its battles with Snakes, even the largest Cobras.

The Felids, consisting of more than thirty-five species, represent the peak of predatory evolution. They are lithe, intelligent killers, equipped with the speed and agility necessary to catch their victims, and sharp claws with which to bring them down.

Their powerful jaws have large canine teeth for tearing flesh and strong, sharp cheek teeth, used for shearing through skin and gristle. All Felids have thirty teeth, a round head, and, except for the Cheetah, retractile claws.

Most Felids, together with Hyaenids and Viverrids, are inhabitants of the tropical regions. Some species of Felids, the Lynx and the Wildcat, for instance, are found in temperate or even quite cold areas. The Tiger, although inhabiting the Bengali jungles, is also comfortable in the cooler mountain forests of Northern China.

Inter-breeding between the various branches of the Felid family is possible, and does sometimes take place. This is particularly true of the European Wildcat and the domestic Cat.

The exact number of Felid sub-species is not known, but the domestic Cat has many variations, and these, together with the wild Felids, are thought by some authorities to bring the total number to more than four hundred.

Genets

(Viverridae)

Eight of the nine species of this animal are found only in Africa, inhabiting the continent from the south to the edge of the Sahara desert. The ninth species, the **Spotted Genet** *(Genetta genetta)*, is present in most of Africa, except the desert, and in Palestine and southern Europe. However, it is seldom seen in Europe, being a shy creature, and somewhat rare. Genets are approximately 90 cm in length and about half of this is tail.

The **Genet** is a solitary, nocturnal

animal, passing the day stretched out on the branch of a tree, camouflaged by its spotted coat, or lying up in a hollow tree, awakening at night to hunt both on the ground and in the trees.

Genets are graceful, elegant in build and very agile and can jump quite long distances. At home both in bushes and in trees, they are sure-footed climbers and stalk their prey rather as a Cat does. Their prey consists of birds, Squirrels and ground-dwelling animals such as Rats and Hares.

Closely related to **Civets** and **Mongooses,** these attractive animals have been known to reach 15 years of age in captivity.

Civets

(Viverridae)

Among the other species of the Viverridae are many forms of the Civet cats.

These animals range in size from the **African Civet** *(Viverra civetta),* which is slightly larger than the Genet, to the near-vegetarian **Binturong** *(Arctictis binturong),* of south-east Asia, which grows to about 4½ ft (1.5 m) in length.

African Civets are less arboreal than Genets, living mainly on the ground. They sleep in abandoned burrows, coming out at night to catch Hares and reptiles, although they eat some fruit.

The **Palm Civets,** forest dwellers of south-east Asia, are slightly smaller than the average and are more arboreal in their habits. They eat fruit and some meat, and are skilful at catching fish.

The **Linsang** *(Poiana),* a handsome, slender relative of the Civets, lives in the West African forests. It is an agile animal, keeping almost exclusively to the tree-tops, and sleeping through the day in a nest high in the branches.

The **Binturong** is the largest of the Civet family. It has long, shaggy hair and is unique among Cats in having a fully prehensile tail.

The Mongooses

(Viverridae)

These skilful hunters are widespread throughout Africa, India, Arabia and south-eastern Europe, the three dozen species being the principal small predators of the tropical regions of the Old World. They have long bodies, with short legs, pointed muzzles and a slender bushy tail. Most species have five clawed digits on each foot. The size varies between the **Egyptian Mongoose** and the **Crab-eating Mongoose,** which are over 3 ft (1 m) long, to the **Dwarf Mongoose** of Africa, which measures about 18 in (46 cm).

These animals are diurnal hunters and are for the most part solitary creatures, with exceptions like the Dwarf Mongoose and **Meerkats** *(Suricata),* which form large colonies in systems of burrows. Their food consists of small mammals, Lizards, birds and insects, while some also eat Frogs, fish and Crabs. A few species will eat fruit. Mongooses are particularly known for their liking for eggs and for snakes.

The Indian, or **Grey Mongoose** is probably the most famous of the species, as a result of its Snake-killing prowess. It will attack a Cobra twice its own size, enraging the Snake by darting at it with teeth bared. When the Cobra strikes, the Mongoose leaps out of range with remarkable speed, eventually seizing the Snake behind the head and shaking it vigorously until it is dead.

The Madagascan species, *Cryptoprocta ferox,* is more Civet-like in appearance and habits, and is the main predator of the island.

The Aardwolf

(Hyaenidae)

This animal, a native of the savannahs of southern and eastern Africa, is not, as the name would suggest, any kind of predator.

The **Aardwolf** was named by the early Dutch settlers, the word meaning 'earth-Wolf'. It does in fact live in burrows, but it is far removed from the real Wolves.

It is a member of the Hyaenidae, but, unlike the Hyaenas, it does not normally eat carrion. It is adapted for a diet of insects and has sparse and simple teeth, unsuitable for chewing larger food, although they have been known to dig up and devour Gerbils. Its only means of defence are anal glands which emit a fluid with an extremely noxious smell which acts as an effective detterent on its would-be attackers.

Aardwolves, with their dark-striped, yellow-grey fur, are nocturnal animals, emerging at night to hunt for food. They are generally solitary or paired creatures, slow and deliberate in their movements.

The Hyaenas

(Hyaenidae)

There are three species of **Hyaenas:** The **Spotted Hyaena** *(Crocuta crocuta),* inhabiting East and South Africa; the **Striped Hyaena** *(Hyaena hyaena),* found in north-east Africa, Arabia and India; and the **Brown Hyaena** *(Hyaena brunnea)* from South Africa. All three species have large heads and powerful jaws and teeth. The ears are quite large and the animals' shoulders are higher than the hindquarters. The male Spotted Hyaenas can measure up to $4\frac{1}{2}$ ft (1.5 m) long, with a weight of 45 lb (90 kg). The females are lighter

◁△ The Meerkat, one of the Mongooses, is found on the dry veldt of South Africa where it feeds mainly on insects. Meerkats live in burrows in small colonies.
△ Mongooses are close relatives of the Cats and are found over most of Africa and in parts of Asia. At one time the Egyptian Mongoose fulfilled the rat-catching role of today's domestic cats and was sometimes even mummified before burial in Ancient Egypt. The most widely known of the Mongooses is probably the Indian or Grey Mongoose, renowned for its ability to kill snakes which it then eats whole.
▷ The close resemblance between the domestic and the wild cat can be seen from this picture of a young Wildcat displaying typical mottled fur.
▷▽ A young Hyaena, unable at this age to fend for itself, its mother bringing it food. The female Hyaena gives birth to three or four cubs once a year. The young are brown and without any markings.
▽ A Spotted Hyaena, found in the bush of East Africa. The low hindquarters, so well developed in the true Cats, are reduced in the Hyaenas but the shoulders are muscular and powerful.

Myths and Magic in Cats

For thousands of years, Cats have been associated with Man. They figure in myths and stories from the East to Scandinavia, where the goddess Freya had her chariot drawn by two Cats.

In Egypt, Cats were objects of worship. The goddess Bubastis is depicted as a woman with the head of a Cat. The animals were as revered as the goddess herself, and to kill a Cat was a capital crime.

Possibly this status stemmed from the days when Egypt was the world's granary and the vermin-killing prowess of the Cat made it a valuable possession.

High value was placed on Cats in Europe also, during the early Dark Ages, for much the same reason; vermin catching ability. Later, however, they became objects of fear and superstition. Old women were burned as witches, and their Cats with them, as their 'familiars', agents of the devil. Even today vestiges of this superstition linger on in stories of 'lucky' black Cats.

The aura of myth and magic is shared by the Great Cats. Arabs once believed the Lion could exorcise evil spirits, and the Indians considered a necklet of Tiger's teeth a powerful charm against the evil eye.

Pumas, Cheetahs, the Lynx, all have their fables and myths.

and smaller, giving birth once a year to three or four brown cubs which are unmarked.

The Spotted Hyaena, also called the **Laughing Hyaena,** sometimes hunts at night in a pack of up to thirty animals, killing about eighty per cent of the animals it eats. During the day, however, it is a scavenger, devouring the left-over carcases from the Lions' kill. Strangely, this role is sometimes reversed, with the Lion driving the Hyaenas away from their kill and becoming the scavenger itself.

The Striped and the Brown Hyaenas are smaller and lighter than the Spotted species, and are not capable of making their own kills. They are nocturnal animals, living in caves or burrows, and are true scavengers. In some areas, they are ruthlessly hunted because of their unpleasant habit of digging up human graves.

Tabby Cats

(Felidae)

The domestic **Cat** has many varieties, sizes and colours. Basically, however, all species stem from the Wildcat. The European species is *Felis silvestris,* and the African, *Felis lybica.*

The **European Wildcat** inhabits remote woodlands in Britain, France

and central and eastern Europe, but it is probable that the domestic Cat is descended from the African species.

The Wildcat looks rather like a large domestic tabby Cat, with a shortish, bushy tail. These animals are very closely related to the domestic Cat, *Felis catus,* and interbreed readily. Usually, Wildcats produce only one litter of kittens a year.

In West Africa, a species of Wildcat called the **Golden Cat,** *(Felis aurata),* is found. This is larger than the other Wildcats, about 47 in (120 cm) long from head to tail-tip.

Wildcats, like most of the Felids, are nocturnal hunters, sleeping through the day in a den. They eat birds and small Rodents and hunt singly.

Jungle Cats and Desert Cats

Felis chaus is found in both Asia and Africa and is one of several species of so-called Jungle Cats, usually inhabiting drier grasslands and scrub. This animal, more widespread in India than in Africa, is present also in Persia and the eastern Mediterranean countries.

Although a Wildcat, the **Jungle Cat** lurks near villages, stealing chickens and other food from the inhabitants. It is less nocturnal than most Cats, and hunts in the morning and evening, preying on birds and small Rodents.

The desert too has its own Cats, *Felis margarita,* found in North Africa and Turkestan; and *Felis manul,* **Pallas' Cat,** which lives in desolate, high-altitude country.

The **Sand Cat** is a small animal, dusty grey-brown with faint tabby markings on its legs. In some respects, this creature and Pallas' Cat are similar. They both have dense, coarse hairs growing on the feet, to assist grip in slippery sand, and their ears are set far apart, probably so that they can peer over rocks without being noticed.

Plain and Marbled Cats

The Asian members of the *Felis* group are in the main quite small animals. The **Marbled Cat** *(Felis marmorata),* is described as a lesser version of the **Clouded Leopard,** having similar variable markings. It is about 2 ft (60 cm), long, with a long tail, and, despite its size, is reputed to be amongst the fiercest of Cats. It lives in the jungle or along river banks in India, Java, and Borneo, nocturnally hunting rodents, birds and Frogs.

The **Malayan Red Cat,** *(Felis planiceps),* sometimes called the **Flat-Headed Cat** is small, about 12 in (30 cm) long, and is a dark red-brown with a white underside. It is similar to the Marbled Cat in habitat and diet, and, although quite rare, is found in Malaya, Borneo and Sumatra.

Very like *planiceps* is the **Borneo Red Cat** *(Felis badia).* It is a little larger, 20 in (50 cm), in length, and with a fairly long tail. It lives and hunts in rocky outcrops on the edges of the jungle, eating small animals and birds.

The Lynxes

The two main species of Lynx are the **Northern lynx** *(Felis lynx),* and the **Caracal** *(Felis caracal).* They are nocturnal hunters, preying on Rodents, birds, Hares and small Deer.

Felis lynx is found in the northern forests of North America, Europe and

Asia, and is a fairly large member of the Felids, about 3 ft (1 m) in length. It has a yellow-brown coat with light spotted markings and a short tail. Its pointed ears are very mobile and are used to indicate the varying moods of the animal. The ears have large tufts of black hair, making any movement more conspicuous.

The Northern Lynx can run quite easily over soft snow, giving it an advantage over more heavily-built animals. Even in deep snow, it can catch and bring down prey larger than itself.

The Caracal inhabits the arid areas of Africa and southern Asia, living mainly on Hares and Rodents. Although small, about 17 in (46 cm) at the shoulder, it is strong and agile, and sometimes manages to include Antelope in its diet.

The Bobcat

The North American continent does not have its own species of Wildcat. In the pattern of wild life, the position is occupied by the **Bay Lynx** and the **Bobcat.**

The Lynx lives in the northern forests whilst the Bobcat is found from southern Canada, through the United States, and into Mexico. It adapts to life in the deserts, the open plains and the forests, preying on Deer, Squirrels and birds.

The Bobcat is smaller than the Lynx and has a slightly longer tail. The ears are not so pointed, and the feet are smaller, since it has not adapted for mobility in snow.

These animals are, like most Cats, hunters by night, lying up during daylight hours in dens, under tree roots or in hollow stumps.

In captivity, Bobcats live for more than fifteen years. In the wild, of course, where life is much harder and more hazardous, they are unlikely to reach this age.

The Spotted Serval

The **Serval** *(Felis serval),* an exclusively African member of the Felidae, stands about 20 in at the shoulder and about 45 in long from head to tail-tip. It is a lithe, slender animal, its shape suggesting speed, and some authorities believe it chases and captures its prey in a similar fashion to the **Cheetah.** The Serval is capable of climbing trees, and does so to escape pursuit and also to catch roosting birds.

Two species of this animal are commonly found, one having a tawny coat with black spots, becoming stripes

◁ △ The coat of the Bobcat, found in parts of Canada and in the United States, is usually a more or less uniform colour but it can be striped or spotted. The Bobcat stands about two feet high at the shoulder and is about three feet long and its normal diet includes birds and small animals but it can bring down a deer by leaping on its back.
△ The Caracal, found principally in East Africa, is easily identified by its characteristic long ear tufts. It is long-legged and a very fast runner, capable of outpacing even a gazelle over a short distance, and an agile hunter.
◁ ▷ The Lynx, a fierce predator of the colder parts of North America, particularly Canada, and the forests of Northern Europe, and much prized for its fur. The Lynx's woolly feet help stop it sinking into the soft snow so often found in its habitat. The feet of the North American Lynx are larger than those of the European type and its hair is longer. the tufted ears and short tail are distinctive.

at the shoulder, and the other having the same tawny colouring, but with a pattern of small spots all over. A rarer, black type has been reported from Kenya, living at high altitudes.

The Serval has very large pointed ears, the bases being so wide that they almost meet across the head. The size of the ears is thought to indicate extremely acute hearing, and it is believed that Servals can, in fact, hear Mole-Rats burrowing at night, whereupon they dig them up for food.

Golden Cats

The beautiful **Golden Cats,** natives of both Africa and Asia, are medium-sized Felids, attaining a length of about 36 in (90 cm) from head to tail.

The African species, *Felis aurata,* is becoming increasingly rare, possibly due to its reputation as a vicious killer and stock raider. It has two phases of colour, sometimes being golden, at other times grey, and it was once thought that the different colours were separate species.

The **Asian Golden Cat,** *(Felis temmincki),* is slightly smaller than its African relation and has a deep red coat and a black and white streaked face. Other variations of colour occur, including grey, stripes and rosettes.

It is found throughout Asia, from Tibet, through India into Sumatra, and is probably nocturnal in habit.

Also an inhabitant of Asia is *Felis viverrina,* the **Fishing Cat.** This animal catches fish by scooping them up with its paw. It is small, about 30 in (76 cm) long, but will attack larger animals, including Sheep, Calves and Dogs.

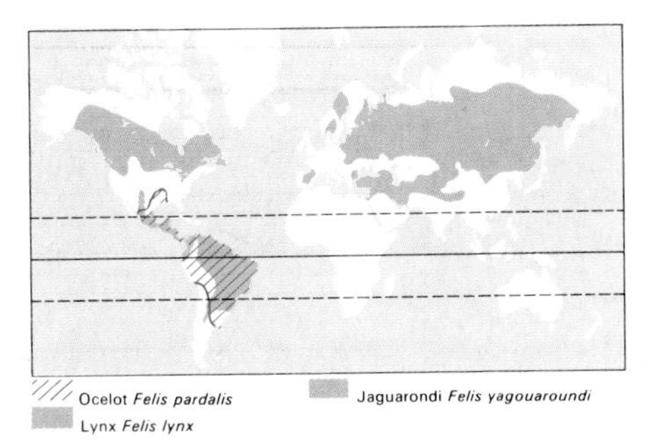

Ocelot *Felis pardalis*
Lynx *Felis lynx*
Jaguarondi *Felis yagouaroundi*

The Ocelots

Ocelots *(Felis pardalis),* are natives of south and central America and are considered by many to be the most beautiful of all the Felids. Their coats, with a tawny-brown background, are spotted and striped with darker colours, many variations being found. These colours, whilst camouflaging the creature in its native forests, have made the animal much prized by trappers.

The Ocelot, about 45 in (120 cm) long, head to tail, is a diurnal hunter, prowling the forest floor for Peccaries, Spiny Rats, and some reptiles. It is an agile climber and sleeps stretched out on a tree branch. In areas where it is intensively hunted, it reverses its life-pattern, becoming nocturnal and sleeping during the day.

Ocelots are sometimes kept as pets, then being given a vegetarian diet. They tame easily and get on well with other domestic animals. In one instance, an Ocelot became a ship's Cat, catching Rats and sleeping in the crew's hammocks. It was lost overboard eventually, trying to catch flying fish which it apparently mistook for birds.

The Leopard Cat

This animal, as the name suggests, has a spotted coat similar to that of the **Leopard.** It is a much smaller creature, however, growing to a length of 45 in (120 cm), from its head to the tip of its tail. The dark spots, solid rather than of the rosette form, are on a yellowish background, and run the length of the body in longitudinal patterns.

The Latin name for the **Leopard Cat** is *Felis bengalensis,* given because the first one ever captured as a live specimen was found swimming in the sea off the coast of Bengal.

Leopard Cats are found over a wide area, from Siberia in the north, throughout south-east Asia to the Philippines, and are the most commonly found Cat in Borneo.

These animals are nocturnal in habits, emerging at night to prey on Hares and small Deer which they capture by dropping on to them, in a manner similar to that of the Leopard. During the day, the Leopard Cats rest in caves and under hollow logs, or sometimes up in the branches of trees.

The Puma

(Felis concolor)

The **Puma,** inhabiting both North and South America has the most widespread latitudinal range, adapting its single species to a range of climates and terrains from the sub-Arctic, through mountains and desert, to the sub-tropical jungles. It is common throughout the entire New World from British Columbia in Canada, down to Patagonia in the south, thriving despite continuous warfare waged against it by Man. It is also known variously as **Cougar, Catamount** and **Mountain Lion,** and eats almost anything from Mice to cattle, its particular favourite being the Deer.

Mountain Pumas are larger than the lowland species, reaching nearly 8 ft (2.5 m) in length.

The Puma has a greyish-brown colouring, merging well with most of the terrain over which it hunts. The cubs, which can be born at any season, and of which there are two or three in a litter, have black spots on a tan background, but these markings disappear after about three or four months. Cubs hunt with their mother until they are about two years old.

The Jaguarundis

Several species of small Cats occur in South America, inhabiting the pampas, jungles, riversides, and the high Andes. Of these, the **Jaguarundi** is perhaps the most interesting, since it appears to resemble certain of the Mustelids (the Otters, Weasels, etc.).

It is dark in colour, ranging from chestnut to black, and has a long body and tail, and short legs. It is quite at home in the water and is known in Mexico as the **Otter-Cat.** Like the Weasel, it hunts by both day and night, eating a variety of prey.

The **Pampas Cat** *(Felis pajeros),* is becoming rare in some parts of South America, due to the onslaught of civilisation. It is nocturnal, living in dense thickets and hunting over grassland.

Geoffroy's Cat, found from the Rio Grande to Patagonia, is larger than the Jaguarundi, about 36 in (90 cm) long. It is a good climber, using the trees for both resting and ambush. On occasion, it enters settlements for poultry.

The Clouded Leopard

The **Clouded Leopard** can be considered either as the largest of the *Felis* genus of Cats or the smallest of the *Panthera.* Since it is difficult to classify, it is often put into a separate genus of its own and called *Neofelis nebulosa.*

Its habitat is quite widespread, the animal being found in eastern Asia from Assam to China and as far south as Malaya. The name Clouded Leopard is derived from the larger patches of colour against a light background on the body and flanks of this creature. Its face is sharply marked and may have a mesmeric effect on its prey. The coat is a yellow-grey dotted with large grey blotches which have dark borders. The centres of these blotches tend to fade in older animals leaving only the dark outlines.

This is a tree-dwelling and hunting animal, living mainly on birds and small mammals.

The Clouded Leopard is a very large Cat, attaining a length of about 25 in (65 cm). Its tail is extraordinarily long often as much as the rest of the animal, and is of great benefit in its tree-dwelling mode of life.

The Lion

Panthera leo, the 'King of the animals', was, until a few hundred years ago, found in southern Europe, as well as North Africa, the northern and southern savannahs, and in India.

Even today, **Lions** thrive in Africa, south of the Sahara, and their numbers are thought to be in the thousands. The Indian sub-species was brought almost to the point of extinction in the early 1900's by British Army officers, armed with sporting rifles. It is reported that by 1908 the numbers of Indian Lions had fallen to thirteen. Since then, a conservation programme has seen the population rise to a total of about two hundred in 1973.

The Lion is the only truly social Cat, living in prides composed of one or two males and several Lionesses. Usually, the Lioness kills the prey, which may be almost any large herbivore, except the adult Rhinoceros, Elephant or Hippopotamus.

The cubs are treated with affection by the entire pride, the male Lions, whose main role appears to be that of protector, taking it upon themselves to teach the youngsters the arts they will need as they mature. Unfortunately, only two or three of a litter of five cubs may survive. After a kill, the male Lions eat first, followed by the Lionesses and finally, the cubs. Because the essential vitamins which Lions need are found in the entrails of the kill which the male Lions eat first, the cubs, feeding last, miss their vitamins.

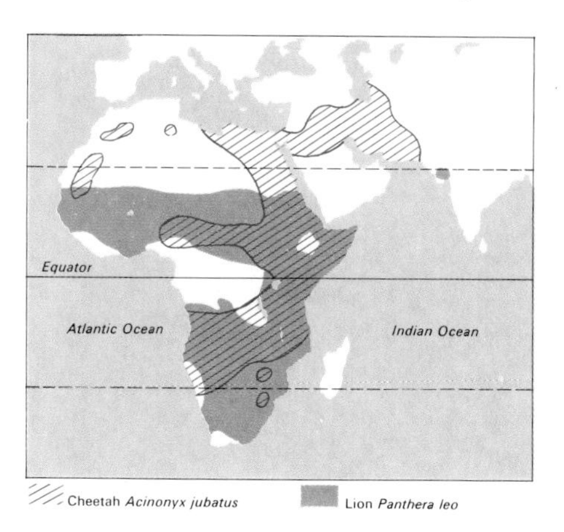

Cheetah *Acinonyx jubatus* Lion *Panthera leo*

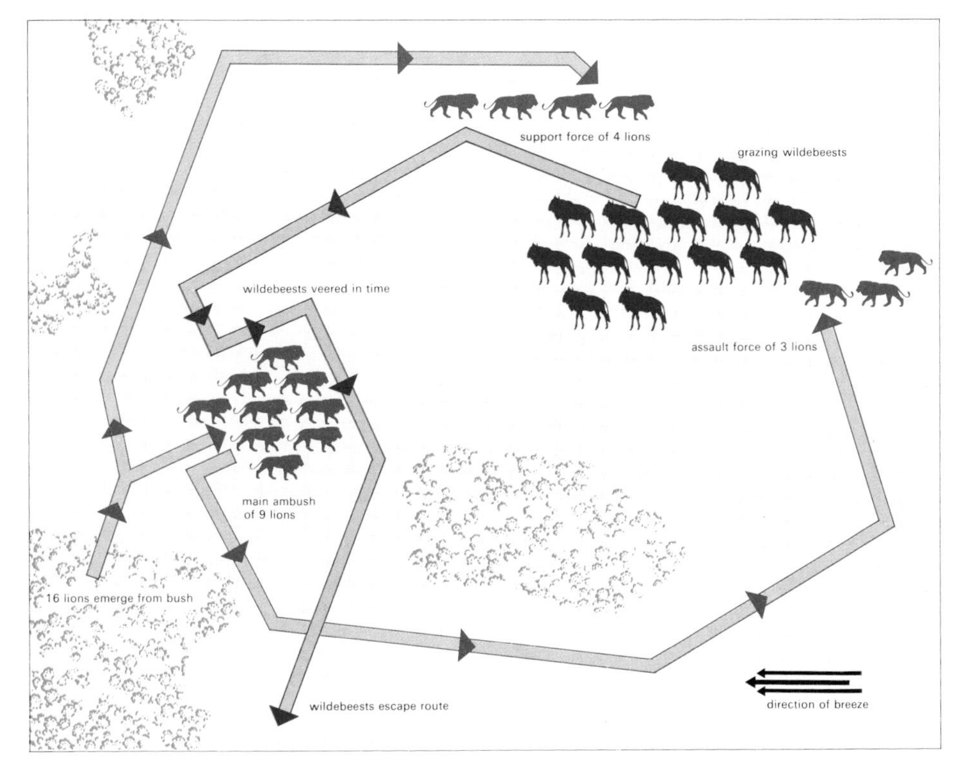

◁ Geoffroy's Cat, one of the beautifully marked black and gold Cats found in South America. The markings help break up the animal's outline in the forests. The prominent whiskers, a characteristic of the Cat family, on nose and face help the Cat find its way through thick bushes in the dark.

▽ A Lioness and her cubs drinking at sunset. Baby Lions have a distinctive mottled pattern to their fur, especially marked on the legs, even at the age of two or three years.

△ A carefully observed attack strategy of a group of Lions on a small herd of Wildebeest. The Lion is the least bloodthirsty of the big Cats, killing only to eat. If it fails to kill its prey a Lion will rarely give chase, though when it does it is capable of wearing down much swifter animals.

Hunting Behaviour

The differing hunting behaviour of each of the species of large Cats has developed as a result of the different terrain which they inhabit.

The Lions, in the borderland between forest and open plain, have cover to conceal them, and the Lioness, who usually makes the kill, uses this cover for a stealthy approach, with a final leap on to the victim. The Lion stands guard against Hyaenas and other carnivores stealing the prey.

Tigers, however, usually hunt alone, although it is not unknown for a pair to work together, one driving the prey towards the ambush of the other. Like the Lions, they are limited by their build to hunting on the ground, silence and great stealth being essential.

The Leopard, as an agile climber, makes most of its kills from the branches of a tree, lying in ambush and dropping down on to the prey as it passes beneath. It does not disdain the stealthy approach though, and will use this method as well as ground ambush.

Cheetahs hunt over open grasslands almost devoid of cover. Initially they use a stealthy approach, but without the final leap. Instead they break from what cover there is, and chase and run down their prey with a great burst of speed.

The Tiger

(Panthera tigris)

The **Tiger** is one of the largest of the Cats, and is found only in Asia. There are different races, differing only in size, colour and markings. The **Manchurian Tiger** for instance, is the largest, and can be over 10 ft in length. Generally, male Tigers are between $8\frac{1}{2}$ ft and 9 ft and weigh between 100 and 140 lb. Female Tigers are smaller and lighter. Tigers prey on herbivores and smaller animals but will eat fish too if necessary. They can swim but, unlike most other Cats, Tigers do not climb well.

Old age or injury may cause a Tiger to attack a human; when this happens, the animal is hunted down and killed, even in conservation areas. It is probable that Tigers were originally animals of the higher, cooler parts of Asia, although they now inhabit the jungles of Bengal. They appear to be uncomfortable in the heat but indifferent to cold and they have a thick winter coat which moults rapidly with the onset of warmer weather.

The last animal census in India revealed that Tigers were becoming rarer, their numbers falling to less than three thousand. They are now under protection, but poaching is prevalent and good Tiger skins bring high prices in the markets of the world. despite conservationist pressure for a ban on their re-sale.

A Bengal Tiger, the only race of Tiger which still numbers more than a few hundred. There is only one species of Tiger but, because it can be found in such a wide range of climates and countries, their is a rich variety of markings and an abundance of adaptations to local conditions, but the stripes, sometimes pale, are always there.

The Leopard

(Panthera pardus)

The **Leopard** is the most widespread of the big Cats, being found in large areas of Africa and India, where it is called the **Panther,** and throughout Asia, to China, Java and Ceylon.

Panthera pardus, although not as big as the Lion or Tiger, is a daring predator. Essentially a tree-dweller, its long tail provides balance whilst it lies in ambush, but it also stalks its prey on the ground, hunting Deer, Antelope, and wild Pigs.

For a long time, it was thought that the Spotted Panther and the black variety were separate species, but it is known now that they are forms of the same species, the two colours sometimes appearing in the same litter. In fact, the Black Panther, which may also have a black tongue and blue eyes, does have spots, but they are visible only under a strong light.

The Jaguar

(Panthera onca)

The **Jaguar,** looking very similar to the Leopard, and about the same size, is the largest of the New World Cats. It probably shares common ancestry with the Leopard but is stockier and has a shorter tail. Its spots are larger and fewer and often have dark centres.

Jaguars range from Mexico, to the forests and plains of Argentina. They are sometimes seen on the open grasslands but on the whole, seem to prefer the forests. They climb well and seem at home in the trees, stalking prey along branches. Unlike most Cats, they swim well too and fish skilfully, flipping fish out with their paws.

Their main food is the Peccary but they also prey upon Sloths, Deer, and even small Alligators.

Like most big Cats, Jaguars will attack cattle and other domestic animals and for this reason are hunted quite vigorously.

The Snow Leopard

(Uncia uncia)

The **Snow Leopard,** also known as the **Ounce,** is found amongst the rocky fastnesses of the Himalayan and Altai mountains. Despite its name it does not live above the snow line but descends to lower levels in winter, following the herds of wild Sheep and Goats upon which it preys. During the summer, however, it may be seen at heights above twelve thousand feet.

In winter, the coat of the Snow Leopard, which is pale fawn or grey-brown with black spots or rosettes, becomes extremely thick and heavy. The animal is well adapted for mountain life, living and breeding in rocky clefts and hunting in the sparse pastures.

When stalking its prey it will creep upon it silently, making one final leap, or it may adopt ambushing methods, much like Leopards.

They are not particularly shy animals, as shown when one prowled all night around a herd of Sheep despite being continuously pelted with stones, and yet they are seldom seen. This may, of course, be due to increasing rarity.

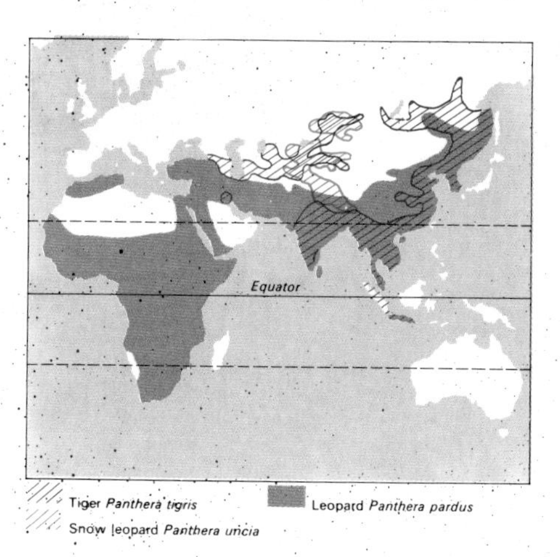

△ A snarling Cheetah shows its stabbing canine teeth and slicing molars.

△ ▷ Leopards spend much of their time resting in trees and do most of their hunting at night. They like to drop out of trees on to their unsuspecting victim though they frequently stalk their prey through thick plant cover. They are the stealthiest of the large Cats. They often maul the hindquarters of larger animals, but their favourite method of killing is to crush the neck of their victims.

▷ Cheetahs, easily recognised by the bold, black line running from the mouth to the corner of the eye, usually move about in twos or in small family groups. Their non-retractable claws, which give them a sure grip, and their powerful, wiry legs help to give them their formidable sprinting speed.

The Cheetah

The **Cheetah** *(Acinonyx jubatus),* lives and hunts in the open grasslands and semi-deserts of Africa and, to a certain extent, south-west Asia. Its prey, small-hoofed animals, Impala and Gazelles, feed well away from any cover so the Cheetah, unable to use stealth, has developed a speed almost unsurpassed in the animal kingdom. It appears that Cheetahs attack only those animals which are moving away from it, and, having caught their prey, kill it by biting through the neck vein.

Cheetahs breed all year round and have litters of between two and five cubs but the survival rate is low—with only 2 or 3 cubs usually surviving the first year.

At rest, the Cheetah looks ungainly and ill-proportioned, its legs too long and its head too small. In motion, these characteristics are seen in their true context, as aids to great speed. Over flat country, these animals have been timed at about sixty miles an hour, but only for four hundred yards or so. If it has not caught its prey inside that distance it gives up and awaits a better opportunity.

The **Indian Cheetah** has, unfortunately, become extinct, the last reported sighting being twenty years ago near Hyderabad.

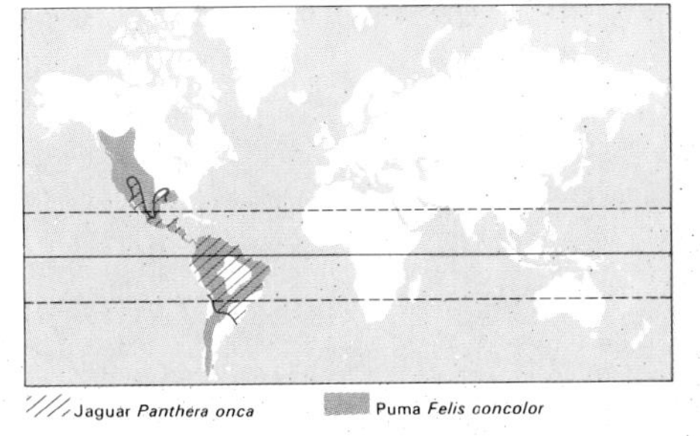

Skilled Hunters and Scavengers

Those skilled hunters, the Wolves; the Bears; the scavenging Jackals; and that favourite film extra, the Coyote, are all members of one super-family.

The super-family, Canoidea, with four families, includes the Dogs, Bears, Raccoons and Badgers etc. Some are entirely carnivorous, others are omnivorous.

The Canidae includes Wolves, Jackals, Foxes, Wild Dogs and Coyotes, all long-legged, long-tailed, runners, rarely able to climb. Their teeth are variable, but all possess the large shearing cheek teeth and have pronounced canine teeth. There may be up to six other cheek teeth, thus the muzzle is long. Some scavenge, but most hunt, using their very keen sense of smell. There are generally five toes on each fore foot and four on the hind.

The Ursidae, the Bear family, presumably from the same ancestral stock as Dogs, are now clearly distinct. They are generally much larger, plantigrade (walking with the whole of the lower surface of the foot on the ground), comparatively slow-movers and with very different teeth. There are no shearing cheek teeth and the molars are designed for crushing and grinding plant material, on which most Bears feed; (the Polar Bear is the exception). Their very short tails are almost hidden and they have five toes on each foot, equipped with strong claws.

The Procyonidae includes the Raccoons and Coatis, Cacomistles, Kinkajous and both types of Pandas. These are mostly small by comparison with Bears and have Dog-like muzzles. They have five toes on each foot, are long-tailed, and some are almost fully herbivorous, living mainly in trees.

The Mustelidae embraces some of the fiercest of all hunters and includes Weasels, Stoats, Mink, Polecats, Ferrets, Martens, the Wolverine, Badgers, Skunks, Ratels and Otters. They are mostly small with long, lithe, muscular bodies on short legs. They have shearing teeth but there are fewer cheek teeth than in Dogs, hence the muzzle is shorter and the jaws more powerful in the bite. Some are digitigrade (walking on the toes), others plantigrade and some have a well-developed scent gland at the base of the tail. One group, the Otters, hunt in water and have appropriate adaptations, such as webbed feet.

Wolves and Wild Dogs

(Canidae)

These are true Dogs and are regarded as the wild ancestors of the domesticated varieties. Wolves once had a wide geographical distribution over Europe, Asia and North America, particularly the colder parts, but they are now very much restricted within these regions.

Wolves, when hunting as a pack, will bring down large animals as prey. They have always been notorious as Man's enemies, taking his flocks and herds, but they have rarely attacked him. There is a variety of size and colour among them.

The grey **Timber Wolf,** *(Canis lupus),* possibly existing as several races, is the largest and is the typical Wolf form. It is like an Alsatian Dog in shape and size but with straight, hanging tail and obliquely set eye pupils. It is, or was, widespread in North America and North Europe, but has now been eliminated from most inhabited areas. Few now remain in Europe and the Wolf was finally exterminated in Britain during the early 18th century.

The **Red Wolf** *(Canis niger),* found only in one or two of the southern United States, is smaller with a coarse, tawny coat.

In India, a form of *Canis lupus* is found in north western parts, whilst *Canis pallipes* inhabits southern areas.

The North American **Coyote** or **Prairie Wolf** *(Canis latrans),* a smaller relative of the **Grey Wolf,** lives in the more open parts and lives chiefly by scavenging. In some areas, it is reported to be invading suburban areas apparently learning to tolerate the artificial environment.

▽ The Black-backed Jackal is one of three species of Jackal found in Africa and in South Africa it is considered a pest. Jackals will hunt down small antelope and in East Africa they are predators of Thomson's Gazelles but it is as scavengers that they are chiefly known.

▷ Dingoes are found only in Australia. They were in all probability taken there by the Aborigines as hunting companions, soon reverting to the wild and being responsible for the extinction of several species of Mammals on mainland Australia.

Another member of the family is the **Australian Dingo** *(Canis dingo)* whose origin in that continent must be traced to its introduction by Man thousands of years ago.

Jackal

(Canidae)

Closely related to Wolves, the smaller **Jackals** resemble them in many ways. They are wild Dogs of the warmer, open parts of the Old World, Asia and Africa, living by scavenging mostly, but sometimes hunting in packs. If they can, they will steal poultry and young animals, stealthily and by night. Some Jackals are particularly noteworthy for their wailing cries at sundown and for the offensive smell produced by secretions of glands at the base of the tail.

The **Common** or **Indian Jackal** *(Canis aureus),* is greyish-yellow with darker hair on its back, standing about 40 cm tall and as long as 60 cm in the body, with a 20 cm tail. It has a sharp muzzle and comparatively large ears. Its range is through the Middle East and North Africa into Asia.

Other members of the Jackal group are the **Black-Backed Jackal** *(Canis mesomelas),* a rusty red, and the greyish **Side-striped Jackal** *(Canis adustus),* both common animals of east and south Africa. The **Simien 'Fox'** *(Canis simensis),* is a large, long-legged Wolf-like animal with thick red fur and a white patch at the base of the tail. In many ways, it is quite distinct from other canids. It is also exceedingly rare and little known, being confined to remote mountains in Ethiopia.

Foxes

(Canidae)

Foxes are distinguishable from Wolves and Dogs mainly by details of skull structure; otherwise they resemble them in looks if not in habits. They are generally nocturnal predators, holing up by day in burrows or 'earths' or some natural underground crevice. They are famed for their cunning in relation to Man who traps and hunts them either for their fur or to control their numbers.

The **Red Fox,** possibly of two species, *(Vulpes vulpes* and *Vulpes fulva),* is the common form of the whole northern hemisphere, living in a great variety of habitats from mountain-sides to forests and farmlands. In Britain it is now also urbanised, scavenging around town dwellings fearlessly. It is a very alert animal, although the more extraordinary tales of 'cunning' are more fiction than fact. Fox fur is valuable, especially the dense winter coat and special varieties like the **Silver Fox** and the **Platinum Fox,** are farmed for their pelts.

There are a number of other Foxes, including the **Corsac** *(Vulpes corsac)* of the Siberian plains, and the pale, cream coloured **Fennec** *(Fennecus zerda),* of the Sahara and other arid parts of North Africa; this being the smallest Fox with disproportionately large ears. The **Grey Fox** *(Urocyon cinereoargenteus),* widespread in North America, is noted for its tree-climbing.

Arctic Fox

(Canidae)

The **Arctic Fox** *(Alopex lagopus)* lives all its life beyond the tree line in the frozen zones of Europe, Asia and North America. In body form, it is intermediate between Dogs and other Foxes. It is comparatively small, up to 80 cm in length overall, with hair on the soles of its feet. In summer, the animal's coat is greyish brown and in winter, pure white. Its beautiful fur is exceptionally thick as protection from the cold (it can survive temperatures of −50°C). A colour variety known as the **Blue Fox** does not turn white in winter. Because of the unusual slate colour, this is an even more valuable pelt and the animals are extensively reared in fur farms. Wild Arctic Foxes feed generally on Lemmings, Squirrels and Mice but will also take fish. The animal is a good burrower and several families may excavate and inhabit a large and complicated system of tunnels. Food is stored in the ground for the worst winter shortages.

Maned 'Wolf'

(Canidae)

This is one of the South American Dog forms *(Chrysocyon jubatus),* locally known as the Guara. It is like a very large long-legged Fox up to 130 cm in length overall, with a shoulder height of up to 32 inches (80 cm). It has a very distinctive appearance and this is further marked by the mane of erectile hair on its neck. The coat is an attractive cinnamon brown in colour, with paler underparts and white-tipped tail. The animal is scarce and inhabits wide areas of south-east South America.

Little is known of its habits but the Guara is a nocturnal hunter and feeds on Mice, fish and plant material, such as sugar cane and fruit. It hunts alone and is reputed to be powerful enough to bring down a Sheep unaided.

Hunting Dogs

(Canidae)

There are a variety of these, widely distributed. They are notorious for their predatory destruction of quite large animals, hunting in packs to effect a kill.

One of the best known is the **African Wild Dog,** or **Cape Hunting Dog** *(Lycaon pictus).* With long limbs, broad head, short muzzle, and heavy body measuring up to 4 feet (130 cm) overall, it is a formidable hunter even when alone. Its blotched, ochre-yellow and black and white colouring is very distinctive and gives it excellent camouflage. It is outlawed by Man for the damage it does to herds.

The smaller, tawny **Red Dog** of Asia, or **Dhole** *(Cuon javanicus),* found from India to Siberia and into Sumatra and Java, is a similar pack hunter, taking Deer and even Buffalo as prey.

The South American **Bush Dog** *(Speothos),* is one of the New World Dog relatives but differs from them in appearance and dentition. It resembles a brown Badger, with a long body, 24 inches (60 cm), and short stubby legs. It lives in the forests of northern South America and is active mainly at night, spending the daytime underground. It seems to be omnivorous in its diet although probably feeds mainly on small Rodents.

African Long-Eared or Bat Eared Fox

(Canidae)

This small, attractive creature *(Otocyon megalotis),* not really a Fox at all, is found on the African plains from Ethiopia to the south. About the size of a Red Fox, it has a grey-brown coat with long black hairs intermingled. Its dentition is unusual for a canid, since it lacks carnassial teeth and has an extra molar on each side. When moving, small groups sometimes travel together, often in daylight. One peculiar feature of the animal's progression is the habit of doubling back on its tracks quite frequently. They are wholly harmless as far as humans are concerned, feeding mostly on Termites, insects and small Rodents.

△ The Cape Hunting Dogs prey mostly on Gazelles, approaching a herd quite openly and selecting their prey when the herd panics and runs, then pursuing it mercilessly for distances of up to five miles. The Indian Wild Dogs, or Dholes, will even attack a Buffalo and are capable of driving off Leopards or Tigers intent on stealing their kill. Wild Dogs are the ancestors of the familiar domestic breeds found all over the world and they are the true dogs. They are savage hunters moving and hunting in packs.

▷ The large ears of the Bat-Eared Fox helps it detect its prey of Mice and other small animals at night.

The Bears

(Ursidae)

The Ursids contain the largest of all existing carnivores with species measuring up to 9 feet (3 metres) long and weighing as much as $\frac{3}{4}$ ton (725 kg). With the exception of the **Polar Bear,** none are truly carnivorous in diet, feeding most commonly on vegetable material although often taking meat too. Fish is taken when the opportunity occurs and honey is sought by almost all of them.

There are seven genera usually described and they are confined in distribution to the northern hemisphere, chiefly the north west of North America and north-east Asia, with the exception of one, the **Spectacled Bear** of South America. None are found in Africa or Australia.

The Spectacled or **Andean Black Bear** *(Tremarctos ornatus),* is one of the smaller Bears, about 3–4 ft (90–120 cm) long and standing 24 in (60 cm) at the shoulder. It is, unfortunately, in danger of dying out as its natural habitat is taken for development by Man and is now quite rare, found only in parts of Ecuador and Peru. It is easily recognisable from the fawn coloured ring around each eye (the 'spectacles'), and whitish muzzle and throat on an otherwise black coat. It climbs well and feeds mainly on fruit gathered from the trees of the Andean foothills.

Black Bears

(Ursidae)

The so-called **Black Bears** are generally large animals, found mostly in North America, and there are several genera included within the broad descriptive name, not all of which are black in colour.

The **American Black Bear** *(Euarctos americanus),* up to 6½ ft (200 cm) long and standing 32 inches (80 cm) at the shoulder, is a North American inhabitant of woodlands where it climbs well, but feeds mostly on the ground on a vegetable diet. It will often approach Man's dwellings to forage and frequently raids dustbins in the American National Parks. Having fed well in summer, the Bear sleeps away most of the winter in its den, and if pregnant, the female produces her young there early in the year. Local strains differ in coat colour and since Bear furs are valuable, the animals have been extensively hunted.

The smaller, more carnivorous, **Asiatic Black Bear** or **Moon Bear** *(Selenarctos),* is found in areas from Iran to Japan and Formosa. A broad crescent of white on its chest, and longer black hair on its shoulders, make it distinctive.

Brown Bears

(Ursidae)

Most **Brown Bears** occur in Europe and Asia and tend to have brown coats, although this may vary from almost black to yellow. They live mostly in forested areas, climb well and feed mainly on vegetable material, but take both honey and fish when possible.

The **European Brown Bear** *(Ursus arctos)* is the typical example, measuring up to 6½ feet (200 cm) long and standing about 3 feet at the shoulder. It once roamed the whole of Europe, including Britain and Asia, differing locally in precise form and colour. It is now much more restricted, being found only in the more remote areas of mountain forest; the last British ones were exterminated several hundred years ago.

The **Kodiak Bear** *(Ursus middendorffi),* a North American relative of Alaska and British Columbia, is the biggest of all Bears with a shoulder height of 5 feet (150 cm) and up to 13 feet (4 metres) tall when fully upright on its hind legs. It is yellow brown in colour and is not nearly as fearsome as its size would indicate. Being almost vegetarian, it will even graze the meadows.

The **Grizzly Bears** also belong to the same genus. The **American Grizzly** *(Ursus horribilis),* now wild only in some National Parks of the United States and small areas of British Columbia and Alaska, is another large Bear, named for its grizzled grey colouration.

Malayan or Sun Bear

(Ursidae)

Otherwise known as the **Bruang** *(Helarctos malayanus),* the **Sun Bear** is the smallest of the Bears, about 4 feet (120 cm) long and 16 inches (40 cm) at the shoulder and weighing up to 90 lbs (40 kg). Its name comes from the patch of white, shaped like a rising sun, on the chest of its otherwise black body. It is a short-haired tree climber of South East Asia found on the Malay Peninsula and in Borneo. Its special feature is the long, extensible tongue and wide, loosely mobile lips which enable it readily to pick up the Termites and grubs on which it mainly feeds. It is also fond of honey, and locally, it is sometimes called the 'honey Bear'.

Sloth Bear

(Ursidae)

A native of Ceylon and India, the **Sloth Bear** *(Melursus ursinus)* is so called from its normally slow moving shuffling progression. It is black and shaggy with a white crescent on the chest and a white muzzle. A large specimen may be 5 feet (150 cm) long and stand 24 inches (60 cm). Its tail, although only about 4 inches (10 cm), is longer than that of most bears. It has a naked face, narrow snout and extremely mobile lips to aid its ant-eating habits.

The Sloth Bear is an excellent climber and ranges through the tree tops very easily, searching for Bees' nests, devouring the honey without regard for the irate owners. Generally it feeds omnivorously on the ground and another favourite food is Termites, which are taken very swiftly through the wide, slack mouth, sucked up with a loud snuffling sound.

The animal shows maternal care for up to three years after the birth of its initially blind and helpless young. When startled, the Sloth Bear can move much more swiftly than its name suggests.

Polar Bear

(Ursidae)

The **Polar Bear** *(Thalarctos maritimus)* is the swimming Bear of the sea, and is one of the largest of all truly carnivorous animals. It may be up to 10 feet (300 cm) long with a shoulder height of 5 feet (150 cm). It inhabits all the Arctic coastal zones of Russia, America and Europe and its fur is white with yellowish fringes in keeping with its snowy background. Its enormous feet, hair-covered on the soles, are adapted to moving over ice. The Polar Bear swims expertly and spends much of its time in the water. It feeds largely on Seals, which it takes largely by stealth.

Males are active all the year but when a female becomes pregnant (which, as is the case with most Bears, occurs every other year), a winter resting place is found in the snow and during this period she gives birth to one or two very small, blind and almost hairless cubs. She gives them maternal care for over a year.

Raccoons, Coatis and others

(Procyonidae)

With the exception of the **Pandas,** the *Procyonids* are all New World animals, long tailed plantigrades most of which climb well.

The **Raccoon** *(Procyon lotor),* is found widely distributed in North America. About 80 cm long overall, with long, coarse black to brown hair, it is easily recognised by its bushy black and white ringed tail and black 'mask'. It is strictly nocturnal, swims well and tends to live near water but makes its home high up in a tree hollow, where it hibernates during the coldest parts of the year. It is omnivorous and often visits gardens, dustbins and rubbish tips looking for food.

The **Ring-tailed Coati** or **Coatimundi** *(Nasua nasua)* is South American and about the same size as the Raccoon but has a longer, flexible snout. It is red-brown to black with yellow coloured underparts and a ringed tail. A number live together in the trees of forested areas, moving at night and during the day. On the ground, the long tail is carried vertically, in the trees it is often used for balance and to grip branches.

A relative, the **Cacomistle** or **Ring-tailed Cat** *(Bassariscus astutus)*, is dark grey to brown with a long, but less distinctly, ringed tail. Slightly smaller than the Raccoon, it is found in the Western United States, Mexico and Panama, where it feeds chiefly on Rodents. Its feet are hair covered on the soles, an adaptation to scrambling over rocky areas.

The **Kinkajou** *(Potos flavus)*, with a body up to 40 cm long, a prehensile tail as long again, and short legs, is an excellent climber and is found in the tropical forests of South and Central America. Its very large, round eyes indicate its nocturnal habits. It is omnivorous and particularly fond of honey.

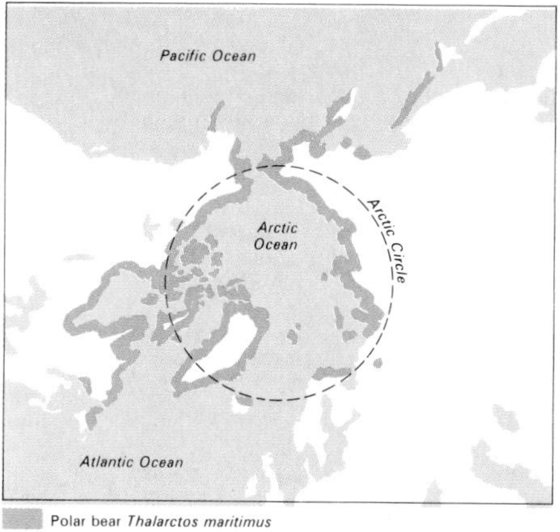

Polar bear *Thalarctos maritimus*

△ The Red Panda has facial markings that are very similar to those of Raccoons and it is a member of the same family. Red Pandas are chiefly nocturnal and spend much of their time in trees.
▷ The Pine Marten is an agile and quick hunter, chasing its prey of Squirrels and birds about in the high branches of trees. It can leap long distances aided by the strong muscles that snap its spine straight as it takes off enabling it to push exceptionally hard against the branch.
◁ ▽ The Polar Bear has the usual cold climate adaptations—a heavy coat, small ears and a smaller ratio of body surface to volume than other bears. It is a patient, stealthy and solitary hunter and preyed upon only by Man.

The **Olingo** *(Bassaricyon)*, of Ecuador and Nicaragua, is very similar to the Kinkajou but is smaller and without the prehensile tail. The Olingo and Kinkajou often live together in the same trees.

Giant and Red Panda

(Procyonidae)

The **Giant Panda** *(Ailuropoda melanoleuca)* of the bamboo forests of western China and Tibet, is a member of the Raccoon group but is more likely to be taken for a Bear. It is a most appealing looking, Bear-like animal, up to 6 ft (180 cm) long and weighing 75 lbs (80 kg). It has a creamy white body with black limbs and ears and a black zone around each eye. On each front foot there is a pad which functions like a thumb to help it to grasp food, chiefly bamboo shoots. Its molar teeth are specially formed as efficient grinders, rather than flesh and bone shearers, but there is no doubt that the Panda is truly descended from carnivorous ancestors. However, its exact relationship to the procyonids or the Bears is still a subject of controversy. Giant Pandas are plantigrade, climb well and form tunnels through the dense bamboo thickets. Not much is known about their habits in the wild, or how many there really are, but they are fully protected animals, though very rare. Two famous Giant Pandas, Chi Chi and An An, were brought together in the London Zoo to mate but this was unsuccessful. There has been reported the birth of several young Pandas in the Peking Zoo, however. Giant Pandas were discovered by Pere David, French missionary, over 100 years ago. Despite familiarity and popularity, live Pandas were not seen outside China until the 1930's. Zoos have still only had a total of about two dozen between them.

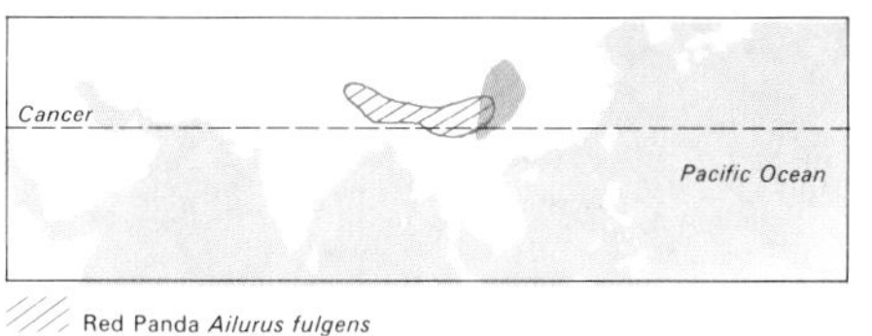

Red Panda *Ailurus fulgens*
Giant Panda *Ailuropoda melanoleuca*

Otherwise known as the **Lesser** or **Common Panda** *(Ailurus fulgens)*, the **Red Panda** is found in forested areas at altitudes up to 15,000 feet in Nepal, the Himalayas and China, occurring in greater numbers than the Giant Panda. It is about 24 inches (60 cm) in body length with a long, indistinctly ringed, bushy tail and a coat of soft, long hair, shiny red on the upper parts and black below. It has short, partly retractable claws. It spends most of its time in trees, active chiefly by night, feeding mainly on fruit and buds but taking eggs and small birds when they are available.

Polecats, Ferrets, Mink, Martens and relatives

The **Polecat** or **Fitch** *(Mustela putorius)*, is a larger form of the Weasel type, measuring up to 24 inches (60 cm) overall. It is found over most of Europe except Ireland. Its coat is dark with long, brown to black guard hair overlying a yellow undercoat. The Polecat hunts very actively, with a preference for wetter places, and will take any small Rodents, Frogs and reptiles that come its way. The **Ferret** *(Mustela furo)* is an albino domesticated form of Polecat used by Rodent trappers.

The **Minks** *(Mustela lutreola)*, the European form, and the **American Mink** *(Mustela vison)*, are closely related to the Polecats and their habits are much the same except that they hunt mostly in water where they are very much at home. The latter is farmed for its valuable fur and escapes have established a thriving population of wild mink on many British rivers. Polecats, Ferrets and Mink all have a similar characteristic smell.

Martens are native to most temperate and sub-tropical areas of the northern hemisphere with one species in Java. Some have a most beautiful golden-brown fur. They feed variously on small animals, fruit, eggs, and will take fish.

The **Pine Marten** *(Martes martes)* is a very agile tree-dweller, now rather scarce in Britain. It is up to 18 in (45 cm) long with a 10-in (25 cm) tail and is found in forested parts of Europe and Asia. *Martes americana* is its North American relative. Also found in parts of Europe, often living in the outbuildings of farms, is the **Beech** or **Stone Marten** *(Martes foina)*. The **Russian Sable** *(Martes zibellina)*, brown or blue-black with white ear tips, is a good climber. It provides one of the most coveted of all furs and is still trapped in large numbers.

Other mustelids are the **Tayra** *(Tayra)*, the **Grison** *(Grison)*, and the **Zorilla** *(Ictonyx)*. The first two are found in South America and the last is a native of Africa. They all have some features in common with both Weasels and Badgers and are sometimes referred to as the 'Weasel-Badgers'.

Weasels and Stoats

(Mustelidae)

These are typical representatives of the Mustelidae, mostly fierce hunters, occurring widely through northern and central parts of Europe to Asia and with relatives in North America.

The **Weasel** *(Mustela nivalis)* is short-legged and small, about 8 in (20 cm) long and with a 2-in (5 cm) tail. Its fur is brown above and white below in both summer and winter, unless the animal lives in a particularly cold area, such as the higher parts of the Alps. It hunts mostly at night over open country, and feeds chiefly on Mice, Rabbits and small birds. The Weasel will swim when necessary.

The Stoat *(Mustela erminea)*, with similar habits, is a little larger, up to 12 in (30 cm) in length with a 10-cm black tipped tail. It lives in old walls and rock crevices. At the onset of winter, as the brown coat is shed, a pure white one is developed, except for the persistent black tail tip. This fur is called ermine, and the black tail tips remain visible on the ermine trimmings seen on ceremonial robes.

In the cases of both Weasels and Stoats, males are somewhat larger than the females, and both produce a strong smell from glands at the base of the tail when angry or alarmed.

Wolverine

(Mustelidae)

The **Wolverine,** otherwise the **Glutton** *(Gulo gulo)*, is the largest of the Weasel family, being up to 30 in (75 cm) long with a 6-in (15 cm) bushy tail and weighing as much as 40–45 lbs. For its size, it is one of the most powerful animals known.

It has comparatively short legs with large feet and sharp claws. Its coat of short wool with long shaggy overhair is black-brown with bands of chestnut at the sides. The fur does not matt or freeze even at very low temperatures.

The animal is an inhabitant of northern regions, parts of Scandinavia and Russia in Europe but more abundant in arctic North America, where it has the reputation of being the most cunning and voracious of predators. It is a solitary burrower and during its largely nocturnal activities it is reputed to kill unaided quite large animals, like Deer and Caribou, and gorge exceptionally heavy meals.

Badgers

The **European** or **Common Badger** *(Meles meles)* is found widely distributed in Europe and a number of very similar species are found in Asia. It is about 30 inches (75 cm) long with a stumpy 6 inch (15 cm) tail and has very distinctive colouring. Its long hair is a mixture of black and white, looking grizzly grey on the body but its head is white with black eye and throat stripes. The

◁ The Wolverine, protected from cold by an ice-shedding coat, hunts by stealth rather than speed but is mainly a scavenger. However, unlike most animals that eat carrion, it will challenge the hunter, usually a Wolf or a Bear, for its entire kill rather than just accept the leavings.
▷ △ The American Skunk, as a mustelid, uses its scent glands, located beneath its tail, to mark out its territory, but also uses them as a form of defence. When it is attacked a skunk will raise its tail as a warning and if this goes unheeded it will then eject two jets of foul-smelling liquid.
▷ The Sea Otter has very thick fur as an adaptation to almost permanently living in the sea.
▽ The European Badger which uses its front feet to dig its 'sett'.

Badger is strictly nocturnal, lives in small groups in a 'sett' which is excavated to some depth, and furnishes some parts of its home with bedding of moss and fern to make comfortable sleeping quarters. It is truly omnivorous, feeding on grubs, larvae, worms, Frogs and other small animals, and sometimes berries. It is in no way harmful to Man but occasionally kills domestic Chickens. Until very recently the Badger was the victim of one of Man's most degrading sports—Badger-baiting—but is now completely protected in Britain.

The **American Badger** (*Taxidea taxus)* is an expert burrower, found in North America, from Canada to Mexico. It resembles the European animal in shape but is smaller, being silver grey in colour. The brown face has a white stripe up the middle which (unlike European Badgers), extends down its back. The **Ratel** *(Mellivora capensis)* of Africa and Asia is a more distant relative, black with a distinctly separate pale grey back.

Skunks

(Mustelidae)

Skunks are the most infamous of the evil-smelling members of the Mustelidae, having the ability to eject an exceptionally offensive oily secretion from the anal glands.

All Skunks are New World forms and the **Common Skunk** *(Mephitis mephitis)* is to be found from the Hudson Bay area to Mexico. It is about the size of a domestic Cat, black with two white stripes on the back, a sharp muzzle and bushy tail. Unless disturbed and molested it is quite harmless, moving unhurriedly around in search of insects and small animals for food, mostly by night, and resting by day in its den or burrow.

Other Skunks are the **Spotted Skunk** *(Spilogale),* also of North America and the **Hog-Nosed Skunk** *(Conepatus)* found from Mexico to as far south as Chile and Patagonia. Its very long naked muzzle is well adapted for rooting in earth in search of grubs and insects for food.

Otters

(Mustelidae)

These are truly water-adapted mustelids, more at home swimming than moving on land. The lithe, close-furred and streamlined body of an **Otter,** with powerful tail, short limbs and broadly webbed hind feet, is admirably suited to its mode of life.

The **Common Otter** *(Lutra lutra)* of Europe and Asia, is up to 24 in (60 cm) long plus a 12-in (30 cm) tail. It has greyish under-fur with stiffer guard hairs of dark brown. Its food is chiefly fish but other small animals may be taken. A den, known as a 'holt' is made under tree roots or in soft earth, with usually a tunnel leading into nearby water. Otters are chiefly nocturnal and move very unobtrusively when following hunting trails from one water stretch to another. There is some anxiety about its survival in Britain where it is still hunted for sport.

The **North American Otter** *(Lutra canadensis)* is larger but has very similar features, as have some other species found in South America, Africa and southern Asia.

The **Sea Otter** *(Enhydra lutris),* is the only member of the Carnivora living entirely in the sea and is completely aquatic, not even coming on land to breed. Its young are born on a bed of floating kelp or seaweed. It is now found only in very restricted areas in the Bering Sea and off the coast of California. Both northern and southern varieties are now completely protected but they were once on the point of extinction through being over-hunted. The Sea Otter will float on its back for long periods and in this position a mother will suckle her baby. Both males and females often adopt the same attitude for feeding, cracking open previously gathered shellfish with small rocks on the platform provided by the chest and belly. The animal will rest and sleep among the bouyant kelp when it requires to do so.

Seals and Walruses the Persecuted Ones

The Seals and Walruses are timid, intelligent creatures. Yet, for hundreds of years, their greatest enemy has been Man. In most of the world's northern waters, various species are still in danger of extinction because Man continues to covet the Seal's fur and Walrus's tusk.

Seals are found in almost every part of the world's oceans, mainly in colder regions, particularly the Arctic and Antarctic. They are well adapted to their harsh environment, being able to store thick blubber under a hairy or furry coated skin.

The order Pinnipedia (fin-footed) contains 32 species of Seals. Within this order there are 3 families: Eared Seals (Otariidae), True Seals (Phocidae), and Walruses (Odobenidae). The 13 species of Otariidae comprise the Fur Seals and the Sealions. These have small projecting ears and are able to rotate their rear flippers forward. In this way they can move quite nimbly on dry land where they spend a good deal of their time. The Phocidae (True Seals), have no outer ears. They can move only clumsily when out of the water. Because their rear flippers only extend backward they have to hump themselves along in an ungainly wriggle. Walruses comprise just one species, with two subspecies, the Pacific Walrus and the Atlantic Walrus, differing only in minor ways. These have no outer ears but can turn their rear flippers forward. They have long, ivory tusks and moustachial whiskers.

Seals are particularly vulnerable at breeding time. Many species gather in enormous numbers at traditional breeding grounds known as 'rookeries'. The breeding pattern for many species is as follows: the males arrive first and the mature bulls, or beachmasters, establish territories on the shore. A few weeks later the females arrive, pregnant from their matings the previous year. Each beachmaster gathers a number of cows into a harem, sometimes as many as 50 or more. The bull fights off challengers and guards his cows until they produce their pups.

Mating takes place again soon after the pups are born. Seal pups grow amazingly quickly on their mothers' very fatty milk and are soon able to withstand their icy environment. Seals vary in size from the 6 metre long Elephant Seals to the 1½ metre Ringed Seals. Seals moult once a year, returning to land to shed their coats.

Fur Seals

(Otariidae)

The **Northern Fur Seal** *(Callorhinus ursinus),* is larger than the **Southern Fur Seal,** the males growing up to 2 metres long and weighing approximately 300 kg. Fur Seals have a small external ear and are distinguished by the ability to turn their back flippers forwards and 'walk', raising their bodies from the ground. The fore flippers are quite large and obviously mobile. Generally, the dense fur is dark brown, although when wet, the females look silvery. The young Fur Seals have this silvery look also. Fur Seals eat fish such as Herring and Salmon but have been known to eat deepsea fish too, indicating that they will dive quite deeply for food.

Although the various species of Fur Seal differ slightly in breeding patterns, the **Kerguelen Seal,** which breeds on islands in the Bering Sea, is typical. Hundreds of thousands of Fur Seals converge on the breeding grounds at the beginning of June. The beachmasters, mature bulls, stake out their claims with a great deal of roaring, writhing and posturing. At the end of June the pregnant cows arrive and join one of the harems. Each produces her pup after a day or two and watches over it carefully for many weeks. In October, the beaches empty as the Seal herds begin their 3,000 mile journey to their winter habitat, off the coasts of Japan or California.

The Southern Fur Seal *(Arctocephalus australis)* is found around the Falkland Islands. Other Fur Seals found in southern waters include the rare **Guadalupe,** the **South African Seal** and the **New Zealand Seal.**

Sealions

The performing Seals that so delight circus and zoo audiences are almost invariably **Californian Sealions** *(Zalophus californianus).* Although many Seals are playful, Sealions seem more so than most and play games even in the wild state. The Californian is the most abundant of the 5 species of Sealions and is found off the southwest coast of North America. They are extremely strong and agile, attaining swimming speeds of approximately 20 to 25 knots as they jet across the surface of the water, and with an ability to jump across rocks and from considerable heights. They are able to do this without suffering injury, by taking the shock on their front flippers and soft cartilaginous ribs.

Sealions have poor eyesight but are good fishermen, eating fish, Squid and occasionally sea birds and Penguins, taking altogether 20 kg of food each day.

Steller's Sealion *(Eumetopias jubatus),* found in the Arctic, is the largest

From Pup to Adult

Grey Seal pups measure about 370 mm when they are born and weigh approximately 15 kg. They are suckled by the mother Seal for about three weeks. Seal milk, far richer than that of a domestic Cow, fattens the pup until it is 3 times its original weight. The pup needs this blubber to keep him alive until he can fend for himself because the mother Seal abandons her baby after three weeks, returning to the sea. The young Seals begin to lose their white coats and their juvenile coat starts to form. They spend their time on the shoreline, learning to swim and feeding on Shrimps and small fish.

The natural curiosity of the Grey Seal and its instincts soon set the young animal on its travels and, within two weeks, it may be 400 miles from the rookery.

By the age of 5 years, Grey Seals are sexually mature and when the next breeding season arrives they are on their way to the breeding grounds. However, at this age, a bull Grey Seal would not be able to hold a territory and presumably they do not mate until they are well past 5 years old.

▽ A Galapagos Sealion lounges with her pups. The Southern Sealions, of which the Galapagos is one, are smaller than the Nothern species and the males have much smaller harems of four or five cows to a bull, as opposed to the 50 females which may form the harem of a Northern beachmaster bull.

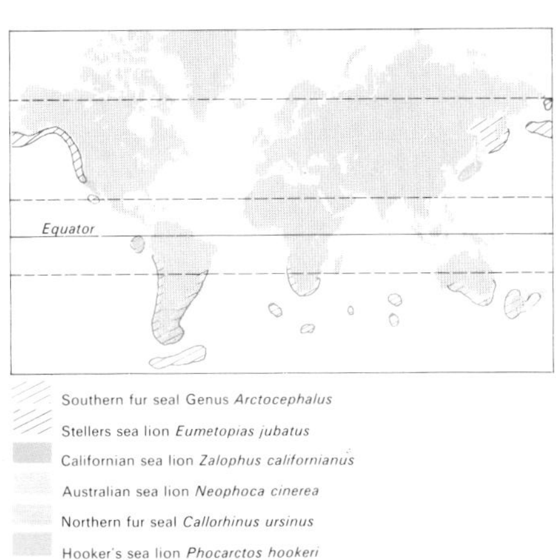

◁ A colony of California Sealions in a typical haunt of a rocky islet just offshore. On some of the animals the characteristic external ears that so obviously separate them from true Seals can easily be seen. Sealions, sometimes called Hair Seals, can also move their hind limbs more freely.

of the eared Seals. The male may reach $3\frac{1}{2}$ metres long and weigh over 11,000 kg. Females, however, reach only a quarter of this size and weight. The maned **Hooker's Sealion** *(Phocarctus hookeri)* is found in considerable numbers round the Auckland Islands south of New Zealand. The other two species of Sealions are the **South American** *(Otaria byronia)* and the **Australian** *(Neophoca cinerea).*

Common, Harp and Ringed Seals

The **Common** or **Harbour Seal** *(Phoca vitulina),* lives around the coasts of Europe, North America and north-eastern Asia. It is comparatively small, reaching just over a metre in length and a weight of approximately 100 kg. It has its habitat on mudbanks and sandbanks, where it also bears its young. Although clumsy on land, the Common Seal is a graceful swimmer and diver. It can stay under water for up to 15 minutes, its nostrils and ears tightly closed, as it searches the sea bed for fish, crabs and mussels.

The **Harp Seal** *(Pagophilus groenlandicus),* also known as the **Greenland** or **Saddleback Seal,** lives on drift ice around the Greenland coast and eastward to the coasts of northern Russia. Harp Seals are gregarious creatures and congregate in enormous numbers at special rookeries. As many as 3 million have been noted, gathered in the Gulf of St. Lawrence, packed 5,000 to the square mile.

The **Ringed Seal** *(Pusa hispida),* the smallest of the ocean Seals, lives in the Arctic, staying close to the shore. It has a grey coat with light coloured rings.

Inland Seals

The two most abundant inland Seals are the **Baikal** *(Pusa sibirica)* and the **Caspian** *(Pusa caspica).* They are both closely related to the Ringed Seal. The Baikal species is the only true freshwater Seal. The nearest sea to Lake Baikal is 1,250 miles away.

Some zoologists believe that the Baikal and Caspian Seals may be the survivors of ancient species of Seals which lived in the waters when they were inlets of the sea. As the waters receded, these early Seals were left behind. The Seal population in the Caspian Sea is about $1\frac{1}{2}$ million, and in Lake Baikal, North Russia, about 70,000.

In the summer, the Baikal Seals gather at the northern end of the lake to breed and mate. In the winter, when the lake is iced over, the cows stay on the ice, but the bulls and the young live in the water, keeping open breathing holes.

Other inland Seals are found in Lake Saimaa in Finland and Lake Ladoga in Russia.

Grey and Bearded Seals

The **Grey** or **Atlantic Seal** *(Halichoerus grypus)* is found in the North Atlantic, the North Sea, the Baltic and well into Arctic waters. It grows to about 3 metres long and weighs about 300 kg. About three-quarters of the world's Grey Seals live around the British coastline, but mostly off the Scottish coasts.

Large numbers of Grey Seals converge on the British rookeries when breeding time comes round. The bulls arrive first, and the cows a little later. The mature bulls set up their harems well in from the sea, while the young bulls stay near the water's edge. The pups are born with a white fleecy coat, which gives way to the juvenile short-haired coat after about three weeks. British Grey Seals bear their young from September to December, but those living in the Baltic and Western Atlantic pup rather later in the spring. Adult seals leave the breeding grounds to feed in the open sea, returning once in the year to moult, before going back to the sea again.

Bearded Seals *(Erignathus barbatus)* live in Arctic regions. They derive their name from their bushy whiskers, which they probably use, like the Walrus, for sensing food on the seabed. They are generally solitary creatures.

Antarctic Seals

The **Leopard Seal** *(Hydrurga leptonyx)* is the fiercest of the Seals. It hunts Penguins in the water or, if its prey is on the ice, will swim below it, shatter the ice with its powerful head and throw the bird into the sea. It will also eat the pups of other species of Seal. The females are larger than the males, measuring nearly 4 metres and weighing up to 500 kg.

The **Weddell Seal** *(Leptonychotes wedelli)* spends the Antarctic winter under the ice. It keeps open breathing holes by chewing at the ice. A Weddell Seal has been known to stay submerged for 43 minutes.

The **Crabeater Seal** *(Lobodon carcinophagus)* in fact eats little but Krill and some small fish. It swims through shoals with its mouth open, squeezing out the water through its unusual cheek teeth, leaving a mouthful of Krill. It is the most abundant of the Antarctic Seals with a world population of up to 5 million.

The **Ross Seal** *(Ommatophoea rossi)* lives a solitary life in the pack ice. Its bulging eyes presumably help it to see in the dark waters below the ice. It is fairly rare, its numbers being estimated at 20,000.

Walruses

(Odobenus rosmarus)

The **Walrus** is found in the Arctic regions round Greenland and Spitsbergen and also in the northern Pacific. They live mainly on ice floes rather than on coastal beaches and their tough hides and thick layer of blubber protect them from the cold. The long ivory tusks, which distinguish the Walruses, are used by the animal for defence and for finding food. It digs up Shellfish and Snails from the seabed with the tusk tips. The Walrus also hauls itself across the ice by its tusks and uses them to lift itself on to ice floes. The prominant whiskers seem to be used to sense where food is likely to be found. A bull Walrus grows to about 3.7 metres in length and attains a weight of over 1,200 kg.

Walruses travel in family groups of one bull with two or three females and a few young. In April, the adults start the long journey northwards, during which the pups are born. In the summer, crowds of Walruses can be seen basking in sunshine, turning rosy red as they dilate their blood vessels to dissipate body heat.

The Disappearing Walrus

Man's persecution of the Walrus has been long and relentless. The Eskimos, of course, have always depended on the animal for their meat, oil, clothing and boat coverings. It was not until Europeans began to slaughter the Walrus for commercial reasons that the animal began to be in danger of extinction. Irresponsible killings and frightening wastage continued until, by 1930, only 100,000 Walruses existed in the world. Today, there are very few of the animals left in the eastern Arctic area and the Bering Sea and it is thought that Soviet Russia is likely to take measures to protect the species. The situation is more satisfactory in the western sector round Greenland and Hudson Bay.

Although the Walrus is no longer found along the coast of Labrador to Nova Scotia, conservation measures are now in force. Of the three main concentrations of Walrus population in the Canadian Arctic, only one area is being overcropped. Most of the 3,000 killed here each year are hunted down by Eskimos who need the meat and oil. But tusk-hunting expeditions are still causing regrettable losses.

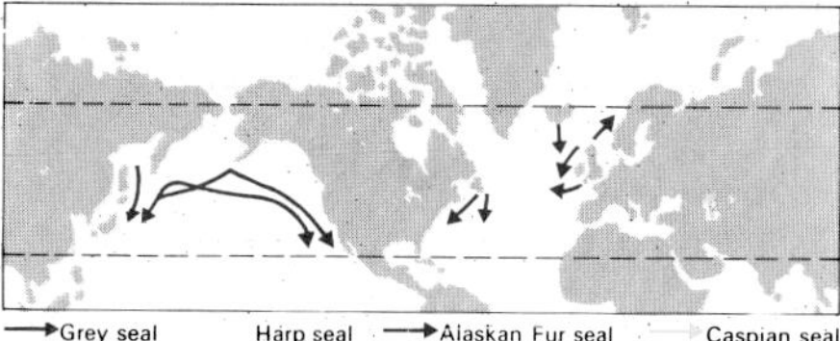

▷ The two maps show the migratory routes of the Northern Sealion, Walrus and seven species of Seal. The Pacific herds of Walrus must clear the Bering Strait before ice closes it in October.

Monk Seals

There are three closely related species of **Monk Seals.** The **Mediterranean Monk Seal** *(Monachus monachus)* is found in the Mediterranean, the Black Sea and around the Canary Islands. Over the last hundred years its numbers have declined from 5,000 to about 500. The **Caribbean Monk Seal** *(Monachus tropicalis)* is either extinct or on the verge of extinction. The last reliable sighting was of two specimens near Jamaica in 1949.

The most interesting feature of the Monk Seals is their tolerance of comparatively high temperatures. In the Leeward Islands, the lowest average temperature is 66°F. And yet the Monk Seal's layer of blubber is no less thick than that of Arctic and Antarctic Seals. It is difficult to see how the animal dissipates its body heat. Another unusual feature is the long-drawn-out pupping season, spread from early January to late May. The cows are very affectionate to their pups and rarely leave them.

▷ The Head of a bull Elephant Seal. The large, flabby nose can be inflated when the animal is excited or aggressive and during the breeding season the bulls are particularly belligerent.
◁ A pair of ponderous Walruses showing the distinctive, long tusks which the animal uses to grub out shellfish from the seabed. The tusks are also used to gain purchase when a Walrus is hauling itself out of the water or moving on ice. The Walruses scientific name means 'ice walker'.

Elephant and Hooded Seals

There are two species of **Elephant Seals,** the **Northern** *(Mirounga angustirostris)*, which breeds on islands off Mexico, and the **Southern** *(Mirounga leonina)*, which breeds in Antarctic regions. Elephant Seals are the largest of all Seals, reaching 6 metres and 4,000 kg. The males have an inflatable proboscis or short trunk. It is a sexual characteristic and also magnifies their roar.

The Southern species is the more numerous. The bulls are belligerent, rearing up when angry. This occurs particularly at the breeding season when fights often break out as a young bull tries to take a cow from a beachmaster's harem. In the summer the Elephant Seals come ashore to moult, a painful process which lasts about 6 weeks. Much of the surface layers of the animal's skin comes away with the hair and, apparently to ease the irritation, the Seals lie sleeping or grumbling in mud wallows, often one on top of another.

The male **Hooded Seal** *(Cystophora cristata)* also has an inflatable facial feature. It can blow its nose up into a bright red bladder—hence its other name, **Bladdernose Seal.** Hooded Seals live on ice floes in the Arctic, sharing breeding grounds with the Harp Seal.

Kindness or Cruelty?

Man has for long hunted Seals for their skins, their meat and their blubber, and sealing was at its most intense during the early 19th century. The southern Fur Seals were much prized for their fine pelts and, at one time, 30 sealing ships were visiting the island of South Georgia in the Antarctic every month, each ship carrying away 50,000 Seal skins. Not surprisingly, the southern Fur Seal became almost extinct. Fortunately, the Antarctic Treaty in 1961 gave protection to the Fur Seals and the similarly threatened Elephant Seal. A limited number of bulls of this species may be killed in any one season, but no pups or cows. Annual quotas are also laid down for Crabeater, Leopard and Weddell Seals, while Ross Seals are completely protected.

In the north, the Elephant Seals are protected, but large numbers of Bearded, Harp and Ringed Seals are killed by sealers and Eskimos. The Hawaiian Monk Seal is fully protected by the United States, after being hunted almost to extinction. Bulgaria, France, Greece, Italy and Yugoslavia similarly protected their Mediterranean species but fishermen regard Seals as enemies and are liable to kill them on sight.

A certain amount of culling may in some cases be beneficial to the world's Seal population. There is evidence that overcrowding causes a rise in infant mortality. The Grey Seal has been totally protected by Britain since 1932 but occasional cullings have been necessary.

The culling operations are generally carried out during the breeding season, at the rookeries when the Seals are ashore. In some countries clubs are used to kill the chosen victims. Many people believe this is cruel, suggesting that the method is used only to prevent damage to the Seal skin. Those who support clubbing say that shooting risks leaving wounded animals to a lingering death. In Britain, culling is carried out by shooting.

△ Sea Elephants, the largest of all the Pinnipeds, are more than twice the size of other seals. The bulls may reach 20 feet or more and can weigh up to four tons. The blubber of a single large Sea Elephant (or Elephant Seal as it is sometimes known) can yield as much as a ton of oil and the animal was almost hunted to extinction before the 1961 Antarctic Treaty gave it protection.

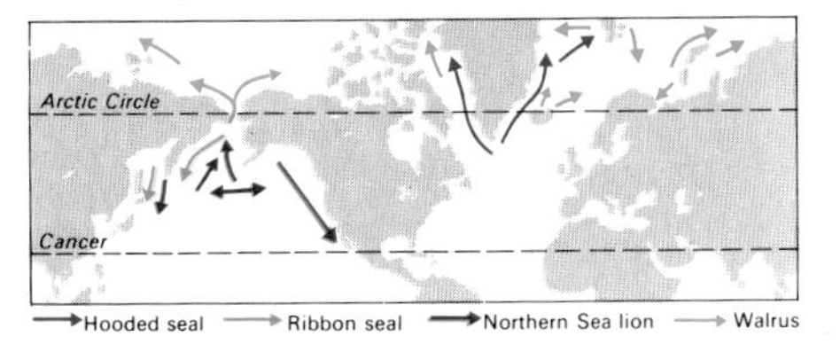

Elephants

The Indian and African are the only two surviving species of the Elephant order which evolved some 25 million years ago and once included the widely known Mammoth and several obscure dwarf forms.

Elephants are classified in the mammalian order *Proboscidea*, the word being used because they possess trunks or very elongated noses. The first members of the order evolved about 25 million years ago, and, as today, the creatures were heavy in bulk and supported on massive legs. The feet of proboscids have always been pentadactyl or based on the five-toed form. Another distinctive feature is seen in the types of teeth. Most commonly, there are ridges on the molar or cheek teeth to aid in the grinding of vegetation, whilst the incisors appear always to have been elongated. The present-day elephants' tusks are extended upper incisor teeth, not canines, as might be supposed.

According to most modern classifications, the order contains three families, two of which are now extinct but known from their fossil remains. The family which contains living members is the Elephantidae but only two genera survive, the African and the Indian Elephants. Extinct members include the Stegodon and the Mammoths and some dwarf forms that lived on eastern Mediterranean islands, including Crete. They stood only one metre high.

The origin of proboscids is obscure, but it is known that by the time of the Miocene period, just over 25 million years ago, great land mammals with proboscid features roamed the earth. These migrated from Africa, through America and Eurasia. Some of the early proboscids were Mastodons but these were never on the direct line of Elephant evolution.

The African Elephant

(Loxodonta africana)

Two races of **African Elephant** survive today, the west African **Forest Elephant** and the **Bush** or **Savannah Elephant,** which ranges across east, central and southern Africa. The African Elephant is generally larger than its Indian counterpart. Height is taken at the shoulder and averages about 8 feet (2.5 metres). The largest specimens stand at about 13 feet (4 metres). The most interesting feature of the Elephant is the drawn out combination of upper lip and nose, called the trunk. Air passes through a pair of nostrils at the tip but the trunk is also used for touching and smelling. Sets of muscles co-ordinate so that the trunk is able to seek out and grasp leafy twigs or bunches of leaves. These are stuffed into the mouth. The tusks, consisting of valuable ivory, are very long, permanently-growing second incisor teeth of the upper jaw. Body heat-regulation is assisted by enormous ear flaps which act as radiators. This is essential as excess heat is not easily lost through the thick skin, which in any case is poorly supplied with sweat glands. Only a little hair is present on the skin surface. Sexual maturity is reached at about the age of 11 years and a single calf is born after a gestation period of a little under two years. Calves are protected in mixed herds which are led by a senior female.

The Indian Elephant

(Elephas maximus)

The **Indian Elephant** is better named the **Asiatic Elephant** as it is found intermittently across India, Burma, Thailand, Malaysia and some of the large Asiatic islands, such as Ceylon and Borneo. There are a number of different strains of the species. On average, it is smaller and lighter than its African relative, the average height at the shoulder being less than 8 feet (2.4 metres). The highest point of the animal is the double-domed head, not the shoulder as is the case in the African elephant. The back is straight or slightly convex compared to the African's saddle-back. Other differences are that the tip of the trunk has only one thin projection which is used for collecting food whereas in the African, upper and lower projections grasp vegetation. The Indian trunk is more heavily marked with annular grooves and the ear lobes are much smaller than those of the African Elephant.

Over the centuries, Man has hunted the Indian (and African) Elephant for the bull's magnificent ivory tusks. Only rarely does a cow carry big tusks. Man has tamed the Indian Elephant and records of his attempts date back two or three thousand years. Besides the animal's use in warfare and ceremonials, troops of Elephants have been trained to move logs or boulders and the animal is considered to be highly intelligent.

The Hyraxes

The **Hyraxes** or **Conies** are small, colonial animals found over a large part of Africa south of the Sahara. However, one species does occur outside Africa, in

Towards the close of the Pliocene period, little more than 2 million years ago, the fossil records show the appearance of the hairy Mammoths and the Elephants. These are classified in the family Elephantidae. Some of the Mammoths were truly massive; a North American species is said to have stood nearly 4.5 metres at the shoulder, with one tusk measuring 5 metres in length. Some woolly Mammoths lived at the time of the last or Pleistocene Ice Age, and besides their thick fur they were protected from the cold by a thick, insulating layer of fat under the skin. Whole, preserved specimens have been found in the frozen tundra of Siberia.

The two surviving genera, *Elephas* the Indian Elephant, and *Loxodonta* the African Elephant, are distinguished by the tusks, the lack of lower incisors and canines, and the characteristic grinding teeth in each half-jaw. These are ridged and covered with a hard layer of enamel. The skin is thick and the hair is sparse. Special skull bones and the trunk characterise the head and large ears function as heat-regulators. The Elephant's brain is bigger than that of any other land mammal and it is of interest that the intelligence and sensory areas are particularly well-developed. Elephants are browsing animals and feed for long periods of the day. The digestive system is adapted to a herbivorous diet.

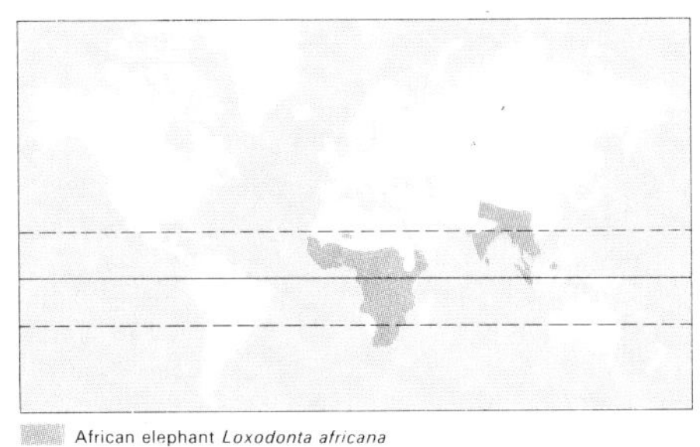

◁△△ The differences between the two species of Elephants can be clearly seen from these two drawings. The Asiatic, or Indian, Elephant which is found in the tropical forest of Southern Asia has a bulbous forehead and very much smaller ears. The African Elephant has a flatter forehead, large ears and two distinct 'lips' at the end of its trunk. The African Elephant has longer tusks which are found in the male and the female while only the male Indian Elephant normally has them.
◁◁ A family group of Indian Elephants feeding in the kind of lush grassland they prefer. Elephants eat a tremendous amount and they are constantly on the move in search of fresh pastures. They usually migrate in the rainy season and they will travel long distances to new feeding grounds, walking in single file, an experienced female in the lead, the male a long way behind.
◁ A group of African Elephants, all fairly young, contemplate the camera. Elephants usually flee when they scent Man, moving away at a fast amble, but herds are sometimes aggressive and rogue elephants will often attack without warning.
▷ Rock Hyraxes are the closest living relatives to Elephants despite the enormous difference in size. Hyraxes weigh about 7 lb and venture only rarely into the open to feed.

parts of Arabia and the near East. Hyraxes are grouped in their own separate order, *(Hyracoidea)*, and although they may at first glance look like Rodents, the nearest relatives are, in fact, Elephants. It should be emphasised, however, that the connection is very remote. Hyraxes are outstanding climbers and jumpers and some scale trees and others, rocks, with great agility. The soles of the feet, which are moist and padded, help with grip. Further grip is provided by toe nails and the curved claw on the second toe on each of the back feet. These small herbivores nibble at grass and shoots and are able to grind their food with cheek teeth. The digestion of cellulose plant material is all important and Hyraxes possess three special digestive tubes or *caeca* which stem from the intestine. The animal's coat is soft and on the back, a ring of hairs surround a number of scent glands.

The **Tree Hyrax** *(Dendrohyrax dorsalis)* is found in tropical forests and feeds at night. The **Ground Hyrax** *(Procavia habessinica)* is a gregarious and active creature which feeds by day. Two or three young form a litter following a gestation of eight months.

The Elephant problem

The Elephant has been hunted by Man since prehistoric times. It may be argued that primitive Man helped maintain some sort of balance between numbers in the herds and the growth or recovery of vegetation in Elephant areas.

The Elephants may spend up to 18 hours a day feeding, and a migratory herd can cause enormous damage to shrubs and trees. Wild Elephants often invade plantations or village crop lands by night. They have a particular liking for bananas, oranges, sugar cane and maize. Various forms of humane deterrent have been tried but the farmers normally end up by shooting the marauding animals.

In an attempt to prevent the unchecked slaughter of big game by the ivory trade and shooting parties, or safaris, game parks and reserves were set up early in the present century. These failed to take into account the overall ecology of any one area. Elephants created havoc and caused destruction of huge stretches of vegetation and the habitat of other creatures. Regrettably, in recent years a number of 'culling' operations have been organised and thousands of Elephants slaughtered. Elephant-ranching is being discussed in order to limit numbers, maintain a healthy stock and protect the restricted areas of the parks and reserves. Despite all attempts to protect them Elephants are still trapped by poachers for their ivory. The methods of killing are usually of the most cruel, inhumane kind.

Inoffensive Sea Cows

The mute and over-hunted Dugong, the sociable Manatee, and the extinct Sea Cow with its sensitive hearing are distant aquatic relatives of the Elephant.

Sea-Cows (order Sirenia) are timid herbivores inhabiting the shallow equatorial waters on either side of the Atlantic and Indian Oceans and the coast of Australia. They are perfectly adapted to their aquatic environment, which accounts for their external resemblance to Whales, Dolphins and Seals. There is little doubt, however, that they are more closely related to the Elephant.

The two modern forms, the Dugong of the Pacific and Indian Oceans and the Manatee of the Atlantic, are different in many respects and represent lines of evolution that have been separate since the Eocene period over 40 million years ago.

The Dugong is a purely marine animal, but the Manatee is also found in the rivers of Africa and America. Both are frequently called Sea Cows.

Steller's Sea-Cow was a large beast discovered in the 18th century and sadly exterminated within 30 years. Sea-Cows have a streamlined body form, a leathery skin with a few hairs and thick blubber. The adult head is comparatively small and the neck is thick. Fleshy, mobile lips are well adapted to take in seaweed and other aquatic vegetation, as is the long convoluted intestine of the creature. It has no hind limbs and the pelvic girdle is vestigial. The large fore-limbs have evolved into paddle-like flippers. Swimming is effected by movement of the body and the broad, horizontally-flattened tail. The bones are characteristically heavy and of high density.

As in Whales and Elephants, the diaphragm of the Sea-Cow is oblique, reaching far back into the body cavity; the lungs contain air sacs and the blood system is adapted to conserve oxygen. Despite these modifications, Sea-Cows cannot remain submerged for much longer than 15 minutes.

The small eyes are placed high on the head and protected by muscular lids. A valve arrangement enables the nostrils to be closed when diving. Although there is no 'ear' visible, simply an orifice a few millimetres wide, Sea-Cows make up for their poor sight by acute hearing.

The snout and teeth of Sea-Cows are distinctive. Molars, up to 20 in number, are simple ridged pegs. Those in front drop out when worn. The well-developed, down-turned snout has large, fleshy, strong but sensitive pads.

△ The Manatee is an inoffensive aquatic animal that lives in shallow, warm coastal waters, both fresh and marine, on both sides of the Atlantic. They eat, sleep and give birth in the water which they never leave. They are less completely adapted to their aquatic life than their close relations, the Dugongs, and still have nails at the tips of their flippers which are supple enough to play a part in the gathering of food. Manatees eat so much vegetation it has been suggested they be used to clear overgrown waterways.

Dugongs have a similarly reduced dentition, but the male's twin upper incisors are developed into tusks.

Little is known of the Sea-Cow's breeding habits. There seems to be no definite breeding season. After a gestation period of 11 months a single pup is born and, occasionally, twins. The Dugong's habit of cradling its young in its flippers and suckling the juvenile at paired breasts at shoulder level, probably gave rise to legends of mermaids.

Sea-Cows are ideally suited to their ecological niche, but they are ill-adapted to coping with the destructive impulses of Man. These harmless creatures are valued all over the world for their flesh and oil-producing blubber. Their exploitation has pushed the surviving members of the order to the verge of extinction. Dugongs are already listed as an endangered species and in some parts of the world are protected by law.

Mermaid Magic

The name for this order, *Sirenia*, refers to the theory that the Dugong gave rise to the legend of mermaids, or Sirens.

Odysseus was supposedly so overcome by the beauty of the Sirens and the loveliness of their song that he ordered his crew to lash him to the mast and block their own ears with wax, lest they all be lured on to the rocks.

The comically ugly, albeit charming, face of the Dugong seems anything but seductive. It is likely, however, that the Dugong's habit of floating on its back and nursing its young at twin breasts may have sufficiently impressed an eager mariner before the animal dived out of sight. As for singing, the normally mute Dugong can manage only an occasional grunt.

The mermaid legend can be traced back to Arab and Greek seamen who would have seen Dugongs in the Red Sea and the Indian seas. Megasthenes, the Greek historian, talks of seeing such creatures near Ceylon. Another Greek writer, Aelian, adopts and expands this information, talking of Lions, Panthers and Rams with fish bodies. These beliefs were accredited by Portuguese and Dutch mariners en route to the east.

But an air of mystery still surrounds the Sea-Cows. Comparatively little is known of their behaviour, habits or history. What is certain is that these strange animals need protection from the hand of Man or they may become, like the mermaid and Steller's Sea-Cow, a part of history.

The Dugong, Big Game of the Ocean

(Dugongidae)

Dugongs can grow to a length of 3 metres and may weigh 270 kg or more. They are brownish or blue-grey in colour, paler on the ventral surfaces. Unlike the Manatee, the tail fluke is notched, the male has tusks and the upper lip has a distinctive disc-like formation, in contrast to the Manatee's bi-lobular lip pad.

Dugongs are said to forage nocturnally, chiefly on marine algae and green seaweeds. Most of their time is spent in shallow waters, but they sometimes pull themselves on to the land to feed. Natives of the Solomon Islands insist that the Dugong roots for shellfish in the sand with its tusks.

Although said to be solitary except at the time of mating, small family groups have been observed, and an extraordinary report of 1895 states that herds of many hundreds were spotted near Brisbane harbour, covering an

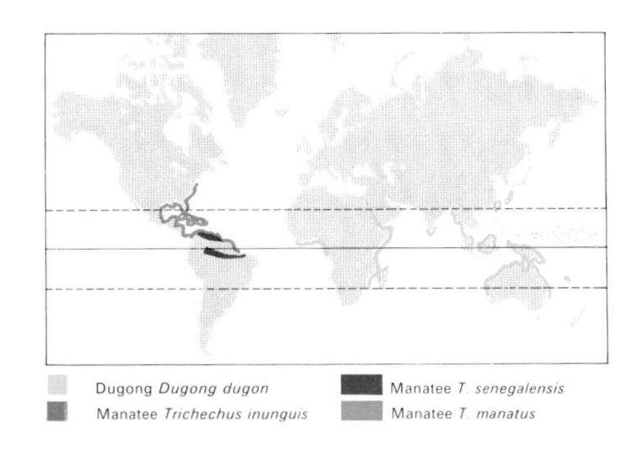

area 3 miles long and 300 yards wide. They were once abundant off Madagascar and 100 years ago were common in the Red Sea.

The Dugong is known as the 'Big Game of the Ocean'. The flesh is said to be tasty and an adult Dugong's body will produce 20 to 25 litres of oil.

Mystery and superstition surrounds these oddly charming animals, and this contributes somewhat to their downfall. In Madagascar, the tusks are ground down and used as an antidote to food poisoning; in China the flesh is much in demand as a supposed aphrodisiac.

Dull-witted and defenceless, the Dugong is an easy target. Indonesians slaughter them on sight along the south coast of Borneo. Aborigines are said to plug the nostrils of the unwary with clay in order to drown them. This would suggest that the Dugong is unable to breathe through the mouth.

Today, the Dugong is protected in many parts of its range, including East Africa, Mozambique and New Caledonia. Most are found around the south coasts of New Guinea and Queensland.

Northern Sea-Cows

(Dugongidae)

The fate of **Steller's Sea-Cow** is a tragic example of Man's senseless destruction of defenceless animals. This giant Sea-Cow was unknown to Western Man until 1741. Within 30 years of the discovery it was virtually extinct.

When the Danish navigator Behring was stranded on an island in the Behring Sea in 1741, he found there vast numbers of these giant but inoffensive beasts. Throughout their ten-month enforced stay, Behring's party fed almost exclusively on them and relied on their oil for heating. Killing them was not difficult: they were toothless, slow and, according to Steller, the naturalist in the party, were unable to dive.

Apart from its cold-water distribution (apparently only in the Behring Sea), Steller's Sea-Cow *(Hydrodamalis)*, probably had similar habits to the Dugong. It was, however, immense. Steller recorded specimens 9 metres in length, 6 metres around the body and weighing as much as 3,630 kg. The head was very small with a down-turned upper jaw; horny chewing pads took the place of teeth. Flippers were comparatively small, covered in bristly hair, and must have been somewhat ineffectual.

Steller's notes state expressly that there was no evidence of a skeletal 'hand' in the flipper; the 'arm' ended abruptly at the extremity of the radius and ulna.

The naked, rugged skin was 2.5 centimetres thick and so tough that the mariners had to use an axe to cut it. Today only a few imperfect skeletons and skins—brown with streaks or spots of white—are preserved.

The story of *Hydrodamalis* may not, however, be over. In 1962, the crew of a Russian whaler observed 6 animals which appeared to be Sea-Cows in shallow water off Cape Navarin, Siberia. They were sighted at a place which does not freeze over in the winter and where sea cabbage and seaweed were abundant. The report has never been verified. At least one naturalist has decried it, but some believe that Steller's Sea-Cow may yet, like the Coelacanth, 'return from the grave'.

Manatees

(Trichechidae)

Unlike the Dugong, **Manatees** are not exclusively marine. They also inhabit estuaries and rivers, and have been found hundreds of miles up the Amazon.

There are three species: *Trichechus manatus* (found in tropical Eastern New World), *Trichechus inunguis* (interior of northern South America) and *Trichechus senegalensis* (tropical West Africa).

The average length of Manatees is 2 to 3 metres and the average weight 200 kg, although specimens of over 250 kg have been recorded.

The tail of the Manatee is spatulate, not notched. Both body and flippers are used in swimming; the flippers each bear three or four vestigial nails. Each half jaw has up to 20 cheek teeth. The upper lip is deeply cleft, and the tip of each half, bearing stout bristles, is independently moveable. This allows a forceps-like movement ideal for feeding. They may eat up to 45 kg of vegetation a day.

Manatees are normally solitary but some travel in small groups and in Florida may congregate in parties of 40 or more, especially in warm spots in cold weather. An apparently sociable animal, they greet each other with muzzle to muzzle 'play'. Manatees make sounds underwater, thought to be for communication.

Females each bear young singly at any time of the year after a gestation period of 152 days. Offspring may remain with the mother for over a year. The greatest threats to Manatees seem to be Man and cold weather.

Hares and Rabbits

As a result of their own versatility the Hares and Rabbits thrive on almost every land mass on Earth. After Man and the Rodents their families are the most successful.

Lagomorph means 'hare-formed', which says little about the structure of these creatures. They are made up of two families; the *Ochotonidae* or Pikas, and the *Leporidae*, known as Rabbits and Hares.

There are fourteen species of Pikas, twelve inhabiting Asia, and two being natives of North America. It is probable that the American species originated in Asia and migrated to Alaska long ago when the two land areas were joined.

The Leporids, of which there are fifty-two species, are found throughout the world except in the Antarctic and Madagascar, although they were not present in Australasia or Chile until introduced by settlers. The European Leporid is thought to have originated in Spain. The name of the country itself is said to be derived from 'Tsapan', which was the Phoenician word for Rabbit.

These animals thrive in all sorts of climates, coping happily with the extreme cold of Siberia and the shimmering heat of the desert. Wherever there is sufficient vegetation these creatures will be found, increasing in numbers as men destroy the predators that are the Lagomorphs' natural enemies.

At one time it was thought that Lagomorphs and Rodents were members of the same order, but it is realised now that the similarities are purely superficial. Close examination reveals differences in skeletal structure that set them apart, and the teeth, although having essentially the same general arrangement in both Lagomorph and Rodent, are dissimilar in one main respect. The Rodent family have one pair of incisors in each jaw, but the Lagomorpha have an extra pair, set behind each pair of upper incisors. Other differences show in the arrangement of the genitals and in the fact that Lagomorpha pass their food through their bodies twice, eating their own faeces in order to extract the maximum nutrition from the available food.

The Pikas live in desolate areas and are not affected by man to any great extent. The Leporids, however, in many parts of the world are considered to be vermin. Attempts at eradication are often made by Man, but the Rabbits and Hares are still thriving.

Myxomatosis

The wild Rabbit has long been considered a pest in a number of countries and means of control have been sought for many years. In 1898, in Uruguay, a disease destroyed an entire stock of laboratory Rabbits. The cause was an unknown virus, eventually traced to immune Brazilian wild Rabbits.

Experiments showed that only Rabbits and a few Hares were susceptible and tests were carried out in Australia in an attempt to control the Rabbits in that country. These were at first unsuccessful, until it was found that the virus was spread by Rabbit-fleas or Mosquitos. Eventually, the Stick-flea was infected and the disease spread with almost total effectiveness.

Myxomatosis was introduced into France by a private citizen who released infected Rabbits in the countryside near Paris. Within a year infected animals were found throughout France despite vigorous efforts to stamp it out.

The disease was introduced accidentally into Britain, despite stringent precautions. The first outbreak was in 1953 at Edenbridge, in Kent. Rabbit fences were erected in an attempt to contain the disease, but by the end of 1954 it was present throughout Britain.

The death-rate from myxomatosis is about 95%, but pockets of immune Rabbits survived, and from these the Rabbit population has once more begun to grow.

△ Baby Rabbits, right, are born underground in burrows. They have their eyes closed and take a week or more to grow their fur. Young Hares, left, are born on the surface without the protection of the burrow and have their eyes open and a well-developed coat from the time they are born.

◁△△ The unusually large ears of this young Black-Tailed Jack Rabbit serve as radiators to disperse excess body heat not required by this common American Hare which is found in the hot climates of Southern USA and Mexico and as far north as the states on the Canadian border.

◁△ The Snowshoe Rabbit or Varying Hare changes colour with the seasons.

△△ A Russian Pika, lacking the large ears and hind feet of Rabbits and Hares, but nonetheless a Lagomorph. Pikas are sometimes known as Mouse-hares and they are found on the mountains of Central Asia, the Himalayas, where they have been found as high as 17,500 feet on Everest, and the Rockies.

Hares and Rabbits

(Leporidae)

These two branches of the Leporids, although similar in appearance, are different in detail and in habits.

European Hares are larger and thinner than **European Rabbits,** and they have longer legs and ears. In spring, the male Hares, known as 'jacks', fight each other by leaping and kicking and boxing with their fore-paws. From this behaviour is derived the saying 'as mad as a March Hare'.

Hares are solitary animals, unlike Rabbits who live in colonies in systems of burrows called warrens. The Hare does not burrow but lies in a shallow hollow called a form.

The baby Hares, or **Leverets,** are born fully furred and sighted. They soon leave their mother, each finding a form of its own, and being visited by the doe only to be fed. In this way, a predator discovering one leveret is likely to miss the others.

Baby Rabbits are born naked and blind, in a special burrow dug by the mother and lined with her own fur. They remain with her until they are quite mature, and even then, stay with the colony.

Whistling Hares

Pikas, the Ochotonid branch of the Lagomorpha, are inhabitants of the bleaker areas of North America and Asia. Of the two North American species, one lives in the Rocky Mountains, the other in Alaska and the Yukon. The colloquial name **Whistling Hare** or **Squeak Rabbit,** given to these animals, describes the high-pitched call made by them when danger threatens.

The twelve Asian species of Pikas live in Siberia, Mongolia and Tibet, and are believed by some authorities to be the remnants of a once widespread single species.

These animals are smaller than Rabbits, with short, rounded ears and an almost non-existent tail. Unlike most other small herbivores of the harsh steppes and mountains, they remain active throughout the winter, often falling prey to Wolves and Eagles.

Pikas are the haymakers of the animal kingdom. During the summer they harvest grass, iris leaves and other plants, and dry them in the sun. The crop is stored in great heaps under overhanging rocks to be used as a winter supply of food and bedding.

Snowshoes and Cottontails

(Leporidae)

The fifty-two species of Leporids belong to a number of genera. In Britain there are two species of Hare, the **Brown Hare,** which is found in the low-lands, and the **Blue,** or **Mountain Hare,** an inhabitant of Scotland and Ireland.

The coat of the Mountain Hare changes colour to white in winter. In this it resembles the **Snowshoe Rabbit** or **Varying Hare** of North America. The Snowshoe gets its name from the bristles that grow around its pads in winter, enabling it to run on soft snow.

European Rabbits of various strains closely resemble each other, but the **Eastern Cottontail** of North America found from Canada to Mexico, differs from them in skull structure.

This Rabbit is not considered a pest like its European cousin since it does not burrow, thereby causing less damage, and it is easier to trap. The North American Hare is known as the **Jack Rabbit.**

Gnawing Mammals

Rodents are the most successful among mammals, making up about one third of all the land forms. They show a wide range of adaptations and there are very efficient burrowers, climbers, runners, jumpers, swimmers and gliders among them. They are essential to the existence of many other animals because as prey they form important links in food chains. They easily make up losses in numbers by their fast breeding.

The order Rodentia includes three sub-orders. The most primitive are Sciuromorphs including the Beavers, Squirrels and Marmots; the Myomorphs and the Hystrichomorphs—the Porcupines, Guinea Pigs and related forms. The common characteristic of all rodents is that they possess persistently growing or open-rooted incisor teeth, one on each side of the mid-line in both upper and lower jaws. These teeth have harder enamel on their outer faces so that they become worn more at the back than at the front, leaving a permanently sharp chisel edge.

Most of them have unspecialised feet, with five fingers and toes. The gnawing habit is associated with a number of skull modifications and the three sub-orders are separated largely on the basis of differences in the musculature of the jaws and the shape of the skull.

Rodents are widely distributed over the earth, from deserts to forests, on or in the ground, or in trees. They have not become adapted to life in the sea, however, although the amphibious Beavers, Musk Rats and others are very much at home in fresh water.

Some are often of considerable nuisance to Man, feeding on his crops, both in the field and in store. Others spread disease-causing parasites. Man expends a great deal of time and energy in trying to control them. Others he finds useful for their fur.

Tree Squirrels

(Sciuromorphs)

The **Red Squirrel** *(Sciurus vulgaris)*, is a tree-liver of Europe and northern Asia. Its colour varies with locality from reddish to dark grey or black. It is about 18 in (45 cm) long, including its bushy tail, used for balance when jumping or climbing or held elevated over the back when resting. It builds a nest or 'drey', large sized for breeding, smaller for sleeping.

The **Grey Squirrel** *(Sciurus carolinensis)*, is a native tree-dweller of North America but has been introduced to many other places, including Great Britain, where it has become a pest. It has been a successful competitor of the Red Squirrel, almost eliminating it. By its bark-gnawing habits it destroys much young timber and is accordingly treated as vermin.

The **Flying Squirrels** *(Petaurista, Glaucomys)*, found in northern parts of Europe, northern Asia and America, have extended their climbing and jumping activity into planing and gliding and perform expertly. A flight membrane, extending along the flanks between front and hind limbs, provides the lift when stretched. The large **Indian Flying Squirrel** *(Petaurista philipensis)*, can glide distances of 150 ft (50 m) or more.

Ground Squirrels

These Sciuroideans are better known as the Marmots, Prairie Dogs, Gophers or Sousliks, and Chipmunks. They exist in greater variety than the tree forms, and most possess cheek pouches. They are generally short-tailed and hibernate in winter.

The **Marmots** or **Woodchucks** *(Marmota)*, are represented in alpine parts of Europe, in northern Asia and North America. They vary in size, up to 35 in (90 cm) long, including the short tail, are good burrowers and sometimes place a guard at the entrance to a family tunnel who, by a shrill whistle, gives the alarm. They lack cheek pouches.

Prairie Dogs *(Cynomys)*, chiefly of the western and northern American plains, are about half this size. They live in colonies in extensive burrows and their warning sound is a bark.

Sousliks *(Citellus)*, of North America, Europe and Asia, up to 12 in (30 cm) long, are also excellent burrowers and seal themselves underground in winter. Some are attractively spotted.

The **Chipmunks** *(Tamias, Eutamias)*, of North America and parts of northern Asia, more lightly built and sometimes showing ability to climb, are noted for their large cheek pouches. They have boldly striped bodies and a relatively large bushy tail for a ground-living squirrel.

◁ Marmots are hibernating animals capable of surviving through low temperatures, though if it does get very cold they wake and shake themselves to warm up. During hibernation they may lose a quarter of their original weight.
▷ The terrestial squirrels, like this Californian Ground Squirrel, lack the silky fur and very bushy tail of the true squirrels.
▽ The American Beaver is a clever engineer, building lodges, dams and canals.

Beavers

(Sciuromorphs)

The two species, *Castor fiber* of North Europe and *Castor canadensis* of North America, are very much alike, being amphibious animals up to 4 ft (120 cm) long, including the broad, flat, scaly tail which is used as a rudder in swimming. The hind feet are webbed and the double claw of each second hind toe is used for cleaning the fur. **Beavers** are now a protected animal in Europe.

Beaver habits are noteworthy. They live in family groups in burrows or 'lodges' with underwater openings, constructed in the banks of rivers and streams, just above a dam built by them from mud, stones, tree branches and trunks. The dam deepens the water and prevents it from freezing to the bottom in winter. Great effort is put into the dam and 'lodge' building and often intelligent behaviour appears to be employed. Beavers are reputed to build canals in order to float large trunks that they cannot carry to the riverside.

Pocket Gophers

Pocket Gophers are so-called from the French word 'gaufre' meaning 'honey-combed', a reference to the maze of tunnels they excavate. They should not be confused with the Ground Squirrels of the same name. Found in North America only, they combine the characters of Squirrels, Beavers and Mouse forms. Some of them have ever-growing cheek teeth and all have cheek pouches. Their fingers are bristly, and this, coupled with enlarged second and third front claws, means that they are equipped with admirable shovels.

In common with many other burrowers, the eyes are small and the external ears rudimentary. The upper incisors are used to loosen earth before it is scooped away and the animals can move through the burrows forwards and backwards without turning round. Pocket Gophers are a nuisance where they burrow in agricultural areas and considerable efforts are made at eradication.

The **Common Pocket Gopher** *(Geomys bursarius)*, about 12 in (30 cm) long overall, is one of a number of species.

Pocket Mice

There are three groups, the **Pocket Mice** *(Perognathus)*, the **Kangaroo Rats** *(Dipodomys)* and the **Spiny Pocket Mice** *(Heteromys, Liomys)*. All are small, mouse-like rodents, found chiefly in North America and with fur-lined external cheek pouches. Some have persistently growing cheek teeth as well as incisors of the same kind. These desert-dwellers are especially interesting for their adaptations to a life without drinking. They have very powerful kidneys which produce concentrated urine, to avoid wasting water.

The typical **Pocket Mouse** is small, about 4 in (10 cm) overall, with a fairly long tail and disproportionately long hind legs, used in jumping. It inhabits desert and plain areas and feeds mainly on seeds and fruit.

The very similar **Kangaroo Rat** is bigger, about 10 in (25 cm) long, including its long tufted tail. It lives in the arid North American zones. There is a strong resemblance to the Old World **Gerbil** and **Jerboa** but it is different in possessing external fur-lined cheek pouches.

The Spiny Pocket Mice of Central America and Trinidad are distinguished by the firmer and longer bristles standing out from the general body hair.

Mountain Beaver

Found only on the Pacific side of the United States of America, the **Sewellel** *(Aplodontia)*, represents the last of a line of primitive rodent ancestors. It is of brownish colour, about 24 in (60 cm) long, including its short tail, and has small rounded ears and small eyes. It is a burrower in damp woodland localities and does not live an aquatic existence as does the true Beaver.

Scale Tails and the Springhare

The **Scale Tails** *(Anomalurus)*, are a very distinct group, similar to the Flying Squirrels in possessing a large flight membrane. They differ in having horny scales in two rows on the underside of the tail. This is not prehensile but the scales are used to grip tree bark, making the animals good climbers. They are found in the African forests only and vary in size from quite small up to 16 in (45 cm) overall, in some species.

The **Springhare** or **Jumping Hare** *(Pedetes caffer)*, found in fairly dry areas of south and east Africa, has a unique appearance, half Kangaroo, half Rabbit. It is about 25–30 in (75 cm) long overall, with its long bushy tail. It has short front legs but by means of its extremely elongated hind legs can travel very fast over quite long distances in leaps of up to 6yd (6 m). It is also an excellent burrower and numerous families will live together in large, composite warrens, living mostly nocturnal lives.

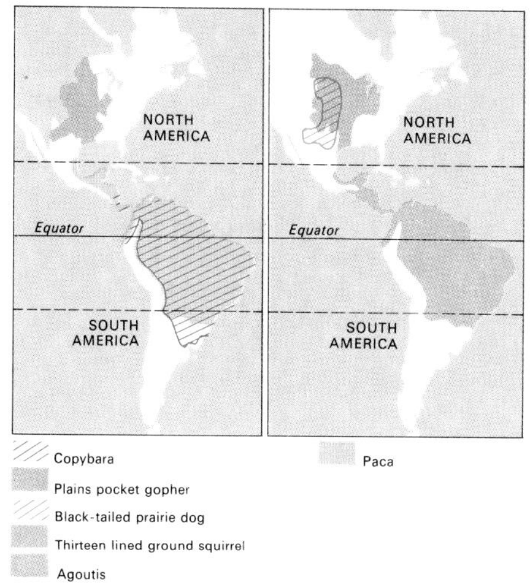

Lemmings

There are two main species, the **Norway Lemming** *(Lemmus lemmus)*, and the **North American Lemming** *(Lemmus trimucronatus)*. A third, the **Collared Lemming** *(Dicrostonyx)*, is found only within the tundra regions of the northern hemisphere. All are typical rodents related to the Voles.

The Lemming may be up to $4\frac{1}{2}$ in (15 cm) long, with a short tail and its feet, the soles of which are hairy, are adapted for digging although it does not burrow extensively. Lemmings are not really gregarious even though they

Legend of the Lemming

Ordinarily each Lemming lives an individual existence and is quite resentful of intruders in its own small domain. But at intervals of about four years, not associated with seasonal or breeding conditions and certainly not a migratory march between localities, large numbers of Lemmings congregate and move together during a so-called 'Lemming year'. They break away from the area of their origin, never to return.

The general direction of movement seems always to be from a higher to a lower elevation—often, therefore, towards the sea—no matter where the starting point. They move chiefly by night, feeding on any available food, preyed upon by larger animals and dying of disease in large numbers. Survivors are not deterred by ordinary obstacles and they will traverse wide stretches of open water without hesitation. None attempt to halt their movement for long even when the sea is

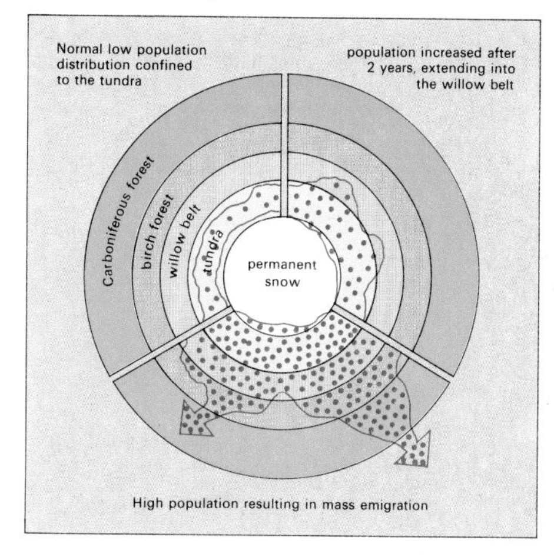

finally reached but every one of them begins to swim away from the shore and continues to do so until it drowns. They will jump from rocks and cliff faces to enter the sea if there is no other way. Some that take this last journey must have been born during the trek since it may take a band of Lemmings up to three years to reach the coast from the starting point. It is interesting that both the Norwegian and North American Lemmings behave in the same way.

Explanations of the behaviour have varied. One embodies the idea that the animals are migrating from the areas of their birth to some place of their ancestry beyond the horizon, in response to an inherited impulse. The place has even been specified as Atlantis. Others hold that it is an anxiety response to the building up of overcrowded and diseased communities that are undoubtedly created at fairly regular intervals.

Whatever the true explanation, the result of a march from any given area enables the growth there of a new and perhaps healthier population.

may be found in large numbers in a small area. They are by nature pugnacious, resenting intruders, and make a peculiar hissing noise when seriously disturbed. They feed almost entirely on grass roots, dwarf birch shoots and lichens and mosses, for which they excavate long tunnels under the turf or snow.

Particularly noteworthy is their mass movement or 'swarming' behaviour, occurring roughly in a four year cycle, during which they move in very large numbers away from their normal feeding and breeding grounds.

True Mice and Rats

These Muroideans are Old World in origin but they have accompanied Man everywhere in his world-wide travels. The typical forms are the **House Mice** *(Mus)*, and the **Rats** *(Rattus)*. They are recognised externally by possessing long snouts with well-developed whiskers, large round eyes naked ears and long, scaly tails.

The best known mouse is probably the **Common House Mouse** *(Mus musculus)*, found everywhere in Man's dwellings or in food stores and even in deep freeze meat carcass stores. It has a body of about 3 in (8 cm), and a tail of the same length. It can run and climb well and is often very destructive. Close relatives are the **Long-tailed** or **Common Field Mouse** *(Apodemus sylvaticus)*, and the tiny **Harvest Mouse** *(Micromys minutus)*. Other true mice are the **African Tree Mice** *(Dendromus)*, good climbers with opposable first toes.

The two best known rats are the **Black Rat** *(Rattus rattus)*, with a 6 in (15 cm) body and a slightly longer tail and the **Brown Rat** *(Rattus norvegicus)*, a little bigger though with a shorter tail. The former has naked ears, a sharper muzzle, slender tail and is generally blackish although the colour may vary to brown. It is more common in the tropics than elsewhere. The Brown Rat, grey to brown in colour is more robust, has larger ears with some hairs, a blunter muzzle and thicker tail. Both are pests and carriers of disease. Black Rats carry the flea that spread the plague (Black Death), in Europe in mediaeval times. In the 18th century, the Brown Rat was introduced into Britain and is now the common rat, *Rattus rattus* being almost superseded. Both Black and Brown Rats swim and climb well, and have defied all attempts by Man to exterminate them.

Among the other true Rats are the burrowing **Bandicoot Rats** *(Bandicota)*, the **African Giant Rats** *(Gricetomys)* over 32 in (80 cm) long; the true **Water Rats** *(Hydromys)*, of Australia and New Guinea have adapted to an aquatic environment.

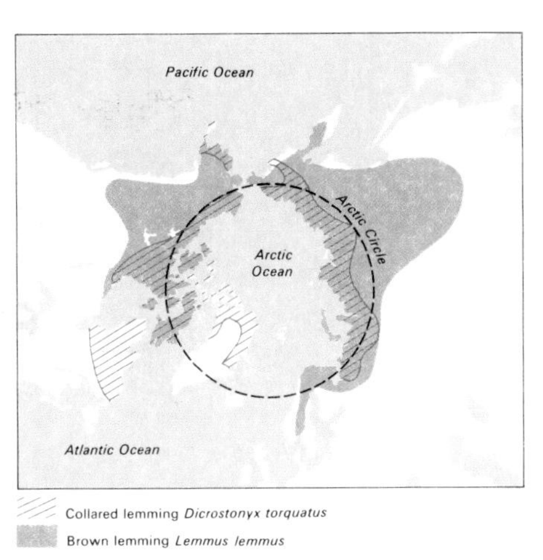

◁ The European or Fat Dormouse is also known as the Edible Dormouse because it used to be eaten in some parts of Europe. Like all Dormice it hibernates in the winter, its body temperature dropping almost to freezing point.

Dormice and Jumping Mice

The **Dormice** differ sufficiently from the true Rats and Mice to be placed in another super-family, the Gliroidea. Externally they resemble Squirrels, with their bushy tails and climbing habits, but their most distinctive features, apart from a very small thumb, are internal. Most are strictly nocturnal and live in or near bushes or shrubs where they build nests for breeding and hibernating purposes.

The rusty-brown **Common** or **Hazel Dormouse** *(Muscardinus avellanarius)*, about 5 in (12 cm) long overall, is found in Great Britain, the Mediterranean areas and the near East. It is an excellent climber. The somewhat larger **European** or **Fat Dormouse** *(Glis glis)*, with very bushy tail, and the **Garden Dormouse** or **Lerot** *(Eliomys quercinus)*, with slender tail ending in a tuft, are found chiefly in central and southern Europe. There are also a number of African species and the **Spiny Dormouse,** distinguished by its tufted tail and spiny or bristly fur, is a native of India.

Jumping Mice and **Jerboas** belong to the super-family Dipoidea. They are mouselike but generally larger with longer tails and disproportionately long hind legs adapted for jumping progression. The North American **Jumping Mice** *(Zapus)*, are nest builders, often underground. They are also called **Kangaroo Mice.** The **Birch Mouse** *(Sicista betulina)*, is a related form found in scattered localities of North Europe and Asia from the Baltic to Siberia.

Jerboas are jumpers from the Arabian and North American desert areas. The **Egyptian Jerboa** *(Jaculus jaculus)*, is about 8 in (20 cm) long in the body with a still longer tail, cylindrical and tufted. The head and eyes are large and the long-whiskered snout is short. The three-toed hind limbs, (Gerbils are five-toed), adapted for jumping, are very long by comparison with the front limbs. The species is gregarious and families live in burrows in sandy soil, coming out to feed only at night. The animal is reputed not to need access to water, its drinking needs being met by the juices of plants.

New World Rats and Mice

(Myomorphs)

Together with the Old World forms, these are the Myormorph rodents that comprise the super-family Muroidea, the richest in number and variety of species of all the mammals. They differ from the Sciuromorphs in their jaw muscle arrangements and internal cheek pouches, in always having a single fused tibia with fibula in the hind leg and having part of the stomach cornified.

Natives of the New World only include the **Rice Rats** *(Oryzomys)*, of North and South America, the **White-footed** or **Deer Mice** *(Peromyscus)*, found from Alaska to Panama, the **Cotton Rats** *(Sigmodon)*, and the **Pack Rats** *(Neotoma)*, chiefly of North America. These latter animals build enormous nests on the ground, using collected brushwood and rubbish. Sometimes the nests can be 6 ft (2 m) in diameter.

The **Vesper Rats** or **'Night Mice'**, *(Nyctomys)*, are squirrel-like and the **Fish-eating Rats** *(Ichthyomys)*, are found in regions from south-west Canada to Ecuador.

Hamsters

There are two well known kinds, the **European Hamster** *(Cricetus cricetus)*, and the **Golden Hamster** *(Mesocricetus auratus)*. Both are short-limbed and short-tailed, with extensive internal cheek pouches and powerful incisor teeth. The former species is larger, up to 12 in (30 cm) overall, nearly twice as big as the Golden Hamster. Both are excellent tunnellers, constructing elaorate, deep burrows including chambers for food storage as well as for nesting and hibernating.

The Golden Hamster, commonly seen as pets, are found wild only in Syria and it is said that all those now in captivity have been bred from a single original pair.

There are other Hamster species occurring in Asia, and a Siberian form turns white in winter. There are others no bigger than house mice. The **South African Hamster** *(Mystromys)*, is the only one to be found south of the equator.

Voles and Gerbils

These are both related to the Lemmings and Hamsters. The vole group contains several forms resembling true rats but can be distinguished by their blunter snouts and comparatively short tails. The largest member, up to 2 ft (60 cm) overall, is the **Musk Rat** or **Musquash** *(Ondatra zibethicus)*, equipped with large webbed hind feet and a long laterally flattened, naked tail. It has aquatic habits and has dense, lustrous fur and the animal is often kept on fur farms. It is an excellent swimmer, found originally in North American habitats, but some have been introduced to Europe. It has become a serious agricultural pest in France but escapees from fur farms in Britain were swiftly exterminated. Its name arises from its musky odour, produced from the secretion of special glands.

Others in the vole group are the small, shy **Field Voles** *(Microtus)*, the **Water Voles** *(Arvicola)*, living near water but burrowing in soft soil and the **Bank Voles** *(Clethrionomys)*, active climbers in banks and hedges. All these are widespread in Europe, Asia and in North America.

The **Gerbils** do not much resemble the voles. They are mostly large-eyed, nocturnal, jumping rodents of Africa and South Asia. The **North African Gerbil** *(Gerbillus aegyptiacus)*, is found burrowing in the desert regions of North Africa. Using exceptionally elongated hind limbs, it can progress very rapidly by leaping. The **Indian Gerbil** *(Tatera indica)*, is more rat-like.

Porcupines and Cavies

(Hystricomorphs)

The **Porcupines** have in common the possession of erectile quills and similar teeth structure but are generally separated into two distinct groups, the Old World Porcupines and the New World Porcupines, on the grounds that their common features have evolved independently from different ancestors.

The former, ground-living kinds are the larger, up to 39 in (100 cm) overall. Their long quills are firmly attached to the skin but can be erected by strong muscles. They are good burrowers. The **Common** or **Crested Porcupine** *(Hystrix cristata)*, is widely distributed in South Asia, Africa and parts of the Mediterranean coast. The **Long-tailed Porcupine** *(Trichys)*, of Malaya, is smaller with shorter spines. The **Brush-tailed Porcupine** *(Atherurus)*, of Africa and Asia, is more rat-like with a long tail ending in a tuft of hollow quills, each resembling a string of flattened beads.

The Porcupines of America are by contrast tree-livers and readily climb using their specially adapted hind feet and tails. The North American **Tree Porcupine** *(Erethizon dorsatum)*, climbs by 'looping' its body, reaching forward then closing up the rear to the front, then gripping with the hinder parts and reaching forward again. Its spines are short, partly hidden by long hairs and loosely fixed in the skin. They make excellent defensive weapons, filling the mouths of attackers that try to bite. The **Brazilian Tree Porcupine** *(Coendou prehensilis)*, uses its unusual prehensile tail in climbing, coiling its

△ The diet of the Water Vole, often incorrectly called the Water Rat, is mainly vegetarian though it does eat fish and crustaceans.
▷△Like all the New World Porcupines the Canadian Porcupine is arborial, the well-developed first toe of its hind foot helping it grip branches. Its long white hairs conceal barbed spines.
▷ The superficial resemblance of the Coypu, sometimes called the Swamp Beaver, to the true Beaver can be seen in this picture.

naked upper side from below the support upwards, opposite to other animals with prehensile tails. The hairy **Tree Porcupine** of Brazil and Paraguay *(Coendou villosus),* rarely leaves the trees even to sleep.

The super-family, Cavioidea are the guinea pig-like rodents, found wild only in South America. They include the largest rodent, the sheep-sized **Capybara.** Generally, they are heavy-bodied, short-tailed (even to rudimentary), and some possess blunt claws almost like miniature hooves.

The true **Cavies** *(Cavia),* are mostly burrowers. The species *Cavia niata* is found in the Andes, whilst *Cavia cutleri* of Peru is the most probable originator of the domestic Cavy or **Guinea Pig** *(Cavia porcellus),* originally bred for eating by natives and now found all over the world in a variety of forms and coat colours. One of its characteristics is a rudimentary tail that does not show externally. By contrast to its relatives it does not burrow to any extent.

The **Mara** or **Patagonian Cavy** *(Dolichotis),* resembles a Hare in its size and shape and tends to run in open country in small herds, although it can burrow. The **Capybara** *(Hydrochoerus),* is a large amphibian form found in jungle marshes, spending a good deal of its time in water. The **Paca** *(Cuniculus),* similar to a large Guinea Pig, but with a striped coat, also inhabits wet places. It is a good swimmer and burrower. Other members of the Cavy group are the **Pacaranas** *(Dinomys),* and the **Agoutis** *(Dasyprocta)* which have elongated limb bones and other adaptations to permit speedy running in the bush and on the forest floor. They appear quite long-legged, especially in comparison with Guinea Pigs.

Chinchillas and Viscachas

These are placed in the super-family Chinchilloidea. They have several distinctive features including ever-growing cheek teeth as well as incisors, peculiar skull characters and the lateral toes on the hind feet are very small or missing to give a three-toed foot. Their tails are long and bushy and the hind limbs are long. They are all South American and well known because of their very valuable and beautiful fur. In most places they have been hunted to extermination but any remaining are protected.

The **Royal Chinchilla** *(Chinchilla laniger),* is silver grey, up to 10 in (25 cm) long in the body with a tail nearly as long again. It is not unlike a Squirrel but lives in loose rocks and crevices. The toes are protected at the tips by soft pads that aid in rock scrambling and there are remnants of claws only. The hairs in the fur are the finest of all known, separate strands being virtually invisible to the naked eye, an adaptation for insulation against the extreme cold of the high altitudes where they live. Occurring up to 18,000 ft (6,000 m) above sea level in the Andes, Chinchillas have the distinction of living higher than almost any other mammal.

The plains **Viscacha** *(Lagostomus maximus),* may be up to 31 in (80 cm) overall with a large head and robust body. A number will make up a colony or viscachera in an extensive warren in the pampas areas of Argentina. The animal's face is remarkable for the two white bands running across it and the patch of long, bushy black hairs under each eye. The fur is also prized. The **Mountain Viscacha,** a rabbit-eared form of the Chinchilla, lives near streams in the Andes foothills.

Other Rodents

Other rodent super-families include the Octodontoidea, Tuco Tucos, the Coypus and Spiny Rats, all of South America, the Bathyergoidea, a group of Mole Rats and the Ctenodactyloidea, the Gundis.

The **Tuco Tucos** *(Ctenomys),* are widespread in South America. They have powerful claws, large incisors and are very pugnacious. They burrow in family groups.

The **Coypu** or **Nutria Rat** *(Myocastor coypus),* is well known because of its brown fur which is known commercially as 'nutria' when the longer guard hairs are plucked. The animal, beaver-like with webbed hind feet, is about 20 in (50 cm) in body length, with a scaly tail about as long again. It is very much at home in water and its burrowing habits cause it to erode river banks. In eastern parts of Britain, Nutria Rats which have escaped from fur farms have become a crop-destroying pest.

The **Mole Rats** are natives of South Europe, Asia and Africa. They are virtually blind with rudimentary eyes only, and spend their lives tunnelling with well adapted teeth and limbs. The European **Mole Rat** *(Spalax microphthalmus),* from South Russia, Yugoslavia and the Middle East, makes large mounds of earth and destroys root crops. The **Naked Mole Rat,** *(Heterocephalus glaber),* from dry parts of East Africa, is a small, blind, almost hairless species, lacking external ears. It never comes above ground.

The **Gundis** *(Ctenodactylus, Pectinator),* North African rodents inhabiting rock crevices in dry areas, are similar in form to Guinea Pigs except for their visible tails.

Horses, Tapirs and Rhinos

Massive but nimble Rhinos, weirdly shaped Tapirs, Horses—thoroughbred and otherwise, Asses and Zebras are all united by the form of their feet.

Many of the world's largest herbivores comprise the order *Perissodactyla;* hoofed, odd-toed mammals in which the original five digits of each foot have been reduced to three or one, most of the weight of the body being supported by the central digit. These animals range from the immense primeval Rhinoceros and its close relative, the equally primitive Tapir to the Asses, Horses and Zebras of the family *Equidae*.

Odd-toed ungulates are divided into two sub-orders: *Ceratomorpha* and *Hippomorpha*, the former being made up of the families Tapiridae and Rhinocerotidae and the latter of the single extant family Equidae.

Tapirs look like a strange mixture of Pig, Donkey and Hippopotamus. They are probably the most ancient members of the order and are unique in having retained four toes on the front feet, while the back feet have only three. All species have a long, prehensile muzzle with sensitive bristles at the tip and a set of rather rudimentary but powerful grinding molars.

The five species of Rhinoceros vary considerably in size and appearance. Among their illustrious ancestors was the largest of all land mammals, the Mongolian *Baluchitherium*, which extended 9 metres in length and 6 metres in height, and the famous woolly Rhinoceros *Coelodonta antiquitatis*, commemorated in ancient cave art. Most early Rhinoceroses were forest dwellers but in the Pleistocene era one branch, *Dicerorhinus etruscus*, moved out into the open plains and became grazers. Another branch entered Africa to evolve along separate lines, dividing into broad-lipped grazers and long-lipped browsers.

All Rhinos have retained tridactyl feet, each digit having a hoof-like covering, though usually vestigial remains of a fourth digit are evident. Dentition is less complete than in other species of odd-toed ungulates and the two African rhinos have no incisors at all.

Horses, Asses and Zebras have only a single digit on each foot. Most members of the Equidae family are swift, intelligent animals, living a peripatetic life in herds on the natural grasslands. Cross breeding is possible between several species and most have proved capable of domestication.

Zebras, which live in the open grasslands of Africa, are typical of the Horse family. Zebras normally live in small family groups and this one contains foals from two successive years, the youngest on the right of the picture and the one from the previous year in the centre. On occasion these family groups will join up to form large herds. The faint brown bands between the bold black stripes are a feature of the South African Zebras and the precise and varying patterns of the stripes can be used to identify the different races of Zebras.

The Wild Horse

(Equus caballus przewalskii)

Przewalski's horse is the only surviving representative in the wild of the original species of **Wild Horse** *(Equus caballus),* from which today's domesticated horse is descended. Named after the Russian explorer, who discovered it in central Asia in 1881, this pale bay or dun coloured horse has a shoulder height of 140 cm, a large, heavy head, erect mane, short ears, dark dorsal stripe, flowing black tail and white muzzle. It emits a shrill whinny.

The total number of pure-bred Przewalski horses in the natural habitat was catastrophically reduced from 2,000 in 1872 to between 20 and 40 in 1972. This was the result of human predation, interbreeding with domestic horses and competition with domestic cattle for grazing land and water. The remaining few are concentrated in the bleak desert area on the borders of Mongolia and Sinkiang, where they graze in small herds. Mating usually takes place in late Spring and a single foal is born after a gestation period of 11 months.

The European counterpart of Przewalski's horse, the **Tarpan,** though long extinct, was genetically re-created by remarkable breeding techniques in Germany during the 1940s.

The Domestic Horse

(Equus caballus)

Of all animals, the horse, with its high intelligence, good memory and ability to obey commands has proved most amenable to domestication.

The earliest horse, *Eohippus* or 'dawn horse', emerged in the humid forests of the Eocene epoch some 50 million years ago. It was a small browsing animal, about 46 cm in height, with four functional toes on the front feet and three on the back feet, like the Tapir of today. It evolved in North America, adapting gradually to life as a grazer on the open plains. Around 2 million years ago, the genus *Equus* appeared, resembling today's Wild Horse. Some of these creatures crossed the land bridge in the region of the Bering Strait, spilling into Eurasia and Africa, where they continued to evolve in a variety of forms. Mysteriously, the line of the horses in North America came to an end.

For early man, horses were primarily a source of food, the first successful attempts at domestication not

▽ Przewalski's Horses, the only true surviving representatives of the original species of Wild Horse still living in the wild. The erect, bristly mane, white muzzle and very long tail are characteristic of this species discovered in 1881.

Przewalski's Breeding Bank

When it became apparent that the stocks of Przewalski Horses were unlikely to survive in their natural habitat, it was decided that this remaining species of Wild Horse would have to be bred in the world's zoos and game reserves until such numbers had been reached that they could safely be reintroduced into the wild.

Two early attempts—at Askania Nova in the Ukraine and Woburn Abbey, Bedfordshire, England—met with complete failure but efforts were renewed, using stock from the Prague stud. The animals bred successfully. By 1956 there were 36 Przewalski horses in captivity and by 1968 there were 153. During 1971, 30 Przewalski foals were reported to have been born in various zoos throughout the world, bringing the total captive population to 206, of which 121 were mares and 85 stallions—descendants of three pairs captured in 1901–2.

The Survival Service Commission of the International Union for the Conservation of Nature has authorised the keeping of stud books for such rare species, and the stud book for Przewalski horses is kept at Prague.

This horse is the most primitive member of the *Equidae* family, having the highest number of chromosomes, and the exercise may help to throw some light on the phylogenetic development of the Horse. Since these Horses often interbreed with other species in the wild, breeding in captivity will also help to preserve the purity of the strain.

being made until around 2500 BC by Asiatic nomads. These peoples invented spurs, curb bits, stirrups and nailed-on horseshoes, several hundred years BC, and the use of the horse for riding, carrying loads, agriculture, warfare and sport spread across Europe.

The Arabs introduced horses to Spain in the year AD 700, and 800 years later, the Spanish conquistadores re-introduced into North America animals that were to be the progenitors of the semi-wild **Mustang.** The immediate ancestry of the domestic horse is complex and confused, consisting of many varieties of wild horses, of which the most dominant were the **Arab,** the **Barb,** the **Turk,** the **Northern Dun.**

Horse Breeding

Selective breeding of horses for improving the stock was first practised by the Arabs who learned, for example, that by crossing a strong horse with a fast runner they could produce a swift and powerful offspring. Specialised breeding over the generations has created 'purpose-built' animals designed for specific tasks.

Saddle horses range from 15 to 16.2 hands (1 hand = 4 in) and weigh from 500 to 700 kg. They are used for a number of sporting activities and for rounding up cattle. The most successful racing horse in the world is said to be the English Thoroughbred, which can trace its pedigree back to three notable stallions: Darley Arabian, Godolphin Arabian and Byerley Turk. In 1689, 1705 and 1728 respectively, these superior **Arab** specimens were crossed with European mares. Racehorse breeding has created a slender, long-legged animal capable of moving at 16 metres per second over a limited distance. The Hunter has been specially developed for its dexterity in negotiating ditches and fences.

Light harness breeds include the American Standard-bred, which measures 15–16 hands and weighs up to

△Brumbies, the Australian name for domestic horses which have returned to the wild.
▷ Common Zebras and, in the background, Wildebeeste which often form mixed herds. The two stallions which are sparring, centre, are typically attempting to bite each other's necks rather than use their hooves. The large herds of Zebras are loose collections of family units usually drawn together by the common need for pastures. Family groups usually consist of a stallion and up to six mares with their foals.

700 kg and the Morgan, both American horses. Coach or heavy harness horses combine speed with strength, though they are not capable of such heavy work as draught horses. The most popular coach horse before the invention of mechanised transport, was the stylish, high-stepping Hackney, developed in eastern England from heavier Thoroughbred stock. Heavy harness horses, which also include English, Yorkshire, German and French coach horses and the Cleveland Bay, range in weight from 550 kg to 750 kg and in height from 14.3 to 16.3 hands.

The powerful draught horses have been bred for weight, muscular power and endurance. These massive animals can weigh up to 1,000 kg and measure more than 17 hands. The Shire, or English farm horse, was one of the earliest breeds of draught horse. Its ancestors were bred to carry heavily armoured medieval knights. The Scottish Clydesdale is a smaller, more agile mixture of Flemish and British blood. The Belgian, bred from the Flemish heavy horse, is strong and docile, while the Suffolk is ideal as farm and dray horse. The French Percheron is unique among draught horses in having Arab and Flemish blood, which has produced a medium-weight, fast-trotting animal.

A horse standing less than 14.2 hands is generally referred to as a *pony.* This smallness of stature is usually a result of a poor environment. The Shetland from the north of Scotland ranks among the smallest ponies weighing as little as 150 kg and standing less than 9 hands high, while the New Forest pony at between 12.2 and 14.2 hands is among the tallest.

Wild Asses

Found only in Central Asia and North East Africa, capable of existing on the roughest terrain and surviving the harshest conditions, the **Wild Ass** is seriously threatened with extinction. Resembling an unstriped zebra, its head over-large for its dun coloured body, it inhabits the plains of Mongolia and the deserts of North East Africa.

The **African Wild Ass** *(Equus asinus),* forages on rough scrubland between Somalia and Ethiopia, where it is becoming increasingly rare. This ancestor of the domesticated donkey can be recognised by its large ears, dorsal stripe, upstanding black mane and black tail tip. It stands just over a metre high at the shoulder.

The smaller **Asiatic Wild Ass** *(Equus hemionus),* still exists in several races, although the Syrian race *(Equus hemonius hemippus)* became extinct

around 1927. Successful attempts at captive breeding have been made in game reserves to prevent the same fate befalling the **Persian Wild Ass** or **Onager** *(Equus hemionus onager).* The pack animal of Biblical times and once widely distributed, it now survives in small numbers on the northern borders of Iran and Afghanistan, many having been ruthlessly slaughtered for meat.

Its close relative, the **Kulun** or **Chigetai** *(Equus hemionus hemionus),* has a population of several thousand, concentrated in the Gobi Desert. The **Indian Wild Ass,** has diminished in numbers as a result of disease and hunting, from nearly 5,000 in 1946 to 870 in 1962, in its native Little Rann of Kutch habitat. The lesser known **Kiang Ass** *(Equus hemionus kiang),* of the Himalayan heights, is equally at risk, although stocks now breeding in captivity are expected to perpetuate the race.

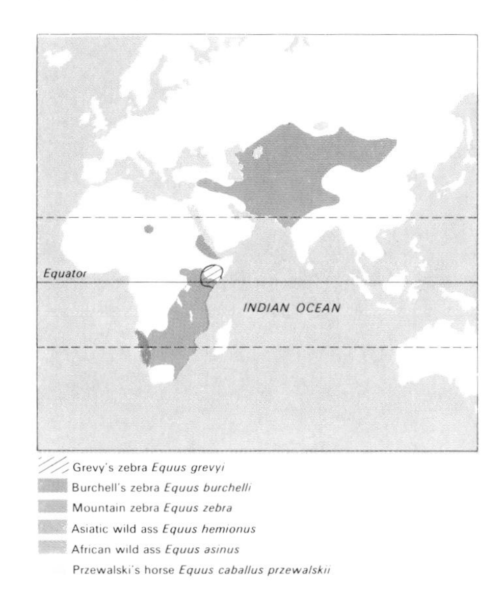

△ This map of the distribution of Horses and Zebras emphasises the rarity of Przewalski's Horses.

Mules and Donkeys

Donkeys are closė relatives of the Horse and descendants of the African Wild Ass. Sure-footed, docile creatures with large asinine ears, they are sturdy, long-lived and resistant to drought but not to cold. She-donkeys were once prized for their rich milk.

The Wild Ass was domesticated several centuries before the birth of Christ, becoming widely used as a saddle and pack animal throughout Egypt, Asia Minor (Turkey). By AD 800 the animal was found in Europe and appeared in North America by the 1800's. Mules are products of complex hybridisation and therefore belong to no definite race.

Selective breeding has resulted in the development of special strains for use in various types of work. The breed established as pack animals in Mexico is known as the **Burro.** Mating between a male Ass and a female Horse produces a **Mule,** whereas a cross between a female Ass and a male horse is a smaller animal called a **Hinny.**

Zebras

The *hippotigris* **(Horse Tiger)** of Roman circuses has been tamed with only moderate success and can be mated with

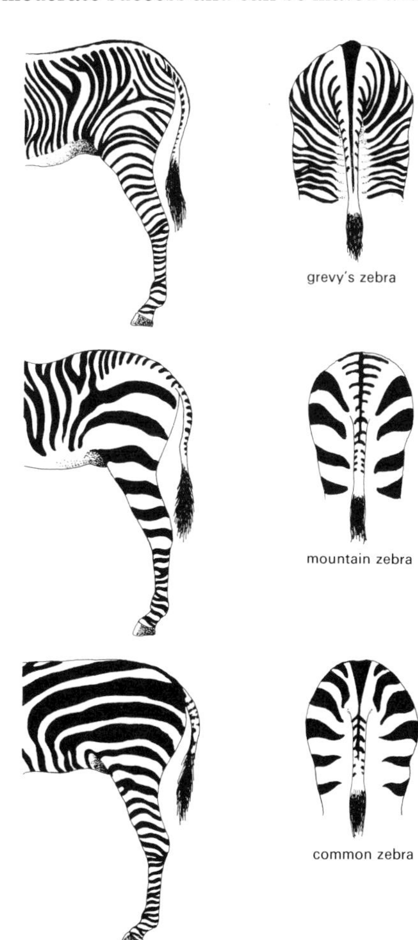

△ Zebras are the most abundant of wild Equidae and the various species can be distinguished by their striping which ranges from the close, almost pin-striped pattern of Grevy's Zebra to the bold striping of the Common and Mountain Zebras.

Ass or Horse, although subsequent offspring have not proved as successful for domestication as the Mule. The various species of wild Zebra in existence inhabit southern and eastern regions of Africa.

Swift, agile animals, Zebras stand between 120 cm and 155 cm and weigh, on average, 350 kg. They have a rather heavy head, large ears and small black hoofs. Their distinctive dark brown or black parallel striping is a form of disruptive camouflage, harmonising with the light and shade of the animal's habitat. The species vary considerably in the size, colour and distribution of the stripes and in general, the most vivid and intensive markings are found on animals in the north, apparently having become paler and more diffuse as the animals evolved and spread south.

Grévy's Zebra *(Equus grévyi),* is a large, lowland species concentrated in Ethiopia, Somalia and northern Kenya.

The stripes are close and narrow, particularly on the hind-quarters, and it gives a loud asinine bray, quite different from the usual Zebra bark.

The **Mountain Zebra** *(Equus zebra),* is smaller and more Donkey-like, having large ears and measuring only about 120 cm high. The striping is narrow but for a unique broad grid-iron pattern on top of the hindquarters. Once abundant throughout southern Africa, indiscriminate hunting has almost led to a complete annihilation of the species. The species was granted legal protection status and in 1937 the Mountain Zebra National Park was established in Cape Province. Numbers of animals in the National Park had increased to 58 in 1967–8, when the total Mountain Zebra population was only 75.

It was not possible to save the **Quagga** *(Equus quagga).* This interesting animal was half horse and half zebra in appearance, its stripes being limited to head and shoulders, leaving the rest of the body a plain fawn colour. This large animal, 140 cm high, once intermingled with the common Zebra on the plains in the extreme south of Africa, in vast numbers. It was completely wiped out by human predation, the last specimen expiring in captivity in 1883.

Burchell's Zebra

From north Ethiopia, to the Orange River in the south, the common Zebra grazes in large, closely-knit family herds, consisting of some half dozen females and their young led by the senior stallion. They move across the open plains and savannas, visiting water holes morning and evening. These gregarious creatures, like other species of Zebra, are accompanied by **Hartebeest, Oryx, Eland, Gnu** and **Ostrich** —a motley association, that provides increased protection in numbers from lurking predators.

At the age of 2 or 3 years young males depart for bachelor herds and young females find their own partners. Foals can be born at any time throughout the year, breeding being timed so that foals are born after the rains when the grass is at its best. The gestation period is 13 months.

Markings vary considerably from race to race. The 'prototype', *Equus burchelli burchelli,* an animal with a striped body and plain legs, is no longer in existence. **Chapman's Zebra** *(Equus burchelli antiquorum),* which is found in Angola and the Transvaal, has plain coloured lower legs and some shadow striping. Farther north, in Rhodesia and southern Zambia, **Selous's Zebra** *(Equus burchelli selousi),* displays close stripes with a little shadow striping. The boldly striped **Boehm's Zebra** *(Equus burchelli böhmi),* ranges from northern Zambia through parts of Ethiopia and the Sudan.

Malayan Tapir

(Tapiridae)

In evolutionary terms, Tapirs are more primitive mammals than members of the *Equidae* family. The ancestors of the Tapir, like those of the Horse, migrated from North America to Eurasia. But subsequent development was minimal and there is little to distinguish the Malayan Tapir from the line that evolved independently in the American continent.

The Malayan Tapir's outstanding feature is its short, bristly, black and white coat. A white patch covers the posterior half of its body, except for the legs. This camouflage makes it difficult for predators to recognise the complete body outline against the light and shade of the forests of Burma, Thailand, the Malayan peninsula and Sumatra where it lives. The Tapir is a largely nocturnal animal and this colouration is particularly effective by moonlight.

This rather solid-looking creature has a shoulder height of only 90 cm to 105 cm. It is a browser, living on low trees, shrubs, fruit and aquatic plants. Tapirs love water and swim well; their sense of smell and hearing are good but their eyesight is weak.

The Tapir's chief enemy is man. In Thailand Tapir meat, called *mu-nam* is much in demand, and even in places where it is not eaten, the forests in which it lives are being destroyed rapidly. Unfortunately, the animal does not always reproduce satisfactorily in captivity. In the wild, offspring are born (usually singly), after a gestation period of 13 months.

South American Tapirs

Of the four Tapirs found in tropical America, three are officially listed as endangered species and are unlikely to survive and the fourth, the **Brazilian** or **Lowland Tapir** *(Tapirus terrestris),* is becoming increasingly rare, despite the fact that it occupies a vast area from Venezuela to the north of Argentina and from Ecuador to the east of Brazil. Shy and solitary, most of its activity is confined to the night hours. The animal's deep brown, velvety coat frequently acquires a layer of dried mud as a protection against biting insects. The average height is about 90 cm.

The **Mountain Tapir** *(Tapirus pinchaque)*, has adapted itself to living in the Andes at altitudes of between 7,000 and 12,000 ft, by developing a much more compact body, with a shoulder height of only 55 cm and a thick, black, woolly coat. This Tapir exists in exceptionally small numbers in Colombia, Ecuador and Peru and is seldom able to survive in captivity.

The other two rare species are **Baird's Tapir** and **Dow's Tapir** *(Tapirus bairdii* and *Tapirus dowi)*. They both live along the central American isthmus from Mexico to Panama. They are threatened primarily by the destruction of their natural habitat—the wet lowland jungle and marshland. Baird's Tapir is the largest of all Tapir species.

Striped Babies

Tapirs seldom produce more than one offspring at a time and to ensure the continuation of the species, it is essential that the vulnerable young Tapirs are protected from such predators as the Tiger in south-eastern Asia and the Jaguar in South America, and in particular from ubiquitous Man.

The mother Tapir invariably finds a quiet secluded hideout in which to give birth, and after a long gestation period of 13 months, baby Tapirs are born, capable of walking within a few minutes of birth. Nature has also provided the young Tapir (Malayan and South American alike), with a remarkable disruptive camouflage in the form of horizontal rows of white spots and stripes. Against a background of low vegetation this renders the animal almost invisible. This protective marking lasts for about a year.

In the case of the Malayan Tapir, the large white patch starts to appear at about 68 days and the spots and stripes begin to fade until finally they disappear, leaving the sharply defined black and white coat of the adult Tapir.

Rhinoceroses

(Rhinocerotidae)

The distribution of Rhinoceroses was once far more widespread than it is today; the many branches and vast numbers of species have dwindled to five: the **Indian, Javan, Sumatran, White,** and **Black Rhinos.** Several of these now face extinction.

For centuries, the animal has been subjected to wholesale slaughter for meat, sport, and for its horn which, particularly in the East, was thought to have aphrodisiac properties. With the colonisation of Africa, India and parts of south-east Asia, Rhino hunting reached catastrophic proportions. Despite the ferocious appearance and pachydermatous hide, the Rhinoceros is not difficult to kill.

The name Rhinoceros comes from the Greek word for 'nose-horned'. The horns are not bone but projections of greatly compressed dermal fibres. Some species have one horn and others two, the smaller one situated behind the other, while some females have only vestigial bumps.

Despite considerable differences in appearance, all species have certain features in common. Having no sweat glands, they rely on external temperature regulation, covering themselves in mud, which acts as a protective layer against the hot sun. All have poor eyesight and acute senses of hearing and smell. Some are extremely territorial in their habits, restricting themselves to a limited area, following regular paths through the undergrowth and keeping to certain wallows. They mate all year round, are not monogamous and do not form a permanent family unit, although the cow guards her offspring solicitously.

The perpetuation of all five species depends on the continuation of international conservation efforts, which now include reintroducing the Rhino into areas where it has been eliminated, translocation to safer environments, and captivity breeding.

Armoured Rhinoceroses

(Rhinoceros unicornis)

Perhaps the most archetypal of all Rhinoceroses is the Great Indian 'armour-plated' species. Its dark greyish-brown exterior consists of large, warty shields divided by deep folds of skin.

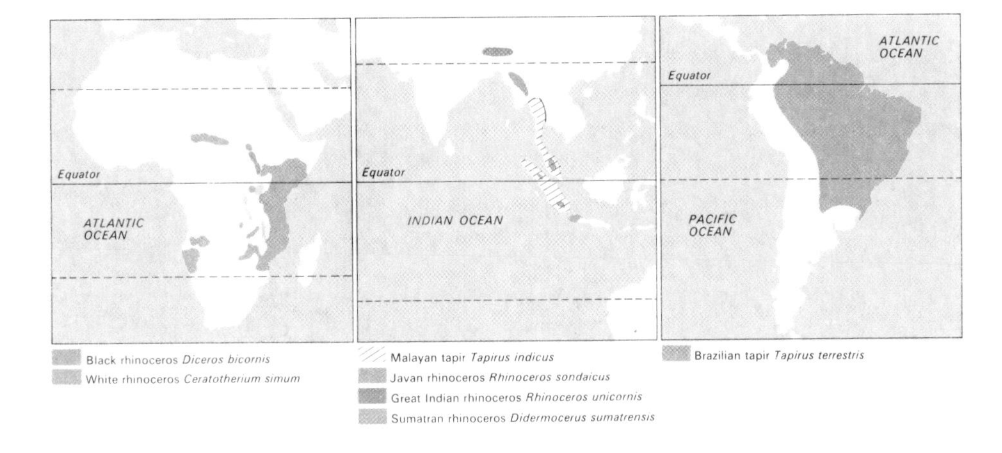

◁ Zebras are characteristically animals of the open grasslands and often Burchell's Zebra, the most common in East Africa, and Grévy's Zebra will form mixed herds. The Mountain Zebra, however, will not and they are now sadly reduced to a small number of animals which roam the Mountain Zebra National Park, founded in 1937 in Cape Province to save them from following the Quagga, half Horse and half Zebra, into extinction.

▽ The South American species of Tapir are brown, the Malayan Tapir, below, is black and white, but the young of all Tapirs are striped in cream and brown to help break up their outline on the forest floor where they live. The remnants of this striping are visible on the baby in this picture. The change of coat pattern and colour takes place over a six- to eight-month period.

There are fringes of hair on ears and tail and a rather blunt single horn may be as long as 60 cm. The armour plating is not as impenetrable as it appears and can easily be pierced by gun or arrow. This is the largest of the Asian species and a mature male measures just over 4.5 metres in length, 2 metres in height and can weigh up to 4,000 kg.

The **Indian Rhinoceros** is a grazer and frequents the grassland and jungle of Nepal, Bengal and Assam, feeding on the long grasses and reeds. In its natural habitat, the Rhino will characteristically be seen carrying jungle **Mynah Birds** *(Aethiops fuscus)*, on its back to act as sentinels. Indian Rhinos are rather irascible creatures. Courtship is accompanied by an elaborate ritual and fights between competing bulls are violent and occasionally fatal. After a gestation period of 16 months a single offspring is born, weighing at least 50 kg.

The **Javan Rhinoceros** *(Rhinoceros sondaicus)*, while bearing a close resemblance to the Indian species, with its dark grey granular skin folded into shields, is in fact considerably smaller and more primitive and has the elongated snout and the shorter molar crowns for grinding tough vegetation of a forest-dwelling browser. The single horn of the male rarely exceeds 25 cm, the female's scarcely develops at all.

The population of this rare animal is now reduced to about a dozen. These remnants are confined to the Udjung Kulon reserve in the extreme west of Java, where they make regular tunnels through the jungle, wallow in mud holes, creeks and even the sea, and browse on trees, shrubs and bamboo.

△ The Great Indian Rhino weighs up to four tons and is found in the swamps at the edge of the forests of Northern India. It is a solitary animal which, like all Rhinos, will charge to frighten an enemy off rather than to actually make an attack. Of all the Rhinos the Great Indian is the most solidly armoured, its thick hide, covered with wart-like granules, falling in stiff and heavy folds.
▽ An adult and two young White Rhinos drinking. The square lip, which differentiates this species from the Black Rhino which has a pointed lip, is clearly visible here.

Sumatran Rhinoceros

(Didemocerus sumatrensis)

Although more widely scattered than its Javan cousin, the **Sumatran Rhino** is also extremely rare, with an estimated total of 170 in 1972. This primitive two-horned species was once found in large numbers throughout the forested regions of south-east Asia, but its proximity to centres where the demand for the magical Rhinoceros horn was greatest, has subjected it to centuries of persecution.

Standing at about 120 cm, measuring approximately 2.8 metres in length, and weighing about 1,000 kg, the Sumatran is the smallest of all species of Rhinoceros. Its coat is smoother than the Indian and Javan rhinos and it has one main fold behind the shoulders. Young animals have a temporary covering of bristly hair. The front and back horns are seldom more than 25 cm and 12.5 cm respectively in the adult bull, while the female's are barely developed. The prehensile upper lip is not as long as in the Javan species and the diet consists of shoots, creepers, bamboo and jungle fruits. As in all Rhinos, the eyes are tiny and sight is poor though senses of smell and hearing are well developed.

Although solitary by habit, the Sumatran Rhino is frequently seen wandering in pairs and the young are kept in close contact with their mother. The Rhinos browse their way through well-trodden tracks and visit wallow holes at least twice a day. They climb and swim well.

There is a colony of 60 Rhinos on the island of Sumatra in the Loser and South Sumatra nature reserves.

White Rhinoceros

(Ceratotherium simum)

The **White Rhinoceros** is the largest land mammal in the world after the Elephant. The adult male may be as tall as 2 metres, 5 metres in length and 3,500 kg in weight. The names 'white' and 'black' as applied to the two African species of Rhinoceros are inappropriate since the coats of both animals are a similar shade of bluish grey. The true coat of the Rhino, however, is seldom seen, being protected from the heat and biting insects by a layer of dried mud, which varies in shade according to the soil colour of the habitat. The names black and white may therefore have arisen from observations of different Rhinos covered in pale and dark coloured mud!

Another explanation is that these designations sprang from the Afrikaans name *wijt renoster* ('wide rhinoceros' an allusion to the species' wide mouth), which became corrupted to *wit renoster* ('white rhinoceros'), and was translated as such into English. The mouth is certainly the White Rhino's main distinguishing feature. It has become adapted for grazing purposes, appearing rather truncated and flat with broad straight lips. The nose is also flattened.

Fossils of *Ceratotherium simum* have been found in many parts of the African continent, but the animal is an easy target and has been completely wiped out in many areas. Today, distribution is concentrated in two distinct zones with a separate northern race *(Ceratotherium simum cottoni)* in Uganda, Zaire and the southern Sudan and another race, *(Ceratotherium simum simum)* some 1,000 miles south in Natal, where it is fortunately now increasing in number.

The placid, docile White Rhino has few enemies, with the exception of the occasional Lion and Crocodile, and of course Man. Among its friends are the small, red-billed Oxpeckers *(Buphagus)* and the cattle Egrets that constantly accompany both species of African Rhino, perching on their large backs and compensating for the Rhino's poor vision by sounding alarm calls whenever they see signs of danger. The White Rhino has a better developed herd instinct than most species and is often found in small mixed groups, numbering sometimes as many as a dozen or so and which can include several adult males. Occasionally fighting breaks out between the males who may also attack unprotected calves.

Black Rhinoceros

(Diceros bicornis)

The **Black Rhinoceros** presents a more ferocious appearance than the square-lipped species. The adult male stands almost 2 metres high, measures well over 3.5 metres in length and weighs about 2,500 kg. The skin on its flanks is creased into parallel folds, its eyebrows, ears and tail have tufts of hair, and its mouth ends in a protruding prehensile lip. This prehensile organ is specially adapted for browsing for it is capable of reaching out and grasping twigs and branches from the trees on which it feeds, while the powerful molars are capable of grinding the woody material.

Like the White Rhinoceros, *Diceros bicornis* sports a pair of very long horns on the top of its nose, one set behind the other. The anterior horn is always much larger than the posterior horn, growing to an average length of 51 cm in the case of the black rhino, although a record length of 153 cm has been found.

The most populous of all Rhinoceros species, Black Rhinos are nevertheless dwindling in number, except where they are protected in national parks or game reserves. Kenya and Tanzania have the largest populations with an estimated total of 5,500–6,500 and the animals are especially numerous near the large lakes. Black Rhinos are more sparsely scattered in Angola, Rhodesia, Botswana, Malawi, Somalia and Ethiopia.

The Black Rhino is less gregarious than the White, usually keeping within strict territorial limits of about 10 square miles and normally living alone. Both species form only temporary mating partners. Mating usually takes place from July to September. A Black Rhinoceros pregnancy can last from 16 to 18 months and twins are unknown. The gestation period of the White Rhinoceros is usually 18 months and twins are uncommon. The calves of both suckle for about two years and are inseparable from their mother for a much longer period. Maturity is reached between 5 and 7 years. The life expectancy of a Black Rhinoceros, in the wild, is estimated at 25 years. In captivity it can live as long as 50–60 years.

The Unpredictable Rhino

Hunters' tales of violent, unexpected charges from fickle Rhinoceroses are not entirely unfounded. The Rhino species differ markedly in temperament but they are not usually aggressive animals. Even the Great Indian, with its reputation for bad temper, has proved docile to the point of carrying men on its back. Most Rhinos, unless provoked or accompanied by vulnerable offspring, will flee in great haste when alarmed. If seriously attacked, they can be violent and dangerous, using their tusks and not their horns as the main weapons.

The Rhino suffers from two disadvantages. Firstly it is extremely myopic. Since it is unable to recognise a stationary object at 24 metres, the animal will very often make an impressive warning charge, using a tree or termite heap as its target, beginning with its head held high, gathering speed up to 30 mph, emitting a loud snort, lowering its horns in a threatening manner, and then veering away at the last moment. The animal is dominated by its acute olfactory sense (it is capable of smelling a human at a distance of 800 metres) and always charges upwind if it scents a possible threat.

The Rhino's second disadvantage is that in comparison with its size it has a very small brain, its lack of intelligence making its behaviour all the more illogical and unpredictable.

The Great Herds

Animals as diverse as the ungainly but agile Hippopotamus and the nimble Deer; the incredibly ugly Wart Hog and the graceful Gazelle; the shy and elusive Okapi and the common Sheep and Goats are all members of the same large order.

Most ungulates are fairly agile creatures and many live in the mountain ranges of the world. They almost all have either horny outgrowths or antlers, ranging from a few inches in length to several feet.

There are altogether about 194 species of eventoed ungulates or *Artiodactyla*. They consist of all the cloven-hooved or even-toed herbivores such as Cattle, Sheep, Antelope, Deer, Pigs and Camels. Most of them are ruminants, that is they chew the cud, digesting it through a complex stomach system. The non-ruminants of the order are Pigs and Hippopotami. They possess the even-toes but not the complex stomachs.

The animals themselves vary greatly in size. The Mouse Deer for instance never exceed 56 cm in height, while the Giraffe grows up to about 6 metres. Some are comic, like the Wart Hog, which shuffles about on its knees when feeding, while others, such as the African Buffalo can be terrifyingly unpredictable in their willingness to attack.

The Wart Hog

The **Bush Pig** *(Potamochoerus porcus)*, is a nocturnal creature which lives in deep thickets and reed beds. They travel in groups of between 4 and 20 and are capable of laying waste a large area of cultivated land.

The coat of the Bush Pig is of short, reddish hair and old males are distinguished by being black with white face patches. Bush Pigs are found mostly in Africa, south of the Sahara, although another type lives on the island of Madagascar.

Like the Bush Pig, the **Wild Boar** *(Sus scrofa)*, is extremely destructive. They are also very ferocious and can become formidable adversaries when hunted. The Wild Boar's narrow, well proportioned body is built for running and fighting and its tusks make daunting weapons.

The Wild Boar has a thick skin and can penetrate seemingly inaccessible thickets. Its scavenging habits brought it into contact with early Man and it is the ancestor of the domestic pig.

Wild Boars range over Europe, northern Africa and southern Asia. Occasionally, they migrate and turn up in regions where they have not been seen for years.

There are many different *Sus* species and these include the **Pygmy Hog**, the **Javan Wild Pig** and the **Bornean Wild Boar.**

Old World Pigs

When the **Wart Hog** *(Phacochoerus aethiopicus)*, is feeding, it sometimes goes down on its knees and shuffles about looking for food. It will eat almost anything although it prefers roots and grass.

The Wart Hog, so-called because it has large, wart-like lumps on its cheeks, is extraordinarily ugly. As well as its warts, it has a misshapen head and grey, almost naked skin.

The Wart Hog is found in open country in most parts of Africa south of the Sahara Desert. It grows up to 155 cm long and can run extremely fast, its tail held high.

The **Forest Hog** *(Hylochoerus meinertzhageni)* was not discovered until 1904 and even now little is known about it. It is like the Wart Hog in appearance and lives in equatorial Africa.

The **Babirusa** *(Babyrousa babyrussa)*, inhabits the damp jungle regions of the Celebes and the Molucca Islands. It is a large, almost hairless Hog with huge curved tusks. Like most Hogs, it is a social animal and travels in groups. Its name means **Pig Deer.**

◁ A brilliant African sunset brightens the normally grey-brown coat of this male Greater Kudu. The spiral horns are not present in the female of this species, amongst the largest of Antelope. Kudu inhabit the drier parts of the East and South African savannah, preferring rocky areas.

New World Peccaries

Peccaries look like small wild Boars and, in keeping with other **Pigs,** they roam the forests in groups. In Spanish, they are known as **Javelinas** and are hunted for sport and meat. However, they can be dangerous and will band together to fight man and any other animal that threatens them.

The **Collared Peccary** *(Tayassu tajacu)*, is so named from the white band round its neck and shoulders. Peccaries are found in South America as far south as Paraguay, and range north to the southwestern states of the United States of America.

As well as its distinctive neck band, the Collared Peccary has a hairless and puffy scent gland on its rump. Males and females of this species tend to be the same size.

The **White-lipped Peccary** *(Tayassu pecari)*, ranges from Paraguay to Mexico. It is similar to the Collared Peccary except that its nose, cheeks and lips are white.

Hippopotamus

(Hippopotamus amphibius)

Despite its ungainly and cumbersome appearance, the **Hippopotamus** *(Hippopotamus amphibius)* is a reasonably agile creature, quite capable of overtaking a running man.

These animals weigh up to 4,000 kg and are much happier in water than on land. Their nostrils and eyes are set high on the head so that their bodies can be totally submerged while swimming. Hippopotamuses can close their nostrils enabling them to remain under water—even walking along the bottom—for several minutes.

The biggest of the species is the **Great African Hippopotamus.** These stand nearly 155 cm high at the shoulder, although their legs are less than 65 cm in length. They live in small herds and spend most of the day dozing.

Hippopotamuses have been hunted so much that they are increasingly rare outside African game reserves. They are dangerous only when provoked.

The **Pygmy Hippopotamus** *(Choe-*

△ A group of Wart Hogs drinking at a water hole. The wart-like lumps on its cheek explain the animals name. The tusks prominent on the animal on the left are greatly extended canine teeth.

▽ When the Hippopotamus takes to the water only its eyes, ears and nostrils, all located near the top of the head, need project above the surface, leaving the animal well submerged and protected.

ropsis liberiensis), as its name suggests, is much smaller than the common Hippo, seldom weighing more than 250 kg. It frequents the forest waterways of Liberia and Sierra Leone.

Hippopotamuses ranged throughout Europe and Asia at one time.

Camels: One Hump or Two

Camels have been used by man since the beginning of civilisation. As pack and saddle animals they have no equal in dry countries, carrying considerable loads over great distances and also providing leather, meat and milk.

There are two species: the **Bactrian** and the **Arabian** Camels. The Bactrian Camel *(Camelus bactrianus)*, is easily distinguished by its two humps. It is also strong and more heavily built than the one-humped Arabian Camel.

Wild Bactrian Camels still roam the Gobi Desert, but for the most part, they have become completely domesticated.

The Arabian Camel *(Camelus dromedarius)*—or **Dromedary**—has only one hump and is wholly domestic. As well as being slighter than the Bactrian, its coat is shorter.

The term Dromedary should be used only when referring to the racing camel. The Arabian baggage camel is the capable beast of burden.

It is a popular misconception that the Camel stores water in its hump, thus enabling it to travel great distances in the desert. Camels are, in fact, able to survive for a long time without water because they have many special physiological adaptations. The hump is a reserve of fat.

Although they are invaluable servants to Man, Camels are bad-tempered and sometimes unpredictable in behaviour.

Llamas

The only way a **Llama** *(Lama glama)*, can defend itself, is by spitting in its enemy's face. Although it has been a domestic animal for a long time it can still be obstinate. When used as a pack-animal, the Llama will refuse to budge if it feels the load is too heavy, and it will often refuse to be loaded unless it is part of a group of other Llamas.

Only male, castrated Llamas are used for this work. The females, together with a few males, are kept in pastures for breeding purposes. They are found in the mountain regions of Chile, Ecuador, Argentina, Bolivia and Peru, where their sure-footedness on the steep paths makes them ideal beasts of burden. In spite of their occasional moods, the Llama is generally willing if the conditions are right. They will carry loads of 45 kg doing 15 km per day, for five days, before needing a rest.

The Llamas' wool is used by Indians for making carpets but it is considered too coarse to be of any commercial value. The leather, however, is prized for its long life.

Llama Relatives

The **Guanaco** *(Lama guaniacoe)*, has the appearance of a long-legged, long-necked sheep.

Guanacos are extremely inquisitive creatures and will examine anything they consider unusual. When this creates a dangerous situation, the male within a group emits a harsh, sheep-like bleat. At this signal, the females move off, butting the young forward, while the male brings up the rear.

It is from the Guanaco that the domestic Llama is descended.

The **Vicuna** *(Lama vicugna)*, is the smallest and neatest of the Llama family. But extensive hunting for its fine wool has also made it one of the rarest. Now it is to be found only in the remote Andes regions above 4,000 metres. It is very timid, a rope stretched across its path being enough to prevent it advancing.

The **Alpaca** *(Lama pacos)*, breeds mostly on the Altiplano by Lake Titicaca on the borders of Peru and Bolivia. This is a domestic animal reared for its wool which is valued almost as much as that of the Vicuna.

It is from the Alpaca wool that the Incas made most of their material.

△ Arabian Camels, or Dromedaries, feeding on coarse vegetation on a dry and rocky hillside. Contrary to popular belief Camels are not able to store water but they use it with great economy, losing little by sweating, which they do not even begin until the temperature has passed 100°F.
▽ Fallow Deer, distinguished by the flat spreading antlers of the bucks, in their summer coat which in winter will be longer and warmer.

The Ruminants

Ruminants—which include nearly all hooved animals—are so-called because they ruminate or 'chew the cud'.

The Ruminant stomach is divided into a complex structure of four separate sections. The first of these sections is the rumen or paunch and it is here that the chewed food first arrives. It remains there, being softened, until the animal is ready to regurgitate it for further chewing. The food then passes into the second section, the reticulum or honeycomb stomach. From there it goes to the psalterium or manyplies before eventually being fully digested in the last section—the rennet.

Ruminants are divided into three groups. These consist of animals with solid horns, animals with hollow horns, and animals with no horns at all. The solid horn animals are mainly **Deer;** hollow horned animals include **Cattle, Sheep, Goats, Antelope,** and **Buffalo,** while the last, the hornless group, include the **Musk Deer** and the **Chinese Water Deer.**

Chevrotains

Chevrotains *(Tragulidae)*, are generally known as Mouse Deer because none of them is more than 36 cm high at the shoulder.

They are the smallest of the Ruminants and are hornless. However, in the males, the upper teeth have developed into small tusks.

All Chevrotains are extremely timid and will hide in hollow logs or among rocks at the slightest hint of danger.

The **Water Chevrotain** is dark brown and covered in white spots and stripes. It lives along river banks and feeds on aquatic plants. The Malayan Chevrotain has a plain coat while the Indian Chevrotain is spotted.

Except during the breeding season, Chevrotains lead a solitary existence.

True Deer

True Deer *(Cervidae)*, are easily distinguished by the fact that the male of the species possesses antlers—solid bony outgrowths on the head—which are shed every year. The new antlers grow very quickly and are at first covered in a soft 'fur', known as velvet. When the antlers are mature, the velvet is rubbed off against trees.

All True Deer are graceful, well-proportioned animals. They can run very fast and are exceedingly agile. The largest of the species grows to almost 2 metres at the shoulder.

The **Fallow Deer** *(Dama dama)*, is multi-coloured, its coat being a mixture of white, cream, silver-grey, sandy and black. The species is distributed all over Europe and has been introduced successfully to parts of the southern hemisphere. During the breeding season the males fight each other for possession of the females.

The **Axis Deer** or **Chital** *(Axis axis)*, is also known as the **Spotted Deer** of India. They are found in large herds in India and Ceylon and have no fear of the water, being strong swimmers. Relatives include the **Hog Deer, Kuhl's Deer** and the **Kalamian Deer.**

There are 13 sub-species of the **Sika Deer** *(Cervus nippon)*. It is an extremely adaptable animal and has been bred in Great Britain and Europe. Its true home, however, is eastern Asia. Sikas are rarely more than 95 cm high and their antlers usually have four points.

Red Deer

During the rutting season, the bellowing of the **Red Deer** *(Cervus elaphus)*, stags, can be heard over several miles as they roar out their defiant challenges to one another.

Fights over the docile hinds can end in death for one of the rivals. Occasionally, they both die from starvation as a result of their antlers locking permanently together. These antlers can grow up to a metre long.

A Red Deer born without the ability to grow antlers is called a Hummel. Without antlers to support, they become strong, heavy Bull-like animals.

The Red Deer has provided man with both sport and food for hundreds of years. It is found all over Europe as well as in parts of north Africa and south-west Asia. They are swift, graceful creatures and can jump as high as 3.5 metres off the ground.

◁ △ The woolly Alpaca, a South American member of the Camel family. Alpaca's are little larger than sheep and are kept in herds for wool and meat.
▽ An Elk, known as a Moose in North America, the largest of all Deer. Mere survival in the long northern winter poses problems for Elks which require up to five tons of vegetable food to last through until spring.
▷ ▽ The cycle of a deer's antler growth and shedding.

American Deer

The North American counterpart of the Red Deer is the **Wapiti** *(Cervus canadensis)*, otherwise known as the **American Elk.** Although related to the Red Deer of Europe, it is larger, 165 cm high to the 135 cm height of the Red Deer; its antlers are also longer.

Whenever the **Mule Deer** *(Odocileus hemionus)*, is protected from hunting, its numbers increase to such a degree that many of the population which ranges across North America, from New Mexico to Alaska, starve during the winter. The name of this animal is taken from their long mule-like ears. They are about 105 cm high and seem to prefer living in wooded country.

The **White-tailed Deer** *(Odocoileus virginianus)*, is so common in North America that it has even been sighted close to New York City. They are destructive to crops and an annual census is taken to decide how their numbers should be controlled.

Instead of having branched antlers, the **Brocket Deer** *(Mazama)*, possesses spikes about 130 mm long. There are four species of this small animal and they occur from Mexico to Argentina.

The **Pudu** *(Pudu)* has spikes too. The Chilean species is the smallest of The American Deer, being only 40 cm at the shoulder. The Pudu of Ecuador are slightly larger.

Moose and Reindeer

The **Alaskan Bull Moose** *(Alces americanus)* is the largest living Deer. It can grow up to 245 cm at the shoulder and weigh as much as 900 kg.

In Asia and northern Europe, the Moose is known as the **Elk.** They are deep brown in colour and have broad, overhanging muzzles. Their hooves are broad, ensuring support when the animal is standing in snow.

The European Elk used to frequent the western part of the continent, but is now confined to Scandinavia and other northern regions. During the last century, the Moose was almost eliminated in America but it is now a protected animal.

The **Reindeer** *(Rangifer tarandus)* used to roam all over the eastern and western hemispheres in herds numbering many thousands. Hunting has reduced this but their population is still considerable.

In the Arctic Circle they are valued in the same way as cattle and indeed the Reindeer is one of only three Deer to have been domesticated successfully, the others being the Elk in Poland and, experimentally in Britain, the Red Deer.

Closely related to the Reindeer is the **Caribou** *(Rangifer arcticus)* of North America. Both Caribou and Reindeer undergo extensive seasonal migrations in search of fresh food supplies and to avoid hordes of irritating insects.

A unique feature of both Caribou and Reindeer is that antlers are present in both sexes.

Musk Deer and Muntjacs

The **Musk Deer** *(Moschus moschiferus)*, is about 50 cm high and lives throughout eastern Asia mainly on high and rocky ground. It has a thick coat well suited to its mountain environment and there are three separate species.

The 'musk' from the animal is sought for the manufacture of perfume and soap. Each male animal yields approximately 28 g of musk, a wax-like substance, secreted in a gland in front of its navel.

Muntjacs *(Muntiacus)*, are also known as **Barking** or **Rib-faced Deer.** The first name arises from their cry which is loud, short and sharp, very like a dog. The slit-openings of the scent glands of the animal give rise to the second name.

There are five species of Muntjacs and 17 sub-species. Among the rarest is the **Hairy-fronted Muntjac.** This is found in the Chekiang Province of China. Other Muntjacs live in the Indian jungles and further east. Both males and females have upper teeth serving almost as tusks.

Père David's Deer

(Elaphurus davidianus)

In 1865, a French missionary and explorer, Père Armand David, saw some Deer in the Imperial Hunting Park of Nan-Hai-Tsue, near Peking. Eventually, a few were brought to Europe where they were known as **Père David's Deer** *(Elaphurus davidianus)*. By 1900, the Deer had become extinct in China and in the zoos of Europe, surviving only on the Duke of Bedford's estate at Woburn in Britain.

A herd was built up from the British survivors and in 1960, breeding pairs were returned to China. There are still no more than 700 Père David's Deer in the world.

▷ Giraffes are so tall that they need to splay their legs wide apart in ungainly fashion to lower their heads to drink. In this position they are most vulnerable to attack by Lions.

▽ Long legs, necks and tongues enable Giraffes to feed, almost exclusively, on tall vegetation.

The Okapi

(Giraffidae)

The **Okapi** *(Okapia johnstoni)* was not discovered until 1900 and was one of the last large land animals to be found by a European.

Its finder, the British naturalist Sir Harry Johnston, thought at first that it was a kind of horse. But a later examination of the teeth and horns, the latter partly skin-covered, established it as a close relative of the **Giraffe.**

The Okapi is a quiet, shy animal which makes hardly any vocal noise. It frequents the dense Congo rain forests where its striking colouring—reddish-brown coat and black and white striped legs—provide excellent camouflage.

The Okapi stands about 155 cm at the shoulder and has the same shaped head as the Giraffe, although its neck is considerably shorter than the Giraffe's, though still somewhat elongated in comparison with the horse.

△◁ The timid Okapi, which hides itself so deeply in the Congo forests that it was only discovered by Europeans in 1900. Like the Giraffe it uses its long tongue to grasp branches and strip off leaves, craning its long neck to reach them. Its teeth and horns are similar to the Giraffe's.

Disruptive Disguise

There are various animals, birds, fish and reptiles, which are helped by nature to blend with their surroundings. Perhaps the best known of these is the Chameleon, an old world Lizard found in Madagascar and tropical Africa.

The Okapi is an excellent example of an animal possessing 'disruptive' colouration. Its coat is black with white markings on the legs, thighs and chin. This enables it to merge comfortably with the dark forest background. The same applies to the Zebra—although they live in open country—with their black or dark brown stripes on a white background.

In birds, feather colouration is caused by physical and pigmentary factors. The physical change is produced by light and pigmentation is affected by bio-chemical substances contained in the feathers themselves.

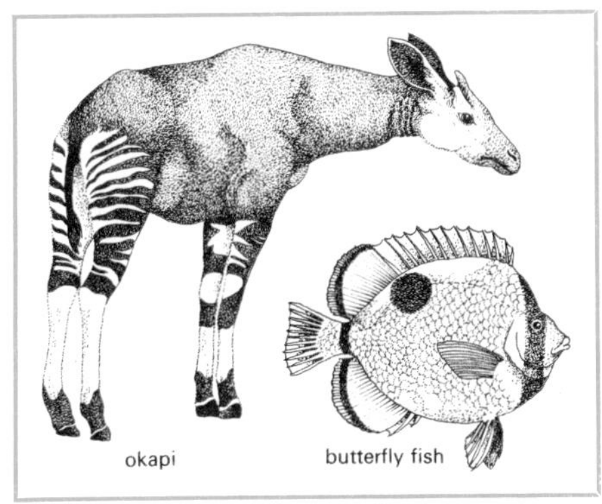

Many animals blend subtly into their surroundings and thus avoid detection by potential attackers, and their disguise is not always dull. The bright colours of the butterfly fish make it difficult to detect amongst coral reefs and the bold stripes of the Okapi break up its outline in the jungle.

The Giraffe

The **Giraffe** *(Giraffa camelopardalis)* is undoubtedly one of the most extraordinary animals alive today. Its long legs enable it to outpace easily the swiftest horse, its sharp hooves can defeat a Lion in combat and its elongated neck allows it to reach leaves in a tree's highest branches.

A Giraffe bull can stand as high as 6 metres and weigh as much as 1,000 kg. Most of their time is spent eating, browsing quietly in small herds. Drinking is a perpetual problem: because of its long neck the Giraffe must straddle its front legs to reach down to water or to the ground.

Giraffes seem happy even during the hottest part of the day. When they sleep, it appears to be only for a few minutes at a time.

Traditionally, Giraffes are thought to be mute but they can produce a low fluttering sound.

Giraffes used to range through Europe and Asia, but now they are limited to Africa.

Ancient Antelope

(Antilocapridae)

The **Pronghorn Antelope** *(Antilocapra americana)* can run at speeds between that of a Horse and a Greyhound and is believed to be the swiftest hooved animal in North America. But in spite of its name, the Pronghorn is not strictly speaking a true Antelope. The Pronghorn is the only hollow-horned ruminant which has branched horns. It is also the only one to shed and renew the horn sheath every year. The males are larger than the females and have a characteristic black spot behind each ear. There is a clump of white hair on the Pronghorn's rump.

Antelopes *(Antilopinae)* are to be found in different habitats—from jungles to desert. They vary in size from that of a large rabbit to an ox. Some species have hollow horns and some have no horns at all.

Antelopes live on grass, shoots and leaves, making the animals unpopular in farming country.

Their herds are much smaller than they once were and several species have become extinct as a result of natural predators and Man's hunting.

A male Nyala Antelope. A rather scarce species of Southern Africa, the Nyala's shaggy mane and long hairs on the underside are rather unusual features for an Antelope.

Spiral-horned Antelopes

The **Nyala** *(Tragelaphus angasi)* is one of the most beautiful members of the Antelope family. It is found in southeast Africa.

The **Bushbuck** *(Tragelaphus scriptus)* is also known as the **Harnessed Antelope** because of the white stripes on its dark red coat. It frequents large areas of Africa, south of the Sahara. The horns, which are tipped with black, have only one spiral turn.

The **Kudu** *(Strepsiceros strepsiceros)* is an imposing animal with long horns and standing 135 cm at the shoulder. It lives mainly in bush country in Africa and takes readily to water if being pursued. They are shy creatures and eat at dawn and dusk rather than during the day. The bulls are much more wary than the females. Kudus form herds of up to 14 animals but never more.

Elands *(Taurotragus oryx)* are quite easy to tame, although attempts to cross them with cattle have been hitherto unsuccessful. The **Giant Eland,** otherwise known as **Lord Derby's Eland,** is the largest living Antelope. It stands 2 metres and weighs up to 1,000 kg. The horns reach over a metre in length and are very useful in breaking down branches in the animal's constant search for food.

The biggest of the Asiatic Antelopes is the **Nylghaie** or **Blue Bull** *(Boselaphus tragocamelus)* from the Indian plains. The horns, found only in the male, are small and curve forward.

Bongos *(Boocerus euryceros)* live in the densest parts of the tropical forests of Africa. When frightened, they move at high speed through almost any obstacle, slipping under those they cannot jump. They are reddish in colour with transverse white stripes.

Cattle

In the many parts of the world where the **Buffalo,** has been domesticated, it has proved a much hardier animal than the **Ox.** Buffaloes are able to withstand extremes of temperature, are happy with a limited variety of food, and seem impervious to insect-carried diseases.

The **Domestic Buffalo,** a tractable placid creature, is used as a beast of burden, and to provide man with milk, meat and hides. Buffalo milk is particularly rich, one of the by-products being genuine oriental yoghurt.

The largest of the Asiatic species is the **Indian** or **Water Buffalo.** This still roams wild in northern India, Borneo, Indo-China and Nepal. There are two other Asiatic types which are related to the Indian Buffalo; the **Tamarau** from the Philippines and the **Wild Dwarf Buffalo** from the Celebes. The latter stand only 1 metre high.

The **African Buffalo** is regarded as a highly dangerous animal. It will charge at the slightest provocation and those who hunt the animal, regard it with respect.

Bison

Although the **Bison** *(Bison bison)* is called a Buffalo in North America, it should not be confused with the true Buffaloes of Asia and Africa.

Bison have enormous heads and are covered in thick shaggy hair over the front part of their bodies. They seem docile but are quick-tempered and can be dangerous if aroused. Their vision is

▷ A family group of Bison, often confusingly called Buffaloes in North America where they once prospered on the open plains but are now confined to a number of special reserves. The taller European Bison, the largest animal in Europe, also exists only under protection.

▽ A herd of African Buffalo seeking shelter from the mid-day sun. Another of their methods of cooling down is to wallow in mud.

Massacre on the Plains

During the mid-19th century, there were between 60 and 70 million **Bison** roaming across the vast plains of North America.

Bison provided the Red Indians with virtually everything they needed in the way of food and clothing and large numbers were killed for this purpose.

Subsequently, the white settlers began to arrive in America and moved westwards. They traded rifles with the Indians for Bison skins but although the Indians could now kill many more of the creatures, this did not make much difference to the total number of the herds.

The almost complete annihilation of the Bison began with the construction of the Union Pacific railway across America. The railway workers had to be fed and the Bison, which teemed everywhere, was killed in greater numbers to provide food. The hunters began to realise the commercial possibilities of shipping Bison hides back to the eastern towns. The practice developed of removing just the tongue—considered a delicacy—and the hide, leaving the carcase to rot.

By the mid-1860s, about 2,000,000 Bison were being killed annually. By 1883, they had almost disappeared. Little remained besides a 10,000 strong herd in North Dakota and towards the end of the same year, this too was completely destroyed.

It was only then that the American national conscience began to stir and the few remaining Bison, numbering 541 in 1889, were carefully preserved. In 1905, the American Bison Society was formed and by 1935 the numbers of the animal in existence had increased to 20,000.

These numbers have recently increased even more and there now seems little danger that the Bison will become extinct.

The European Bison suffered the same catastrophic depredations. Huge herds once wandered throughout Western Europe but these were reduced to a few hundred head by the 18th century. In 1932, a society was formed to ensure their preservation and the numbers are now on the increase again.

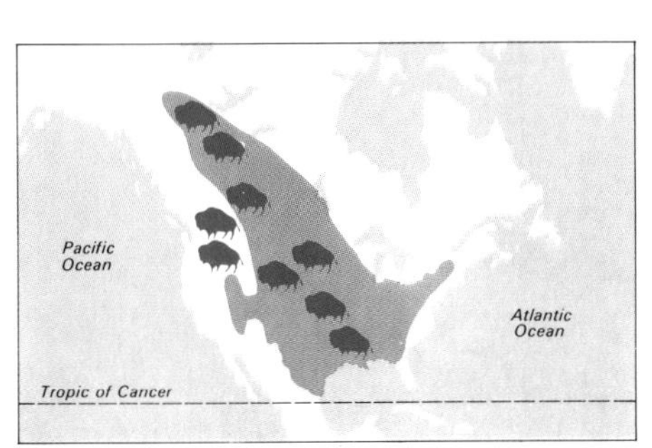

poor but the animal has a keenly developed sense of smell and hearing—particularly useful if Wolves or other predators are about.

The Bison is the largest animal in North America. It stands up to a full 2 metres at the shoulder and can weigh as much as 1,500 kg. During the winter the Bison grows a substantial new coat. This falls off in patches with the arrival of spring and the growth of its summer coat.

The **European Bison** *(Bison bonasus)*, is closely related to the North American Bison but is smaller and more graceful. It is also much less shaggy. There was a time when these animals were found all over Europe, but they were gradually eliminated and it is only now, as the result of strict controls and protection, that they are on the increase.

The Duikers

There are thought to be 17 separate Duiker species. The best known are the **Black Fronted Duiker,** the **Blue Duiker, Cape Duiker** and the **Yellow-backed Duiker.**

Duikers inhabit forest and thicket country south of the Sahara. They have bare muzzles and a small tuft of hair between their short horns which makes them instantly recognisable. Their fur is short and either reddish brown or grey.

None of the Duikers is more than 76 cm. high. They feed at night on fruit and seeds and generally lead a fairly solitary existence. Their horns point back in line with their face, which accounts to some extent for the fact that they make hardly any noise when pushing through even the thickest undergrowth.

In 1933, the **Jentinks Duiker** was listed as an endangered species. But it was not accorded special protection and Jentink's Duiker, along with others of this small breed of African Antelope, is still sought for its valuable skin.

Bucks, Kobs and Lechwe

(Reduncinae)

If the **Reedbuck** is alarmed it makes its escape only at the last possible moment and, as it does so, emits a curious high-pitched whistle.

These are medium-sized animals with simple, curved ringed horns. There are three types of Reedbuck: the **Common Reedbuck** *(Redunca arundinum),* which inhabits the Cape Province area; the **Bohor,** which is smaller than either the Common Reedbuck or the **Mountain Reedbuck** measuring about 72 cm high. The latter frequents the mountain areas from Cape Province to Natal, north-eastern Tanzania to Ethiopia and the Cameroons The Mountain Reedbucks are much more sociable animals than the others of the species and move in groups of 20 or more.

Outside of the breeding season, the **Kob** is, like the Mountain Reedbuck, also a fairly sociable creature.

The **Lechwe,** *(Onotragus)* is another kind of Antelope belonging to the same general group of *Reduncinae.* Lechwes enjoy an unusual degree of safety from predators and something of a monopoly on their feeding grounds, largely because they feed belly-deep on flooded grasses and reeds.

A group of Common Waterbuck at dawn. The male Waterbuck will mark out quite a large area of territory, sometimes as much as 500 acres, and mate with any receptive females that enter it.

Roans and Sables

The ears of the **Roan Antelope** *(Hippotragus equinus)*, are almost as long as its head. It is a large animal, sometimes as high as 155 cm and has horns reaching as far as 117 cm

Except for the rutting and calving season, the Roan live singly or in pairs. When choosing a mate, it carries out the selection with great ceremony, rubbing the other's hind legs and sniffing its sexual organs.

The **Sable Antelope** *(Hippotragus niger)* is also large, and closely related to the Roan. It possesses splendid, curved horns, a black-marked face and a fine mane on its shoulders and neck.

Sable Antelopes live in groups that can number several dozen, each male pairing off with each of three or four females during the mating season. As soon as the young males reach puberty they are expelled from the group and wander off to form their own band.

Sable Antelope are found in parts of the Transvaal, the Congo and Kenya. The best-known of the species is the **Giant Sable.** The males are black and the females, golden-chestnut.

The Oryx

It is thought that the **Oryx** can exist indefinitely without water, absorbing whatever moisture it needs from desert plants. Certainly it seems quite happy in remarkably high temperatures—up to 40°C.

The most common Oryx is the **Gemsbok** *(Oryx gazella)*. It stands just over a metre high and is fawn with a black stripe and black garters on the forelegs, which are white from the knee down.

The rarest is the **Arabian Oryx** *(Oryx leucoryx)*. This, in fact, is so rare that but for the determined efforts of a group of dedicated conservationists in the early 1960's, it would soon be extinct. The Arabian Oryx is slightly smaller than the Gemsbok and is white with brown legs.

The **Scimitar-horned Oryx** *(Oryx algazel)* is so called because of the shape of its horns, which are long, approximately 1 metre in length, and gracefully curved.

A large **Addax** *(Addax nasomaculatus)* can weigh as much as 220 kg, rather more than its Oryx relatives. It lives in the deserts of North Africa and, like the Oryx, survives a long time without water. It is comparatively rare—only about 5,000 are believed to remain in the wild—and lives in small herds.

Save the Oryx

In December 1960, a motorised hunting expedition, from Qatar on the Persian Gulf, discovered that there were only about 48 Arabian Oryx left in existence. For centuries, killing and eating an Oryx was believed to enhance Arab potency, and, as a result, the animals' numbers were reduced to this frightening figure.

Determined to save the animal before it became extinct, conservation groups began planning a campaign, but, even as they talked, a further 16 of the graceful animals were killed off.

'Operation Oryx' as the campaign came to be called, went into action immediately and managed to capture four animals (one of which died from an earlier bullet wound). The remaining three were shipped first to Isiola in Kenya, and from there to the Phoenix Zoo in Arizona in the United States.

King Saud of Saudi Arabia then announced that he wanted to present two breeding pairs of the Arabian Oryx to the World Wildlife Fund; the king apparently owned a private herd of 13 Oryx. These four were added to those already at Phoenix, and by the mid-1970s the herd numbered over 300.

At least two herds of the Oryx have now been re-established in the Middle East—one in Qatar and the other in the Oman and there is encouraging evidence that the Arabs are now taking a serious interest in ensuring their survival.

△ Hornshape in different kinds of even-toed ungulates. The sturdy horns of Sheep and Goats are often used for sparring between males. The longer more delicate horns serve more for visual recognition of social status.

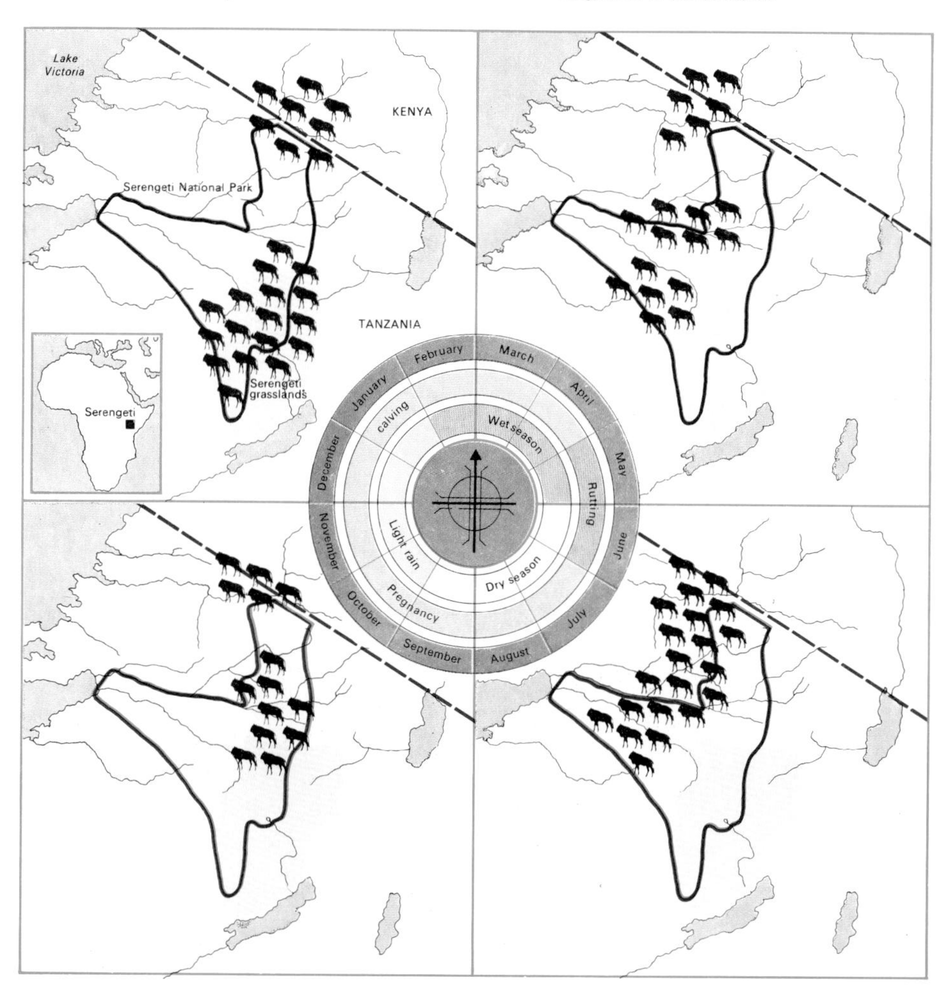

▷▷The migration of Wildebeeste on the Serengeti plains. The herds move anti-clockwise, reaching each area when the forage is in peak condition. Their arrival at the south eastern grasslands coincides with the birth of the young. They move to the north and spend several months in dispersed groups and mating takes place prior to the return journey so that the females again give birth in the best areas.

Hartebeest and Gnu

With its long, gloomy face, lumpy forehead and v-shaped horns, the **Hartebeest** *(Alcelaphus)* is an extremely odd looking animal. There are several species, two of which, the **Bubal** and the **Red,** are now unfortunately extinct. Another, **Swayne's,** is fast approaching extinction.

The **Common Hartebeest** is, as its name suggests, the most prolific and widespread. Other species are the **Western Hartebeest; Jackson's;** the **Tora** and **Kongoni,** or **Coke's Hartebeest.**

△ A group of Gemsbok at a water hole in the kind of dry, semi-desert country to which they are so well adapted.
▽ Wildebeeste on the floor of the Ngorongoro Crater when the grass is in peak condition.

Hartebeests inhabit the open plains and desert country of East Africa, where they exist solely on a grass diet.

The greatest enemy of the **Gnu,** or **Wildebeest,** is the Lion and to avoid this strong predator, the Gnu has developed a keen sense of smell and sight.

There are two species: The **White-tailed Gnu** or **Black Wildebeest** *(Connochaetes gnou),* which is now reduced to nothing more than a few well-protected herds; and the **Brindled Gnu** or **Blue Wildebeest** *(Connochaetes taurinus),* the common species of the East African plains.

A 1963 survey of the massive Serengeti Plain on the Tanzanian side of Lake Victoria revealed more than 300,000 Wildebeeste which have been known to congregate in herds of more than 10,000 as the dry season advances and the grazing areas are reduced.

Dainty Rock Antelopes

During the rutting season, male **Klipspringers** *(Oreotragus oreotragus)* stand watch on high rock pinnacles with their legs brought very close together, blunt-pointed hooves almost touching.

They are solitary animals, found mainly in the rocky regions of South and East Africa. Their height is about 50 cm and they have rough, short coats and short pointed horns placed wide apart on a thickset head.

Other very small Antelope include the **Suni** *(Nesotragus moschatus)* from East Africa; the **Steinbok** *(Raphicerus campestris)* from South Africa and the **Grysbok** *(Raphicerus melanotis),* also from South Africa.

None of these dainty little creatures are more than 56 cm high but in spite of their diminutive size they share many of the characteristics of other, larger Antelope.

Dik-Diks and Royals

The **Dik-Dik** *(Madoqua, Rhynchotragus),* of which there are six species, are small Antelope frequenting the eastern and southern plains of Africa. These animals weigh between 3 kg and 5 kg and stand up to 40 cm. Their muzzles are long and curve down like a little trunk.

Dik-Diks live mostly by themselves or in pairs but occasionally they will form small family groups.

The **Royal,** or **Pygmy Antelope** *(Neotragus pygmaeus)* is the smallest of all living ruminants. It stands no more than 30 cm high—about the size of a Hare. It is a nocturnal and solitary animal and lives in West Africa.

Other Antelopes, only very slightly bigger than the Royal include **Bates' Dwarf Antelope** and the **Blue Duiker.**

Impala and Blackbuck

The **Impala** *(Aepyceros melampus)* is quite capable of jumping as much as 10 metres for the nearest cover if alarmed. It rarely ventures far beyond the safety of trees and bushes, but keeps a very wary eye open for predators when it does so.

Found in large parts of East Africa, the Impala is a gregarious animal and moves in herds numbering several hundreds. In the summer, they tend to divide into much smaller family groups, one male to about 20 females.

Impala are about 1 metre high and reddish-brown in colour with a buff underbelly. Only the males possess horns, which are long and lyrate.

The **Blackbuck** *(Antilope cervicapra)* an Indian Antelope, prefers open grass country to woodland. They are extremely swift animals and could probably outrun a Greyhound, but they cannot outrun the Cheetah. Previously, Cheetahs were used to hunt the Blackbuck and, as a result, the herds have now been drastically reduced in number.

Speedy Gazelles

Gazelles *(Gazella)* are to be found throughout India, Central and Western Asia and North Africa and there are many different species.

They are beautiful, graceful creatures, fawn and white in colour with long, elegant ringed horns. Gazelles vary in size: one of the smallest, the **Dorcas,** is only 60 cm high. Other species include the **Mhor Gazelle, Chinkara** or **Indian Gazelle, Grant's Gazelle,** the **Korin Gazelle** and **Thomson's Gazelle.**

A Gazelle can run at speeds of up to 40 miles an hour for 15 minutes but hunting has greatly reduced the size of the herds, in spite of the animal's speed.

The **Gerenuk** *(Litocranius walleri),* sometimes known as **Waller's Gazelle,** has a curious long neck. Because of this it is sometimes called the **Giraffe-necked Gazelle.** When feeding from bushes or the lower branches of trees, the Gerenuk stands erect on its hind legs and stretches its neck to reach food inaccessible to shorter-necked gazelles.

Once, vast herds of **Springbok** *(Antidorcas marsupialis),* roamed all over South Africa. Now their numbers have decreased to those found mainly in the Angola and Kalahari regions.

Large-nosed Saiga

The **Saiga** *(Saiga tatarica)* looks rather like a Sheep. It has a woolly coat and a long, round protruberant nose. This nose is very mobile and serves the Saiga extremely well. Not only does each nostril contain a sac of mucous mem-

branes which warms inhaled breath but the position of the nose enables the animal to avoid breathing in dust when moving quickly.

Saigas are about 75 cm high and the males weigh up to 55 kg. They used to frequent parts of Europe and fossilised remains have been found there but the animal is now confined to the Volga steppes and the Caspian Sea regions. Another, smaller, Saiga type lives in the Gobi desert.

A close relative of the Saiga is the **Tibetan** or **Chiru Antelope.** This is a much more handsome creature with long, almost straight horns. It lives on the high, Tibetan plateaux singly or in small groups between 20 and 25.

▽ Two rival male Impala sparring to establish social dominance. At maturity the young males challenge herd leaders and when they are sucessful gain harems. The fighting usually takes the form of pushing contests.

△◁ The bold markings of the Bontebok serve for social recognition within the herd.
▷△Male Ibex using their heavy horns in a trial of strength. These are mountain Ibex which are found on high, rocky precipices in Central Europe and though short and squat they are extremely agile.

Goats and Chamois

The **Rocky Mountain Goat** *(Oreamnos americanus)* lives well above the tree-line in the remote mountain areas of North America. It is a thick-set, muscular animal and is the only ruminant which keeps its white coat during the entire year.

Both the males and females look alike, both having beards and short, black horns. They move about in small groups, feeding mainly on alpine plants. It is only rarely that the Mountain Goat comes down from the heights—and then only to the alpine meadow level.

The **Chamois** *(Rupicapra rupicapra)* is a small, agile creature which lives in herds of up to 50 high in the mountain ranges of Europe. Its coat is beige-grey in the summer and black during winter. Both sexes have horns which are set close together and straight, except for a curve towards the tip.

The males live alone for most of the year, some following the herd at a discreet distance. They are fierce fighters in the rutting season.

Gorals *(Naemorhedus)* are known to live well above 15,000 feet in the Himalayas but at somewhat less exalted heights elsewhere, (China, Burma and Korea). They have rough, woolly coats, a thick mane and short, straight horns.

A close relative to the Goral is the **Serow** *(Capricornis)*. These are also mountain animals and are to be found in south-west China, Sumatra and Malaya. One species lives in Japan and Formosa. They are solitary creatures, rarely found in groups of more than four or five.

Musk-Ox

The **Musk-Ox** *(Ovibos moschatus)* is a massively-built, aggressive creature weighing up to 400 kg. When threatened by predators such as Wolves, the adults bunch together in seemingly impenetrable strength. This may protect the Musk-Ox from Wolves but it does not seem to be of much use against the modern Eskimo, who hunts the Musk-Ox for food and hides. At one time it was feared that the Musk-Ox might become extinct and the Canadian and Danish governments established two reserves; one is in Greenland and the other between the Great Slave Lake and Hudson Bay.

Musk-Oxen are extremely hardy creatures and seem able to exist in the most difficult conditions. They live mainly on lichen and moss which they have to dig out from under the snow. There is no shelter—not even trees—and the Musk-Ox must rely on its thick, shaggy coat to keep out the cold. They live in herds of up to 100 and the cows give birth to single calves.

The **Takin** is closely related to the Musk-Ox. It too has short legs and is heavily-built. It lives in the densely wooded areas of south-west China and the Himalayas.

Tahrs, Goats and Ibex

Tahrs *(Hemitragus)* form part of the Goat-Antelope group. There are several species: the **Himalayan Tahr,** which has a brown coat and lives in wooded and rocky places up to 10,000 feet; the **Nilgiri Tahr** from the Indian peninsula and the **Arabian Tahr** of the Oman.

They all have short horns and are very like Goats except that the male Tahrs exude a different smell.

Goats are related closely to Sheep but have differently shaped horns, narrower heads and beards. There are many species, the main ones of which are the **Wild Goat** *(Capra hircus)* and the **Ibexes**.

The **Wild Goat** lives in the mountain areas of Asia Minor and is believed to be the forerunner of the domestic goat. The largest is the **Markhor,** a shaggy creature with flat, spiral horns.

There are seven types of **Ibex.** They are all magnificently agile, mountain creatures and differ from the Wild Goat in the shape of their head and horns. The **Alpine Ibex** is fairly typical: it is a squat, short-legged animal weighing up to 20 kg. At one time they became almost extinct but have now been successfully re-introduced to many of the areas they once frequented.

Sheep

Domestic Sheep *(Ovis aries)* have been used by Man for centuries to provide him with food and clothing. They are hardy, easily-managed creatures and can be kept on open grazing country in enormous numbers.

The **Merino,** which originated in Spain but is now found all over the world, is the best of the wool-producing Sheep; its wool is long and fine. The Merino has been used as the basis of various other breeds of Sheep.

One of the best wool-producing Sheep is New Zealand's **Corriedale.** A cross between the **Lincoln** and the Merino, a single Corriedale yields as much as 10 kg of wool in a year.

Wild sheep are found all over the world and there are 37 different species. The smallest is the **Moufflon** group, the **European Wild Sheep.** The **Barbary Sheep** is the furthest removed from the domestic breeds and the **Bighorn** or **Rocky Mountain Sheep** *(Ovis canadensis)* is the only wild Sheep in North America. The largest is the **Argali** from Central Asia. Other species include the **Red Sheep,** the **Urial, Snow Sheep,** and **Thinhorn Sheep.**

A group of Musk-Ox, found in the Arctic regions of Europe, which when threatened form into a dense mass with the young in the middle and their horns facing outwards for protection.

Index

T

U

V

W

Z

Picture Credits

The publishers wish to thank the following photographers and organisations who have supplied photographs for this book. Acknowledgements to the Australian Information Service and the South African Tourist Board have been abbreviated to AIS and SATOUR respectively.

1 Sea Otter; P. Morris. **2–3** Giraffe; Commonwealth Institute. **4–5** White-handed Gibbon; J. Mackinnon. **6–7** Repeats **8** Coral; H. Angel. **9** Scarce Swallowtail; Heather Angel. **10–11** Flamingos; H. Angel. **12** Bottle-nosed Dolphin; H. Angel. **13** Stag's Horn Coral; H. Angel. **14–15** Elephant Herd; The South African Tourist Board. **16–17** Pearl Bordered Fritillary; H. Angel. **18** Fossilised Sea Lily; Pic on Tour. **18** Fossils of Trilobites; P. Morris. **19** Ammonites; P. Morris. **19** Peripatus; P. Morris. **20** Amoeba; NHPA (Banta). **21** Radiolaria; NHPA (Banta). **22** Sea Anemones; H. Angel. **23** A Hyroid; British Museum (Natural History). **24** Dahlia Anemone; H. Angel. **24** Lions Mane Jellyfish; H. Angel. **26** Lugworm; H. Angel. **27** Peacock worm; H. Angel. **27** Fresh Water Flukes; H. Angel. **28** Snails; H. Angel. **28** Sea Slug; H. Angel. **29** Open Scallop; H. Angel. **30** Sectioned Nautilus; H. Angel. **31** Common Octopus; H. Angel. **32** Goose Barnacle; H. Angel. **34** Fiddler Crab; Australian Information Service. **35** Crawfish; Angling Photo Service. **36** Soldier Crabs; AIS. **36** Millipede; H. Angel. **36** Centipede; H. Angel. **37** Birdeating spider; P. Morris. **37** Trapdoor Spider; AIS. **37** Black Widow; P. Morris. **38** Orange knee'd Tarantula; P. Morris. **39** Scorpion; AIS. **39** Sea Spider; H. Angel. **39** Sea Spider; H. Angel. **40** Common Blues; H. Angel. **41** Red-eyed Damsel Fly; M. Chinery. **42** Leaf Long-Horned Grasshopper; AIS. **43** Anthill; AIS. **43** American Cockroach; M. Chinery. **44** Stick Insect; AIS. **45** Cicada Feeding; M. Chinery. **45** Stink Bug; M. Chinery. **45** Green Lacewing; M. Chinery. **46** Witchetty Grub; AIS. **47** Great Diving Beetles; H. Angel. **47** Stag Beetle; H. Angel. **47** Longhorn Beetle; M. Chinery. **48–49** Wanderer Butterfly; AIS. **50** Privet Hawk Moths; H. Angel. **50** Swallowtail Butterfly; M. Chinery. **51** Indian Moon Moth; P. Smart. **51** Assam Silkmoth Caterpillar; H. Angel. **52** Honey Bees; H. Angel. **52** Crane Fly; H. Angel. **53** Bulldog Ant; AIS. **54** Hoverfly; H. Angel. **54** Geometrid Caterpillar; H. Angel. **54** Meadow Grasshopper; H. Angel. **56** Sunstar; H. Angel. **57** Sea Cucumber; H. Angel. **58** Edible Sea Urchin; H. Angel. **58** Aristotles Lantern; H. Angel. **58** Brittle Stars; H. Angel. **60–61** Stripey Surgeon Fish; H. Angel. **62** Coelacanth; Pic on Tour. **63** Fossil Port Jackson Shark; Pic on Tour. **63** Star Sea Squirt; H. Angel. **63** Sea Squirts; H. Angel. **64** River Lamprey; H. Angel. **65** Brook Lampreys; H. Angel. **65** Hagfish; H. Angel. **66** Egg Cases of Dogfish; H. Angel. **67** White tipped shark; H. Angel. **68** Shark with Remora; P. Morris. **69** Ray; P. Morris. **69** Skate Ray; P. Morris. **70–71** Batfish; H. Angel. **71** Salmon Alevins; H. Angel. **73** Piranha Fish; P. Morris. **73** Pike; H. Angel. **74** Catfish; H. Angel. **75** Sea Horse; H. Angel. **75** Moray Eels; H. Angel. **75** Elvers; H. Angel. **76** Butterfly Fish; P. Morris. **77** Mud-skipper; AIS. **78** Scorpion Fish; P. Morris. **79** Plaice; H. Angel. **79** Lungfish; P. Morris. **79** Trigger-fish; P. Morris. **80–81** Green-eyed Tree Frog; AIS. **84** Salamander; P. Morris. **85** Caecilian; P. Morris. **85** Newts; H. Angel. **86** Axolotl; P. Morris. **87** Marbled Newt; H. Angel. **88** Common Frogs Spawning; H. Angel. **88** Toad; H. Angel. **88–89** Bullfrogs; AIS. **89** Frog Tadpoles (series); H. Angel. **91** Common Toad; H. Angel. **91** Tree Frog; H. Angel. **91** Spade Foot Toad Burrowing; H. Angel. **92** Green Tree Frog; H. Angel. **92–93** Land Iguana; H. Angel. **94** Ichthyosaur; P. Morris. **95** Crocodile; The Commonwealth Institute. **95** Iguana; H. Angel. **100** Crocodiles; SATOUR. **101** Newly Hatched Crocodiles; AIS. **101** Saltwater Crocodile; AIS. **102** Eastern Water Dragon; AIS. **103** Leaf Tailed Gecko; AIS. **103** Varigated Gecko; AIS. **104** Boyd's Forest Dragon; AIS. **105** Frilled Lizard; AIS. **105** Chamaelion; P. Morris. **106** Marine Iguanas; H. Angel. **107** Sand Goanna; AIS. **107** Water Skink; AIS. **107** Slow-Worm; H. Angel. **108** Grass Snake; H. Angel. **109** Grass Snake swimming; Angling Photo Service. **110** Egg-eating Snake; P. Morris. **110** Brown Tree Snake; AIS. **110** Tiger Snake; AIS. **112** Adder; H. Angel. **113** Russell's Viper; P. Morris. **113** Western Desert Rattle-snake; P. Morris. **114–115** Flamingos in flight; H. Angel. **116** Moor hen; NHPA (Stephen Dalton). **117** Archaeopteryx; P. Morris. **117** Archaeopteryx fossil; P. Morris. **117** Crowned Crane; H. Angel. **118** King Penguin; Pic on Tour. **118–119** Penguin Colony; J. Boswell. **119** Penguins; J. Boswell. **120** Cassowary; P. Morris. **120** Emus; AIS. **121** Kiwi; P. Morris. **122–123** Nesting Colony; H. Angel. **125** Pelicans; SATOUR. **125** Cormorants; SATOUR. **125** Double Crested Cormorant; H. Angel. **125** Little Bittern; AIS. **126** White Stork; H. Angel. **126** Sacred Ibis; SATOUR. **129** Galapagos Hawk; H. Angel. **129** African Fish Eagle; P. Morris. **129** Golden Eagle; G. Kinns. **130** Osprey; NHPA (J. Tallon). **131** Vultures; H. Angel. **131** Peregrine Falcon; AIS. **133** Helmeted Guinea fowl; H. Angel. **133** Peacock; H. Angel. **134** Demoiselle Crane; Angling Photo Service. **135** Crested Cranes; SATOUR. **135** Sunbitten; Brazilian Embassy. **144** Touraco; P. Morris. **144** Cuckoo; P. Burton. **145** Roadrunner; P. Morris. **146** Galapagos Short Eared Owl; H. Angel. **147** Snowy Owl; P. Morris. **147** Eastern Grass Owl; AIS. **148** Pigmy Kingfisher; P. Morris. **149** Malchite Sunbird; P. Morris. **150** Red-Billed Hornbill; H. Angel. **151** White-throated Night Jar; AIS. **151** Night Jar Chicks; H. Angel. **151** Potoo; P. Morris. **152** Barbet; H. Angel. **153** Great Spotted Woodpecker; E. Herbert. **153** Toucan; P. Morris. **154** Lyre Bird; AIS. **155** Cock of the Rock; P. Morris. **155** Swallow; E. Herbert. **156** Great Tit; E. Herbert. **157** Long Tailed Tit; E. Herbert. **157** Tailor Bird; AIS. **158** Weaver Bird; P. Morris. **159** Silvereye; AIS. **160–161** Cheetah; SATOUR. **162** Giraffe; H. Angel. **163** Pilot Whale; P. Morris. **164** Platypus; AIS. **165** Spiny Anteater; AIS. **165** Baby Echidna; AIS. **165** Spiny Anteater swimming; AIS. **166** South American Opposum; Brazilian Embassy. **167** Eastern Native Cat; AIS. **167** Native Cat; AIS. **167** Tasmanian Devil; AIS. **169** Koala; AIS. **169** Greater Glider Possum; AIS. **169** Pigmy Glider; AIS. **169** Hairy Nosed Wombat; AIS. **170** Scrub Wallaby; AIS. **170–171** Kangaroo and Lake; AIS. **171** Kangaroo and young; AIS. **171** Baby Opposum; AIS. **172** Spiney Tenrec; P. Morris. **172** Aardvark; P. Morris. **173** Hedgehog; H. Angel. **174** Elephant Shrew; P. Morris. **175** Mole; H. Angel. **175** White-toothed Shrew; P. Morris. **177** Long-nosed Bat; P. Morris. **177** Flying Fox; H. Angel. **178** Horseshoe Bat; P. Morris. **178** Greater Horseshoe Bats; J. Hooper. **178** Baby Serotine Bat; P. Morris. **179** Greater Horseshoe Bat; J. Hooper. **180** White-handed Gibbon; J. MacKinnon. **181** Tree Shrew; P. Morris. **182** Dwarf Lemur; R. D. Martin—Pic on Tour. **182** Slow Loris; J. MacKinnon. **182** Bush Baby; R. D. Martin—Pic on Tour. **183** Sifaka; R. D. Martin—Pic on Tour. **184** Squirrel Monkey; P. Morris. **185** Wooley Monkey; P. Morris. **186** Baboons; H. Angel. **186** Olive Baboons; SATOUR. **186** Rhesus Monkeys; NHPA (I. Polunin). **187** Macaques; NHPA (I. Polunin). **188** Langur; NHPA (E. H. Rao). **188** Vervet monkey; H. Angel. **189** Proboscis Monkey; J. MacKinnon. **190** White Gibbon; P. Morris. **190–191** White-browed Gibbon; J. MacKinnon. **191** Orang Utan; P. Morris. **191** Orang Utan in Trees; J. MacKinnon. **192** Chimp; J. MacKinnon. **194** Minke Whale; The Whale Research Unit of the National Institute of Oceanography. **196** Killer Whales; P. Morris. **197** Pilot Whales; H. Angel. **197** White-sided Dolphin; P. Morris. **198** Tamandua; Bruce Coleman (D. and J. Bartlett). **199** Sloth; P. Morris. **200** Three-toed Sloth; Bruce Coleman (F. Erize). **200** Pangolin; NHPA (E. G. Rao). **201** Sloth's Tail; P. Morris. **202–203** Lion; SATOUR. **204** Meerkat; P. Morris. **204** Mongoose; SATOUR. **204** Hyena; A. J. Sutcliffe. **204** Hyena Cub; A. J. Sutcliffe. **205** African Wild Cat; SATOUR. **206** Bobcat; P. Morris. **206** Canada Lynx; National Film Board of Canada (E. Cesar). **206** Lynx; National Film Board of Canada (E. Cesar). **207** Caracal; SATOUR. **208** Geoffroys Cat; AIS. **209** Lioness; SATOUR. **210** Tiger; P. Morris. **210** Cheetah; SATOUR. **211** Cheetahs; SATOUR. **211** Leopard; SATOUR. **212** Black Backed Jackal; H. Angel. **213** Dingo; AIS. **214–215** Bateared Fox; P. Morris. **214** Cape Hunting Dog; SATOUR. **216** Polar Bear; H. Angel. **216–217** Red Panda; Pic on Tour. **218** Wolverine; National Film Board of Canada. **218** Badger; H. Angel. **219** Sea Otter; P. Morris. **219** Skunk; National Film Board of Canada. **220** Seal Colony; AIS. **221** Galapagos Sea Lion; H. Angel. **222** Walrus; National Film Board of Canada. **223** Sea Elephants; AIS. **223** Bull Elephant Seal; J. Boswell. **224** Indian Elephant; Government of India Tourist Office. **224–225** Elephant; SATOUR. **225** Rock Hyraxes; H. Angel. **226** Manatee; Bruce Coleman (M. Freeman). **228** Juvenile Jack Rabbit; P. Morris. **229** Snow-shoe Rabbit; National Film Board of Canada. **229** Russian Pika; P. Morris. **236** Zebra; SATOUR. **237** Przewalski's Horse; P. Morris. **238** Wild Horses; AIS. **238–239** Zebra at Water; SATOUR. **240** Zebra; SATOUR. **241** Tapir; Common-wealth Institute. **242** Rhino; Government of India Tourist Office. **242–243** White Rhino; SATOUR. **244** Kudu Bull; SATOUR. **245** Warthog; SATOUR. **245** Hippos; SATOUR. **246** Fallow Deer; H. Angel. **246** Camel; P. Morris. **247** Elk; National Film Board of Canada. **247** Alpaca; P. Morris. **248** Giraffe; SATOUR. **248** Okapi; P. Morris. **248** Giraffe; SATOUR. **250** Herd of Buffalo; SATOUR. **250** Bison; National Film Board of Canada. **251** Water Buck; SATOUR. **253** Oryx; SATOUR. **254** Bonte Bok; SATOUR. **254** Impala; SATOUR. **255** Muskox; National Film Board of Canada. **255** Ibex fighting; A. J. Sutcliffe. **253** Wildebeest; A. J. Sutcliffe.

Acknowledgements

The publishers wish to acknowledge their indebtedness to the following books which were consulted for reference or as a source for illustrations.

Highly Developed Eyes—**A History of Fishes** by J. R. Norman, second edition by P. H. Greenwood (Benn) and **Invertebrate Zoology** by Robert D. Barnes (W. B. Saunders).
How bees give directions—**The Dancing Bees** by Karl von Frisch (Methuen).
Mark of the true crocodile—**Schildkroten, Krokodile, Bruckenechsen** by Dr. H. Wermuth, and Professor Dr. R. Mertens (Veb Gustav Fischer Verlag, Jena, Berlin).
Legless bodies moved by waves of muscular contraction—**Animal Locomotion** by Sir James Gray (Weidenfeld and Nicolson).
The Seas by Sir Frederick S. Russell and Sir Maurice Yonge (Warne); **The Whale** edited by Dr Leonard Harrison Matthews (Allen and Unwin; Tre Tryckare, Cagner & Co., Gothenburg), American Museum of Natural History; and **On the Osteology of the Cachalot or Sperm-whale** by W. F. Flower; **Transactions of the Zoological Society of London**, Vol. 6.
Atlas of Animal Migration by Cathy Jarman © Aldus Books Limited, London 1972.
The Living World of Animals published by The Reader's Digest Association Limited.
Book of British Birds published by Drive Publications Limited.
The World of Birds by James Fisher and Roger Tory Peterson published by Macdonald.
The population of Whales based on a diagram in **Animals of the Antarctic** by Stonehouse published by Peter Lowe (Eurobook Ltd.).
Snakes of Southern Africa by V. F. M. Fitzsimons published by Macdonald.
A History of Fishes by J. R. Norman revised by P. H. Greenwood published by Ernest Benn Ltd.